AMS/MAA | TEXTBOOKS

VOL **41**

Journey into Discrete Mathematics

Owen D. Byer, Deirdre L. Smeltzer,
and Kenneth L. Wantz

Providence, Rhode Island

2010 *Mathematics Subject Classification*. Primary 97K20, 97K30, 97F60, 97N70, 97E30.

For additional information and updates on this book, visit
www.ams.org/bookpages/text-41

Library of Congress Cataloging-in-Publication Data

Names: Byer, Owen, author. | Smeltzer, Deirdre L., author. | Wantz, Kenneth L., 1965- author.
Title: Journey into discrete mathematics / Owen D. Byer, Deirdre L. Smeltzer, Kenneth L. Wantz.
Description: Providence, Rhode Island : MAA Press, an imprint of the American Mathematical Society,
    [2018] | Series: AMS/MAA textbooks ; volume 41 | Includes bibliographical references and index.
Identifiers: LCCN 2018023584 | ISBN 9781470446963 (alk. paper)
Subjects: LCSH: Set theory. | Number theory. | Mathematical analysis. | Combinatorial analysis. | AMS:
    Mathematics education – Combinatorics, graph theory, probability theory, statistics – Combinatorics.
    msc | Mathematics education – Combinatorics, graph theory, probability theory, statistics – Graph the-
    ory. msc | Mathematics education – Arithmetic, number theory – Number theory. msc | Mathematics
    education – Numerical mathematics – Discrete mathematics. msc | Mathematics education – Founda-
    tions of mathematics – Logic. msc
Classification: LCC QA248 .B9657 2018 | DDC 511/.1–dc23
LC record available at https://lccn.loc.gov/2018023584

Contents

Contents

Preface

What Is Discrete Mathematics?

The word *discrete* in mathematics is in contrast to the word *continuous*. For example, the set of integers is discrete, while the set of real numbers is continuous. Thus, *discrete mathematics* describes a collection of branches of mathematics with the common characteristic that they focus on the study of things consisting of separate, irreducible, often finite parts. Although largely neglected in typical precollege mathematics curricula, discrete mathematics is essential for developing logic and problem-solving abilities. Questions located within the realm of discrete mathematics naturally invite creativity and innovative thinking that go beyond formulas. Furthermore, the cultivation of logical thinking forms a necessary foundation for proof-writing. For these reasons, discrete mathematics is critical for undergraduate study of both mathematics and computer science.

Goals of the Book

Simply stated, the goal of *Journey into Discrete Mathematics* is to nurture the development of skills needed to learn and do mathematics. These skills include the ability to read, write, and appreciate a good mathematical proof, as well as a basic fluency with core mathematical topics such as sets, relations and functions, graph theory, and number theory. The content and the corresponding requisite mathematical thinking are appropriate for students in computer science and other problem-solving disciplines, but the content presentation and the nature of the problem sets reinforce the primary goal of training mathematicians. Throughout the book, we emphasize the language of mathematics and the essentials of proof-writing, and we underscore that the process is very important in mathematics.

Entry-level discrete mathematics serves as an excellent gateway to upper-level mathematics by priming students' minds for upper-level concepts. *Journey into Discrete Mathematics* is designed for use in the first noncalculus course of a mathematics major, employing a writing style that models a high degree of mathematical accuracy while maintaining accessibility for early college students. For example, the treatment of inclusion-exclusion provides both informal and technically precise explanations. Ultimately, the goal behind this approach is communication: we want to model and teach students to communicate both accurately and clearly.

Journey into Discrete Mathematics utilizes problems and examples to lay the foundation for concepts to be encountered in future mathematics courses. For example, the chapter on relations and functions introduces students to definitions such as one-to-one and onto; several problems in Chapter 4 guide students through definitions of continuity using nested quantifiers; the treatment of greatest common divisor foreshadows finding the GCD of polynomial functions; the binomial and multinomial theorems are presented as tools for combinatorial counting; and Euler's totient function and Fermat's Little Theorem are important number-theoretic concepts that students will see again in an abstract algebra course. The homework questions are divided into sections according to difficulty, spanning the gamut from routine to quite challenging. The first section generally includes exercises that are more routine or computational, meant to give students a chance to practice given techniques, while the latter sections generally consist of problems that require creativity, synthesis of multiple concepts, or proofs.

This book takes the time to describe the origins of important discrete math topics as well as the connections between concepts. The treatment of matrices references Arthur Cayley's first use of matrices; the introduction of Fibonacci numbers is placed within historical context; the work on inductive thinking and proof by induction exhibits care for making connections with deductive thinking, the Well-Ordering Principle, and other mathematical concepts. Inspirational quotes throughout the book and the incorporation of the first names of mathematicians in examples and exercises (with a corresponding summary list providing a brief biography for each one mentioned) contribute to familiarizing students with the names of key figures within discrete mathematics.

Features of the Book

Convince Me Chapter. This opening chapter contains a selection of interesting, nonstandard problems of varying degrees of difficulty. Readers are invited to think creatively and argue persuasively as they work to find solutions. This process cultivates an understanding of the importance of making a good mathematical argument, while setting the tone for the problem-solving nature of the book. Moreover, many of the solutions in this chapter foreshadow the mathematical techniques and theorems that will be encountered later in the book.

Hook Problems. In the manner of the *Convince Me* problems, each chapter begins with an intriguing and challenging problem intended to capture the reader's interest. Each hook problem can be solved using techniques to be developed in the chapter and usually reappears later in the chapter, either as an example or as a homework problem.

Presentation of Logic. Chapters 3 and 4 of the book combine the topics of sets, logic, and proof-writing in a distinctive way. This approach helps to highlight the high level of congruity between concepts such as DeMorgan's Laws for sets and logic, membership tables and truth tables, logical operators and set operators. The chapter on logic and proof-writing appears early in the book to help students bridge the gap between intuitive thinking and the formal presentation of an argument, both of which are necessary in mathematics.

First Thoughts and Further Thoughts. Solutions to many examples in the book are preceded by "First Thoughts", describing the initial thought process that one might engage in when first considering a new problem. This is intended to be both helpful and reassuring to students who might be intimidated by seeing final polished proofs and assuming that "real" mathematicians can produce these immediately, without intermediary struggles or failed attempts. First Thoughts help train students in the ways that mathematicians actually operate. Similarly, "Further Thoughts" often follow a solution in order to provide additional insight about, alternative approaches to, or extensions of the given solution; once again, the goal is to cultivate a spirit of *doing mathematics*.

Advanced Topic Chapters. Several core topics (counting, number theory, and graph theory) are addressed twice in this book: first in an introductory chapter covering standard content and later in a chapter with extended optional (often, but not always, more advanced) material. This provides instructors with flexibility to customize the course, depending on their particular goals, or expand beyond a typical first course in discrete mathematics.

Course Outline

This book is designed to be used as a stand-alone text for a three-credit or four-credit discrete mathematics course for average to above average math majors who are learning to write proofs; however, since there is more material in this book than can be covered in a single semester, instructors will have to make some choices. For students who have already had an introduction-to-proofs course, select portions of the first six chapters of the book can be covered rather quickly, and the last half of the book can serve as a main text for a junior-level course in combinatorics. If this book is supplemented with a few extra topics (such as probability, solving recurrence relations, or finite-state machines), then there is enough material for a two-semester sequence in discrete mathematics.

With that goal in mind, we suggest one design for such a three-credit course. The second column in Table 0.1 on page x lists the core sections we believe should be covered. We estimate that the core sections can be covered in about thirty-four fifty-minute lectures. The remaining class periods could be used for review days, testing days, and optional sections from the third column. The first column of the table lists sections and chapters containing material that is essential for students to know before covering corresponding sections of the middle two columns. The middle columns contain material that is used in sections and chapters listed in the fourth column, though they may not be absolute prerequisites. For example, although matrices (first addressed in Section 2.3) also appear in Chapter 9 (Graph Theory), one need not study Section 2.3 in order to be able to understand the essential components of Chapter 9.

Table 0.1. Section priorities and interdependencies

Prerequisite sections and chapters for →	Core section	Optional section	→ Material used in
	1.1–1.2		
	2.1		
	2.2		Chapters 4, 5, 6, 7, 12
		2.3	5.1, 5.2, Chapter 9
	3.1–3.6		Chapter 4
		3.7	
2.2, 3.1–3.6	4.1–4.2		proofs throughout book
4.2		4.3	
2.2, 3.1–3.6	4.4–4.5		proofs throughout book
2.2	5.1		5.2, 5.3, 9.1
2.2	5.2		7.4, 9.1
2.2	5.3		
		5.4	6.1, 6.2, Chapter 11
		6.1, 6.2	
4.2, 5.3	6.3–6.4		proofs throughout later chapters
		6.5	
2.2, 6.3	7.1		12.1
2.2, 6.3	7.2		Chapter 12
2.2		7.3	12.4
2.2, 6.3	7.4		
2.2		7.5–7.6	12.4
	8.1–8.3		Chapters 9 and 11, 13.2
	8.4		11.3, 11.4
		8.5	
8.3	9.1		Chapter 10, 13.4
	9.2		
		9.3	13.1, 13.2
	9.4		13.1, 13.2
6.4		9.5	
Chapter 4		10.1–10.2	
3.2, 3.4, 3.6, 8.3	11.1		
2.2	11.2		
		11.3	11.5
Chapter 8		11.4	
Chapter 8, 11.3		11.5	
7.1		12.1	
2.2		12.2	
7.4		12.3	
7.3, 12.3		12.4	
7.4, 12.3		12.5	
9.1, 9.4		13.1	
6.4, 9.1, 9.3, 9.4		13.2	
9.1, 9.4		13.3	
9.1, 9.4		13.4	

Acknowledgments

Many years of effort went into the conception, writing, and publication of this book. We thank our ever-supportive families, including spouses Barbara, Sherwyn, and Beth, for their encouragement and understanding through the entire endeavor. We thank our students for bearing with us through early rough drafts and for pointing out numerous errors. We give special thanks to Daniel Showalter for his feedback from careful reading of numerous parts of the book. Our greatest thanks goes to Felix Lazebnik, who was not only an inspiring discrete math teacher and mentor for two of us, but who provided significant vision and content for the book. In particular, Felix contributed heavily to Section 2.2, Section 6.5, Chapter 7, and Chapter 10. Finally, we thank MAA and AMS, particularly Stephen Kennedy, Stan Seltzer, and Christine Thivierge, for their excellent work in facilitating the publication process.

1

Convince Me!

Obvious is the most dangerous word in mathematics.
—E. T. Bell (1883–1960)

How did you learn how to ride a bicycle? If you learned as a child, you may have first ridden a tricycle and then a bicycle with training wheels before attempting to balance on a two-wheeler. Although you probably first understood how a bicycle works by watching someone else use one, and perhaps an adult ran along behind with a hand on the back of the bike during your first tentative rides, the primary way in which one becomes an expert bicyclist is through practice.

The same is true of mathematics. You cannot learn to do mathematics simply by watching someone else do mathematics. To become adept, you must be willing to practice, sometimes failing and sometimes succeeding, even if only partially. In this chapter, we encourage you to "dive in" to mathematics by presenting you with problems that are easy to understand and of variable difficulty to solve.

You may already have an idea of what mathematical problems are, but we hope to expand your view. Some of the challenges that we pose in the next section may not even seem like math! A "mathematical problem" isn't necessarily the same as an "exercise"; an exercise is for practicing a known procedure, whereas a problem is something to be solved when the approach may not be obvious.

Improving skill in problem solving may require learning to recognize patterns, developing theoretical notions, and determining how best to use a variety of mathematical tools and techniques. The particular focus of this book is on solving problems in discrete mathematics—which has some differences from solving problems in, say, calculus. Understanding the language of discrete mathematics, including symbols and definitions, will be necessary for understanding the concepts; just as familiarity with common abbreviations (e.g., "BRB") allows for quick everyday communication, so the language and symbolism of mathematics will allow for quicker communication of mathematical ideas. In discrete mathematics, this will be the language of sets, logic, functions, enumeration, and graphs, and it will be introduced systematically as needed.

In mathematics, simply knowing how to solve a problem is rarely satisfying. A second important step is to persuade others that your solution is correct. Thus, another invaluable purpose of this book is to develop skills in constructing mathematical proofs. We will demonstrate and discuss a variety of types of proof, illustrating sound methods for convincing others of a correct solution. As with problem solving, though, the development of proof-writing expertise requires practice, practice, and more practice. Without further ado, let us begin!

1.1 Opening Problems

Although the problems below are quite varied in nature, each one has a solution that does not require high-level mathematics. We urge you to select several of interest to solve. Try to give a convincing argument as to why your solution is correct.

(1) In a game for two players, seventeen coins lie on a table. The players take turns, at each turn removing between one and four coins from the table. The player who takes the last coin loses. If you are going to play this game, ask yourself if you prefer to go first or if you should ask your friend to go first—or does it matter?

As an example, suppose you go first and take four coins (leaving thirteen coins on the table), your friend goes second and takes four coins (leaving nine coins), you then take two coins (leaving seven coins), your friend takes one coin (leaving six coins). Now things look bad for you: you take two coins and your friend wisely removes three more, forcing you to take the last coin and lose the game.

Could you have done something different so that you would have won the game? It might be hard to find a solution immediately. Some may have insight and see the strategy right away, but for most people, finding the solution requires previous experience or lots of experimenting; this is what most mathematicians would do when faced with this problem.

Additional experimentation should lead you to conjecture that the player who goes first can always win; that is, no matter how the second player plays the game, if the first player follows the proper strategy, she can always win. Give a convincing argument of this conjecture.

(2) Must there be two people in Los Angeles who have the same number of hairs on their heads? You may say that the answer is clearly "yes" since there are surely at least two completely bald people in L.A. Let's agree to disregard the people who are completely bald. Now, what is your answer?

(3) What is the last digit of 3^{1776}? We believe most of you can answer this question by finding a pattern for the last digit of 3^1, 3^2, 3^3, and so on, looking for a pattern. Having done that, what about finding the last *two* digits of 3^{1776}?

(4) Consider a single-elimination tournament, which starts with $n \geq 1$ teams. In the first round, all teams are divided into pairs if n is even, and the winner in each pair passes to the next round (no ties). If n is odd, then one random (lucky) team passes to the next round without playing. The second round proceeds similarly. At the end, only one team is left—the winner. Find a *simple* formula for the total number of games played in the tournament.

(5) Suppose you have access to an unlimited supply of 3-cent and 5-cent stamps. If you need to send an envelope that requires 47 cents, can you create the exact postage using these stamps? What other postage amounts can you pay by using combinations of these stamps? Describe as many as possible.

(6) Can you create a 3×3 rectangular array of numbers (repeat numbers permitted) such that the sum of numbers in every row is 10 and the sum of numbers in every column is 10? Can you do the same for a 3×4 array of numbers?

(7) Which is larger, 3^{400} or 4^{300}?

(8) You are a participant in a game show in which there are three doors to choose between. Behind each door is a different amount of money. You choose a door and are shown the amount behind it. If you wish, you may take the money and the game ends. If you decline, you may choose another door and see the amount behind that door, but you forfeit the opportunity to accept the amount behind the previous door. You may accept the prize behind the second door; if you decline, then you must take the amount behind the third door, sight unseen. If you play this game optimally, what is the probability that you will win the largest amount of money possible?

(9) Consider a simple game for two players in which the players take turns placing coins on a round table. The coins may not overlap and may not be moved once they are placed. The loser is the first person to be unable to place a coin on the table. Is there a winning strategy? If so, which player has the advantage, the one going first or the one going second?

(10) There are seven cups on a table—all standing upside down. You are allowed to turn over any four of them in one move. Is it possible to eventually have all of the cups right-side up by repeating this move?

. (11) A large chocolate bar is composed of 40 smaller squares in a 5 by 8 grid. How many cuts will it take to cut it into those 40 squares if one must always slice along grid lines and the knife may pass through only one chunk of chocolate per cut (i.e., one cannot line up several previously sliced chunks and cut them simultaneously).

(12) Represent the number 1,492 as a sum of two positive integers whose product is the greatest. Next try to maximize the product if you are permitted to use three positive integers. Finally, maximize the product if any number of positive integers is permitted. (It may be helpful to experiment with a much smaller number than 1,492.)

(13) A group of people sits in a circle, each holding an even number of pieces of candy. Each person simultaneously gives half of his or her candy to the person on the right. Any person who ends up with an odd number of pieces of candy selects one piece from a large bowl of candy in the center of the circle, so that once again all persons have an even number of pieces. Show that after enough iterations of this procedure all persons will have the same number of pieces of candy. Is this statement also true if at each iteration each person gives half of his or her candy to each person to the left and right?

(14) Students in an elementary school classroom are seated in five rows, with each row having six desks. From each row, a tallest student is chosen; then, a shortest of

these five students is chosen (call this student A). Now, from each column, a short-est student is chosen; then, a tallest of these six students is chosen (call this student B). Who is taller, A or B? Can they be the same height? Can A and B be the same person?

(15) Given a group of 21 people, show that there are as many ways to select an odd num-ber of them as there are ways to select an even number of them. (Each "way" is determined by the people chosen, not by the number of people chosen. So, choos-ing Al and Sal counts as a different selection than choosing Hal and Cal.)

(16) Each of two bright math students, Terence and Srinivasa, has a number on his fore-head. Each can see the number on the other person's forehead, but not the number on his own. They are told that the two numbers are consecutive positive integers. They start the conversion with Terence saying, "I don't know my number." Srini-vasa thinks for a moment and replies, "I don't know my number either." They then repeat this exact same conversation, for a total of 53 times, at which point Terence exclaims, "Hey, I just figured out my number!" Srinivasa immediately follows, "Yeah? Me too!" What are the possible numbers on Terence's forehead?

(17) Consider a group of 100 prisoners. They are told they will be given a collective chance at a pardon if they can work together to solve the following challenge.

At the appointed time all 100 prisoners will come into a large room to gather in a circle. Each prisoner will have a hat placed on his head as he enters, and it will be one of seven colors (red, orange, yellow, green, blue, indigo, or violet). Each prisoner will be able to see all hats except his own. One at a time, each prisoner will conjecture out loud the color of his or her own hat. If at most one of the 100 prisoners makes an incorrect conjecture, they will all go free; otherwise they will all be executed. The prisoners can plan a strategy ahead of time, but once they are brought into the room, there will be no communication, other than that each prisoner can hear all of the other guesses. No clues can be given based on eye contact, positioning of the prisoners, timing of responses, etc. What strategy could the prisoners adopt to best guarantee survival? Under this strategy, what is the probability of survival?

1.2 Solutions

(1) According to the rules of the game, whichever player is faced with only one coin on the table will lose. In fact, once there were only six coins left and it was your turn, you were bound to lose: if you took n (with $1 \le n \le 4$) coins, your friend would take $5 - n$ of them, leaving just one coin. Therefore, each player's goal should be to leave his opponent with six coins on his turn.

Arguing backwards from there, if you can leave your opponent with eleven coins, if he takes n of them ($1 \le n \le 4$), then you can always take $5 - n$ of them, which will be a number between 1 and 4, as required. Then it will be his turn with six coins remaining. Similarly, whoever's turn it is with sixteen coins remaining will lose if the other player always selects $5 - n$ coins immediately after n coins are taken.

In general, if a player is, at any turn, faced with $5k + 1$ coins (for any integer k), he will lose if his opponent plays strategically. Therefore, no matter how many coins are initially on the table, the first player can always win if it is possible to use the first turn to remove an appropriate number of coins so that $5k + 1$ coins remain for his opponent for some integer k. If both players follow the proper strategy, the first player will lose only when there are $5k + 1$ coins on the table at the start of the game for some integer k.

Can you generalize the strategy to the case where initially m coins are placed on the table and each player may remove between 1 and t coins on his turn?

(2) The population of Los Angeles is about 4 million people. It is possible for them all to have a different number of hairs on their heads only if the human head could have 4 million hairs on it. How many hairs could fit on a human head? Suppose that half of a typical human head is covered with hair. If the radius of the head is 6 inches and it is assumed to be spherical, half of the surface area would be $(1/2)4\pi6^2 = 72\pi < 250$ square inches. How many hairs are there per square inch? Assuming the follicles are 20 per linear inch would give 400 per square inch and about 100,000 hairs on a human head.[1] This is nowhere near 4 million, so we can safely conclude that it must be the case that two residents of Los Angeles have the same number of hairs on their heads.

(3) The consecutive powers of 3 are 3, 9, 27, 81, 243, 729, etc., so the last digits seem to follow the pattern 3, 9, 7, 1, and then the cycle repeats. Indeed, multiplying each digit in that sequence by 3 gives the next one in the sequence, wrapping around. Every fourth power of 3, therefore, has 1 as its last digit, and since 1,776 is divisible by 4, $3^{1,776}$ has 1 as its last digit.

Now, can you figure out what the last two digits of $3^{1,776}$ would be? While the above method works, we will find a faster solution in Chapter 7 that uses properties of modular arithmetic.

(4) As is often the case, this is best approached by way of small cases. Suppose there are only four teams (call them A, B, C, and D) participating in the tournament. In round 1, A competes against B and C competes against D. Assume that A and C are the winners of their respective games; these teams then compete against each other. Regardless of whether A or C wins, the total number of games played will be 3.

Can you easily see how this generalizes? If not, try diagramming the situation for some other values of n. You should see that if there are n teams competing, the number of games played is always $n - 1$. In fact, this is sensible: in each game, exactly one team is eliminated, and all teams but one (the tournament winner) will be eliminated at some point.

(5) There are many ways to obtain 47 cents, including nine 3-cent stamps and four 5-cent stamps. One cannot obtain a total of 7 cents, as can readily be checked. However, $8 = 2(4)$, $9 = 3(3)$, and $10 = 2(5)$. Note that 11-, 12-, and 13-cent amounts can be obtained by adding one 3-cent stamp to each of the previous configurations, respectively. Similarly one can continue adding 3-cent stamps to these arrangements to obtain any integer amount greater than 13 cents. This idea will be explored more formally in Chapter 6 (Induction).

(6) A 3×3 grid with a 10 in each of the three positions on one of the diagonals and a 0 in all other positions will meet the requirements given.

---

[1] A quick Internet search confirms that this is a reasonable approximation.

It is not possible in a 3×4 rectangular array, however. If each column sum is 10 and there are four columns, then the total of all the entries in the array will be 120. On the other hand, if each row sum is 10 and there are three rows, the total of all of the entries in the array will be 90. This is a contradiction, so such a grid is not possible.

(7) Note that $3^{400} = (3^4)^{100}$ and $4^{300} = (4^3)^{100}$. Since $3^4 = 81$ is larger than $4^3 = 64$, we see that $3^{400} > 4^{300}$.

(8) You should pick any door, see the amount behind it, and decline it. Then select another door and compare its value to the first one. If it is higher, keep it. If it is not, take the amount behind the third door. Label the three amounts as Low, Medium, and High, and note that there are six corresponding permutations of L, M, and H. You should then verify that the above strategy will enable you to select the largest amount in half of them. Two such orderings are (M, L, H) and (M, H, L). An ordering in which you will not obtain the largest amount is (H, M, L).

(9) The first person can win if she places her coin in the exact center of the table. After that point, she always takes her turn by placing a coin diametrically opposite to the coin placed by the second player. The space will always be available, so she will always be able to move.

(10) Think about the possible numbers of glasses that can be right-side up at any given time. We will discuss a formal solution to this problem in Chapter 10.

(11) Every time a cut is made, one chunk of chocolate becomes two chunks of chocolate. Thus, no matter in which order the pieces are cut, it will take 39 cuts to break the original bar into 40 squares.

(12) The maximum value of ab if $a + b = 1{,}492$ is 746^2, when $a = b = 746$. Of course, one can try other values of a and b in a futile effort to exceed this product, but a short proof will show an improvement is impossible. If $a = 746-x$ and $b = 746+x$, then $ab = 746^2 - x^2$. Clearly, the product will be maximum when $x = 0$, which means $a = b = 746$.

This solution will generalize to $a = b = n/2$ if $1{,}492$ is replaced by n. Of course, if n is not even, you must modify a and b appropriately since we are required to use integers. What does this generalization tell you about how to maximize the product of a set of more than two positive integers whose sum is $1{,}492$? Try with three integers whose sum is 10.

(13) Consider what happens to the difference between the greatest and fewest number of candies held by anyone as the rounds progress. A formal approach for solving problems like this will be discussed in Chapter 10.

(14) Certainly, it is possible that A and B may be the same person. For example, this will happen if the students are seated in order of height, with the shortest student in the desk that is in Row 1 and Column 1, the next shortest student in the Row 1, Column 2 spot, and so on, with the tallest student in the desk that's in Row 5, Column 6.

Let $h(A)$ and $h(B)$ be the heights of students A and B, respectively. Then, in the general case, $h(A) \geq h(B)$. To see this, for each i, define t_i to be the height of the tallest student in Row i; then $h(A) = \min\{t_i : 1 \leq i \leq 5\}$. Similarly, define s_j to be the height of the shortest student in Column j; then, $h(B) = \max\{s_j : 1 \leq j \leq 6\}$.

Clearly, if A and B are in the same row, $h(A) \geq h(B)$. By definition, $h(B) = s_k$ for some k, and B is a shortest student in Column k. Likewise, $h(A) = t_l$ for some l, and A is a tallest student in Row l. Let $h(l, k)$ be the height of the student in Row l and Column k. It follows that $h(A) \geq h(l, k) \geq h(B)$.

(15) If a group of n persons is chosen, then $21 - n$ persons remain. Notice that n and $21 - n$ have different parities (one must be even and one must be odd). Therefore, for each group with an odd number of persons, there is a corresponding group with an even number of persons, and vice versa. These two sets of groups are therefore in one-to-one correspondence, so the sets have the same size.

(16) Try the problem with a smaller number of rounds than 53 (say, two rounds). What must Terence's number be if he immediately knew his number after seeing Srinivasa's number? What can you conclude if Terence could figure out his number as soon as Srinivasa first acknowledged that he didn't know his own number? We will see this problem again in Chapter 6 where we learn to use induction as a proof technique.

(17) Assign each color a number and have the first person add up all the "numbers" he sees. We will return to this problem when we discuss modular arithmetic in Chapter 7.

Think of an integer between 41 and 59, inclusive. Subtract 25 from your number and write down the resulting 2-digit number. Now subtract 50 from your original number and square the result to form another 2-digit number, using a zero as the first digit if necessary. Append this 2-digit number to the right side of the one you wrote down earlier, forming a 4-digit number. This 4-digit number should be the square of the original number. Why does this work? Can you amend the method to work for numbers outside of the 41–59 range?

2

Mini-Theories

*To many, mathematics is a collection of theorems. For me, mathematics is a
collection of examples; a theorem is a statement about a collection of examples and
the purpose of proving theorems is to classify and explain the examples....*

—John B. Conway (1937–)

In the first chapter, *Convince Me!*, we gave many examples of convincing arguments in
solving mathematical problems. In this chapter we continue our journey into mathematics in a more formal way by exploring several familiar mathematical theories, developing them from the ground up, in order to serve as a microcosm for how mathematical theories are created. By a mathematical theory, we mean a collection of undefined terms, a set of **axioms** (statements that are accepted as being true), definitions,
theorems—all of which relate to a given mathematical topic—and the results which
can be derived from them. The topics we undertake in this chapter, properties of real
numbers, divisibility of integers, and matrices, will likely not be new to the reader. This
leaves us free to focus our attention on the importance of definitions and the development of the theories.

2.1 Introduction

We begin with definitions and properties of real numbers that are commonly used in
mathematics. The reader should not dismiss the importance of the "elementary" notions given here, as they will be used to build mathematical theory in later sections of
the book.

   The following notation for sets of numbers is commonly used, other than perhaps
$[n]$ to denote the first n natural numbers.

   - $\mathbb{Z} = \{..., -3, -2, -1, 0, 1, 2, 3, ...\}$ is the set of **integers**.
   - $\mathbb{N} = \{1, 2, 3, ...\}$ is the set of positive integers, also known as the set of **natural**
     numbers.

- $[n] = \{1, 2, 3, \ldots, n\}$.
- $\mathbb{N}_0 = \{0, 1, 2, 3, \ldots\}$ is the set of nonnegative integers, also known as the set of **whole** numbers.
- $\mathbb{Q}$ is the set of **rational** numbers. Its elements can be represented by fractions $\frac{m}{n}$ or m/n, where $m, n \in \mathbb{Z}$ and $n \neq 0$. Notice that such a representation is not unique:

$$3/5 = (-6)/(-10) = 60/100.$$

  Alternatively, the elements of $\mathbb{Q}$ can be represented by periodic decimals. Such a representation, again, is not unique:

$$\frac{23{,}917}{1{,}000} = 23\frac{917}{1{,}000} = 23.917 = 23.917000\ldots = 23.916999\ldots.$$

- $\mathbb{R}$ is the set of all **real** numbers. Its elements can be represented by decimals, both periodic (e.g., $3.\overline{45} = 3.45454545\ldots$) and nonperiodic (e.g., $0.101001000100001\ldots$).

Nineteenth century German mathematician Leopold Kronecker (1823–1891) is quoted as saying, "God created the integers; all else is the work of mankind" [**20**]. In fact, the above sets of numbers form nested subsets: for example, the natural numbers are contained within the integers, the integers are contained within the rational numbers, and the rational numbers are contained within the real numbers. We will not go through the rigorous process of demonstrating how one constructs the reals from the integers (though it can be done). Rather, while acknowledging that the domain of discourse of discrete mathematics is almost always the integers, we accept that the following basic "laws" apply to all real numbers.

Property 2.1. *Let $\mathbb{S}$ represent any one of $\mathbb{Z}$, $\mathbb{Q}$, or $\mathbb{R}$. The following properties hold for all a, b, and c in $\mathbb{S}$.*

(1) *Closure: The sum, difference, and product of two numbers in $\mathbb{S}$ is also in $\mathbb{S}$.*

(2) *Commutative laws: $a + b = b + a$ and $ab = ba$.*

(3) *Associative laws: $(a + b) + c = a + (b + c)$ and $(ab)c = a(bc)$.*

(4) *Distributive law: $a(b + c) = ab + ac$.*

(5) *Additive and Multiplicative Identities: $0 + a = a$ and $1 \cdot a = a$.*

(6) *Additive Inverse: There is an additive inverse in $\mathbb{S}$, denoted $-a$, such that $a + (-a) = 0$.*

(7) *Multiplicative Inverse: If $\mathbb{S} = \mathbb{Q}$ or $\mathbb{S} = \mathbb{R}$ and $a \neq 0$, there is a multiplicative inverse in $\mathbb{S}$, denoted a^{-1}, such that $a \cdot a^{-1} = 1$.*

(8) *Exponential laws: $(a^n)^m = a^{nm}$, $(ab)^n = a^n b^n$, and $a^n a^m = a^{n+m}$ for all real numbers m and n.*

Notice that we are not attempting to prove the above properties; in that sense, we are taking the statements in Property 2.1 as axioms. The statements in the following theorem may seem as obvious as the ones just listed, but we will see that each one is actually a consequence of one or more parts of Property 2.1.

Theorem 2.2. (1) *For any real number a, the additive inverse of a is unique.*

(2) *The number 0 is the unique additive identity for $\mathbb{R}$.*

(3) *For any real number $a \neq 0$, the multiplicative inverse of a is unique.*

(4) *The number 1 is the unique multiplicative identity for $\mathbb{R}$.*

(5) *For any real number a, $-(-a) = a$.*

(6) *For any real numbers a and b, $-(a + b) = (-a) + (-b)$.*

(7) *For any real number a, $a \cdot 0 = 0$.*

(8) *For any real number a, $-1 \cdot a = -a$.*

(9) *For any real numbers a and b, $(-a)(-b) = ab$.*

(10) *For any real number a, if $a \neq 0$ and $ab = ac$, then $b = c$.*

(11) *Zero product property: For any real numbers a and b, if $ab = 0$, then $a = 0$ or $b = 0$.*

Proof. We provide a guided proof of each property. In each proof there is a series of "Why?" questions. In lieu of completing a problem set for this section, the reader should supply a reason (usually a property from Property 2.1 or from an already proven part of this theorem) that justifies each statement in the proof preceding a "Why?" query.

(1) To prove the uniqueness of the inverse, assume that b and c are both additive inverses of a. Then

$$
\begin{aligned}
b &= b + 0 && \text{(a. Why?)} \\
&= b + (a + c) && \text{(b. Why?)} \\
&= (b + a) + c && \text{(c. Why?)} \\
&= 0 + c && \text{(d. Why?)} \\
&= c. && \text{(e. Why?)}
\end{aligned}
$$

Therefore, $b = c$, and a has a unique additive inverse.

(2) To show that there is only one additive identity, we let e be an additive identity and prove $e = 0$. If e is an additive identity, then $e + a = a$ for all a (a. Why?). Then for any given a, $e + a + (-a) = 0$ (b. Why?). It follows that $e + (a + (-a)) = 0$ (c. Why?). Therefore, $e + 0 = 0$, so $e = 0$ (d. Why?). Thus, 0 is the unique additive identity.

(3) To prove uniqueness, assume that b and c are both multiplicative inverses of a. Then $ab = ac = 1$ (a. Why?) and $ab = ba$ (b. Why?). It follows that

$$
\begin{aligned}
b &= b(1) && \text{(c. Why?)} \\
&= b(aa^{-1}) && \text{(d. Why?)} \\
&= (ba)(a^{-1}) && \text{(e. Why?)} \\
&= (ba)(c) && \text{(f. Why?)} \\
&= 1 \cdot c = c. && \text{(g. Why?)}
\end{aligned}
$$

Thus, $b = c$ and the multiplicative inverse of a is unique.

(4) Assume that e is a multiplicative identity and let a be any nonzero number. Then a has a multiplicative inverse, a^{-1}. It follows that

$$e = e \cdot 1 \qquad \text{(a. Why?)}$$
$$= e(a \cdot a^{-1}) \qquad \text{(b. Why?)}$$
$$= (e \cdot a)a^{-1} \qquad \text{(c. Why?)}$$
$$= a \cdot a^{-1} \qquad \text{(d. Why?)}$$
$$= 1. \qquad \text{(e. Why?)}$$

This proves 1 is the unique multiplicative identity.

(5) $-a$ has a unique additive inverse (a. Why?), denoted $-(-a)$, and $-(-a) + -a = 0$ (b. Why?). On the other hand $a + (-a) = 0$ (c. Why?). Therefore $-(-a)$ and a are both additive inverses of $-a$. This proves that $-(-a) = a$ (d. Why?).

(6) It suffices to show that $(-a) + (-b)$ is an additive inverse of $a + b$ (a. Why?). Indeed, $(a + b) + ((-a) + (-b)) = (a + (-a)) + (b + (-b))$ (b. Why?). This sum is 0 (c. Why?), which proves that $(-a) + (-b)$ is an inverse of $a + b$ (d. Why?).

(7) $a \cdot 0 = a(0 + 0)$ (a. Why?), which equals $a \cdot 0 + a \cdot 0$ (b. Why?). By adding $-(a \cdot 0)$ to both sides of $a \cdot 0 = a \cdot 0 + a \cdot 0$, we obtain $0 = a \cdot 0$ (c. Why?).

(8) It suffices to prove that $-1 \cdot a$ is the additive inverse of a (a. Why?). We have $a + (-1 \cdot a) = 1 \cdot a + (-1) \cdot a$ (b. Why?), which equals $(1 + (-1)) \cdot a$ (c. Why?). This value is $0 \cdot a = 0$ (d. Why?), which proves that $-1 \cdot a$ is the additive inverse of a (e. Why?).

(9)

$$(-a)(-b) = (-1 \cdot a)(-1 \cdot b) \qquad \text{(a. Why?)}$$
$$= (-1)(a)(-1)(b) \qquad \text{(b. Why?)}$$
$$= (-1)(-1)(a)(b) \qquad \text{(c. Why?)}$$
$$= -(-1)(ab) \qquad \text{(d. Why?)}$$
$$= 1(ab) = ab. \qquad \text{(e. Why?)}$$

(10) Since $a \neq 0$, a has a real-valued multiplicative inverse, a^{-1}. By assumption, $ab = ac$; thus, $a^{-1}(ab) = a^{-1}(ac)$, and therefore $(a^{-1}a)b = (a^{-1}a)c$ (a. Why?). It follows that $1 \cdot b = 1 \cdot c$ (b. Why?), so $b = c$ (c. Why?).

(11) If $b = 0$, we are done. If $b \neq 0$, then b has a real-valued multiplicative inverse, b^{-1}. Then

$$a = a \cdot 1 \qquad \text{(a. Why?)}$$
$$= a(bb^{-1}) \qquad \text{(b. Why?)}$$
$$= (ab)(b^{-1}) \qquad \text{(c. Why?)}$$
$$= 0 \cdot b^{-1} = 0. \qquad \text{(d. Why?)}$$

Thus, either $a = 0$ or $b = 0$, which completes the proof. $\square$

We close the section with a discussion of two commonly used functions in discrete mathematics. For any x, the **absolute value** of x, denoted $|x|$, is defined as

$$|x| = \begin{cases} x & \text{if } x \geq 0, \\ -x & \text{if } x < 0. \end{cases}$$

For example, $|5| = 5$, $|0| = 0$, and $|-7| = -(-7) = 7$. Note that $|x|$ can be viewed as the distance on the real number line from the number x to the origin, 0. Clearly, for any a, $-|a| \leq |a|$, and for any real numbers a and b, $|a| \leq |b|$ if and only if $-|b| \leq a \leq |b|$. Furthermore, the distance on the real number line between any a and b is $|a - b| = |b - a|$; see Figure 2.1 for an illustration of this in the case when $a < 0 < b$.

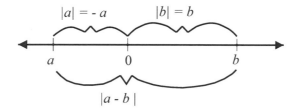

Figure 2.1

This background provides us with two additional properties. Each one follows from the definition of operations on real numbers, by considering various cases for a and b.

Theorem 2.3. (1) *Absolute Value Product:* $|ab| = |a||b|$.
(2) *Triangle Inequality:* $|a + b| \leq |a| + |b|$.

While real numbers are the dominant domain of discourse in many areas of mathematics, most questions in discrete mathematics yield integral answers. In some cases, a real number must be rounded (up or down) to obtain an integer. The **floor function** of a real number x gives the largest integer less than or equal to x; it is denoted $\lfloor x \rfloor$. Therefore,

$$\lfloor 2.9 \rfloor = 2, \ \lfloor \pi \rfloor = 3, \ \lfloor -4.2 \rfloor = -5, \text{ and } \lfloor 6 \rfloor = 6.$$

Notice that if $x > 0$, $\lfloor x \rfloor$ can be obtained by truncating the decimal part of x or by rounding down to the nearest integer. If $x < 0$, truncating does not produce an integer less than x; the floor of a negative noninteger is one less than the truncation of x.

As an application, consider the problem of finding the number of multiples of seven that lie between 1 and 100. One could list the successive multiples of seven and then count them, but a moment of reflection should lead to the answer of $\left\lfloor \frac{100}{7} \right\rfloor = \lfloor 14.29 \rfloor = 14$.

The "round up" counterpart to the floor function is the **ceiling function**, which will be explored in the exercises.

2.2 Divisibility of Integers

As we noted in Section 2.1, integers are featured prominently in any discrete math book. In fact, for the remainder of this book, unless otherwise stated,

> letters $a, b, c, \ldots$ will be used to represent integers.

We begin our discussion about integers with a simple statement:

> *The sum of two even integers is an even integer.*

Undoubtedly, the reader can distinguish between even and odd integers and is quite certain that this statement is true. How would one *prove* it, though, based on a definition of even? Many people would define even integers as those ending in 0, 2, 4, 6, or 8. With this definition, proving that the sum of two even integers is even would require considering all possible pairs of numbers ending in the digits 0, 2, 4, 6, or 8 and demonstrating that in all cases, the sum also ends in 0, 2, 4, 6, or 8. While such a proof is valid, it is quite tedious and would not be the method of choice of most mathematicians.

Rather, a mathematician would prefer to define a number n to be **even** if it can be written as the product of 2 and some integer. Seeing that the integers that satisfy this definition must be those that end in 0, 2, 4, 6, or 8 is not difficult; at the same time, this definition will be far more useful in proofs. To wit, here is a proof that the sum of any two even integers is an even integer.

Proof. Let m and n be any two even integers. Then, by definition, $n = 2j$ and $m = 2k$ for some integers j and k, in which case $n + m = 2j + 2k = 2(j + k)$. Since $j + k$ is an integer (by the closure property), we have demonstrated that the sum of n and m can be written as the product of 2 and some integer. Therefore $n + m$ is even. $\square$

Hopefully the above example underscores that a good definition must be both accurate and useful. This definition of "even" is actually a special case within the broader notion of divisibility.

For two integers a and b, $b \neq 0$, if there exists an integer q such that $a = bq$, then we say that b **divides** a or a is **divisible** by b, and denote this relationship by writing $b|a$. If $a = bq$, then a is called a **multiple** of b, b is called a **divisor** of a, and q is called the **quotient** of the division of a by b. For example, $5|(-15)$ since $-15 = 5 \cdot (-3)$, 2 is a divisor of 20 since $20 = 2 \cdot 10$, and 0 is a multiple of 5 since $0 = 5 \cdot 0$.

Our use of "the" when referring to "the quotient" above is justified by the fact that if such q exists, it is unique. To see this, note that if $a = bq_1 = bq_2$ for some q_1 and q_2, then $b(q_1 - q_2) = 0$. Since we have assumed $b \neq 0$, then by the zero product property, $q_1 - q_2 = 0$. This proves that $q_1 = q_2$, so the quotient is unique.

Why do we need to restrict b from being zero when we say $b|a$? The reason is the following. The equality $a = 0 \cdot q$ implies $a = 0$; therefore the only number a which seems to allow division by 0 is 0 itself. But $0 = 0 \cdot q$ is correct for every q, which means that the quotient of the division of 0 by 0 could be any number. This proves to be too inconvenient when properties of integers (as well as rational or real numbers) are discussed, and therefore the division by zero is not defined at all.

We defined an even integer to be an integer divisible by 2; i.e., n is even if $n = 2k$ for some integer k. In a similar manner, we define an integer n to be **odd** if $n = 2j + 1$ for some integer j. It may seem obvious that no integer can be both even and odd; in a Section 4.5 exercise, you are asked to prove this.

Though not as elementary as "even" and "odd", the concept of a prime number is also familiar, but one which we wish to define formally here. A positive integer $p \neq 1$ is called a **prime number**, or **prime**, if it is divisible only by ± 1 and $\pm p$. The first nine primes are 2, 3, 5, 7, 11, 13, 17, 19, and 23. The number 2 is the only even prime; more generally, a prime number p is the only prime divisible by p. A positive integer with more than two positive divisors is called a **composite number**, or **composite**. The integer 1 is therefore the only positive integer that is neither prime nor composite.

The first result of the following theorem has already been proven. Proofs of the remaining parts of the theorem are asked for in upcoming problem sets.

Theorem 2.4. (1) *The sum of any two even integers is an even integer.*

(2) *The sum of any two odd integers is an even integer.*

(3) *The sum of an even integer and an odd integer is an odd integer.*

(4) *The product of two odd integers is an odd integer.*

(5) *The product of an even integer and any other integer is an even integer.*

Part (1) of Theorem 2.4 is the specific case with $c = 2$ in part (3) of Theorem 2.5, which lists several important properties related to the division of integers. Though most of these properties may look familiar or seem obvious—and we are no doubt being redundant in saying this—mathematical rigor includes being able to prove such statements. Short proofs like these are important for learning to apply definitions, and they help build mathematical understanding. The reader should study them thoroughly.

Theorem 2.5. *For all integers $a, b, c, x,$ and y,*

(1) *if $b|a$, then $b|ca$;*

(2) *if $c|b$ and $b|a$, then $c|a$;*

(3) *if $c|a$ and $c|b$, then $c|(a + b)$ and $c|(a - b)$;*

(4) *if $c|a$ and $c|b$, then $c|(xa + yb)$;*

(5) *if $a \neq 0$, then a and $-a$ are divisors of a;*

(6) *1 and -1 are divisors of a;*

(7) *if $b|a$ and $a \neq 0$, then $|b| \leq |a|$; and*

(8) *a nonzero integer has a finite number of divisors, whereas zero is divisible by any nonzero number.*

Remark. At this point in the book, we have not provided the reader with the tools for proving statements such as those in Theorem 2.5. Therefore, we encourage reading through the proofs below to gain as much understanding and insight as possible without undue concern for the formal structure. Proof techniques will be addressed in a formal manner in Section 4.5.   ◇

Proof. (1) We need to show that $b|ca$, i.e., that there exists an integer q such that $ca = bq$. If $b|a$, then $a = bq_1$ for some integer q_1. Then $ca = c(bq_1) = b(cq_1)$. Since cq_1 is an integer by closure, setting $q = cq_1$, we obtain that $b|ca$.

(2) We have to show that $c|a$, i.e., that there exists an integer q such that $a = cq$. If $c|b$ and $b|a$, then $b = cq_1$ and $a = bq_2$ for some integers q_1 and q_2. Then $a = bq_2 = (cq_1)q_2 = c(q_1q_2)$. Setting q equal to the integer q_1q_2, we find $a = qc$, which completes the proof.

(3) We prove the statement for $a + b$, as the proof for $a - b$ is nearly identical. We have to show that $c|(a + b)$, i.e., that there exists an integer q such that $a + b = cq$. Since $c|a$ and $c|b$, there are integers q_1 and q_2 such that $a = cq_1$ and $b = cq_2$. Then $a + b = cq_1 + cq_2 = c(q_1 + q_2)$. Since $q_1 + q_2$ is an integer, setting $q = q_1 + q_2$, we obtain $a + b = cq$.

(4) Before we start our proof, we point out that this statement is a generalization of the previous one: taking $x = y = 1$, we obtain $c|(a + b)$, and taking $x = 1, y = -1$, we get $c|(a - b)$. We present two proofs of (4), one based on (3) and (1) and another that is independent of this theorem.

Proof 1. Since $c|a$ and $c|b$, from (1) we see that $c|xa$ and $c|yb$. It then follows from (3) that $c|(xa + yb)$.   □

Proof 2. We have to show that $c|(xa + yb)$, i.e., that there exists an integer q such that $xa + yb = qc$. Since $c|a$ and $c|b$, there are integers q_1, q_2 such that $a = cq_1$ and $b = cq_2$. Then $xa + yb = x(cq_1) + y(cq_2) = c(xq_1 + yq_2)$. Since $xq_1 + yq_2$ is an integer, setting $q = xq_1 + yq_2$ shows that $xa + yb = cq$.   □

Further Thoughts. Since (4) implies (3) and since the second proof of (4) is independent of (3), one might ask why we bothered to prove (3) at all. The answer is twofold. First, development of a mathematical theory most often follows an "inductive" path, i.e., a generalization from particular cases to a general conclusion. On the other hand, having (3) proven enabled us to construct the first proof of (4).   ◇

(5) Since $a = a \cdot 1$ and $-a = a(-1)$, the statement follows. (Both 1 and -1 are integers.)

(6) Since for every integer a, $a = 1 \cdot a = (-1) \cdot (-a)$, the statement follows.

(7) By definition, $b|a$ implies that $a = bq$ for some integer q, and therefore $|a| = |b||q|$. Since $a \neq 0$, $q \neq 0$, and therefore $|q| \geq 1$. Hence $|a| = |b||q| \geq |b|$, giving $|b| \leq |a|$.

(8) If $b|a$ and $a \neq 0$, then (7) gives $|b| \leq |a|$. Thus b is an integer in the set $\{-a, -a+1, \dots, -1, 1, \dots, a-1, a\}$. Therefore a nonzero integer a has at most $2|a|$ divisors, and this proves the first statement. The second statement is obvious, since $0 = b \cdot 0$ for any b.   □

Regardless of how basic the statements of Theorem 2.5 appear, in the right hands they become powerful tools, which can be used to establish many interesting and not-so-obvious facts about integers. This is not always easy and several attempts are often needed to find (and write) a valid proof. Below we give several examples of simple applications. We assume that the integers are represented in base ten, so the term "digit" refers to an integer from 0 to 9.

Example 1. Take a 2-digit integer, switch the digits, and subtract the obtained number from the original one. Prove that the difference will always be divisible by 9.

First Thoughts. If your initial reaction is one of disbelief, you should experiment. Two examples include $83 - 38 = 45$ and $75 - 57 = 18$; both 45 and 18 are divisible by 9. The statement seems to be true, though the examples may not give an indication of why. We should consider a general way to represent an integer in terms of its digits. In the base-10 system, the number 83, with digits 8 and 3, can be represented as $83 = 8(10) + 3$. We can generalize this representation to *any* 2-digit number.   ◇

Solution. Let N be an arbitrary 2-digit number. Then $N = 10a + b$ for some digits a and b. After the digits are reversed, we obtain a number $M = 10b + a$. Then $N - M = (10a + b) - (10b + a) = 9a - 9b = 9(a - b)$. Since $a - b$ is an integer, $9|(N - M)$, so the proof is complete. $\square$

The following problem is similar to Problem 5 in Chapter 1.

Example 2. Is it possible to pay an exact total of $100,674 when buying only $12 items and $32 items?

First Thoughts. In general, if you have no idea how to begin to answer a question, it is a good idea to try either a simpler or slightly different problem, and such an approach could be helpful here. For example, would it be possible to pay an exact total of $100,673 when buying only $12 items and $32 items? Hopefully, you will recognize that if one buys only items whose individual costs are even, an odd total like $100,673 cannot be obtained. What insight can that give to the original problem, though, since 100,674 is even? A slight rephrasing will help: if one buys only items whose costs are divisible by $k = 2$, then a total that is not divisible by $k = 2$ cannot be obtained. But the same statement is true if k is replaced by any integer! Now, can you find an appropriate value of k to help with the given problem?   ◇

Solution. The answer is "No." To show this we assume the contrary and let integers x and y represent the number of $12 items and $32, respectively. Then the total price is $12x + 32y = \$100,674$. Since $4|12$ and $4|32$, then $4|(12x + 32y) = \$100,674$ (according to Theorem 2.5 (4)). But 4 does not divide 100,674 (check it!). This contradiction proves that a total of $100,674 cannot be obtained. $\square$

The following theorem, known by several names, is used by middle school students and number theorists alike. It simply says that integers can be divided to give a unique quotient and remainder.

Theorem 2.6 (Division Theorem or Division with Remainder Theorem or Division Algorithm). *For any two integers a and b, $b \neq 0$, there exists a unique pair of integers q and r, $0 \leq r < |b|$, such that $a = qb + r$.*

Here, we give some examples; a proof of the theorem will be given in Section 6.5, where the method of mathematical induction is used.

(1) If $a = 20$ and $b = 6$, then $q = 3$ and $r = 2$, since $20 = 3 \cdot 6 + 2$ and $0 \leq 2 < 6$.

(2) If $a = -20$ and $b = 6$, then $q = -4$ and $r = 4$, since $-20 = (-4) \cdot 6 + 4$ and $0 \leq 4 < 6$. (Notice that $q = -3$ and $r = -2$ do not meet the conditions of the theorem, since $r < 0$.)

(3) If $a = 20$ and $b = -6$, then $q = -3$ and $r = 2$, since $20 = (-3)(-6) + 2$ and $0 \leq 2 < 6$.

(4) If $a = 0$ and $b = 7$, then $q = r = 0$, since $0 = 0 \cdot 7 + 0$ and $0 \leq 0 < 7$.

If $a = qb + r$ and $0 \leq r < |b|$, then we will continue calling q the **quotient** upon dividing a by b and will refer to r as the **remainder** upon dividing a by b. Another way of denoting the relationships between a and b is with modular arithmetic and congruences. These will be explored more fully in Chapter 7, but for now we simply write $a \operatorname{div} b = q$ and $a \operatorname{mod} b = r$ to denote that q is the quotient and r is the remainder in the Division Theorem when a is divided by b. This notation is commonly used when writing pseudocode for programs or algorithms. In the above examples, we have

(1) $20 \operatorname{div} 6 = 3$ and $20 \operatorname{mod} 6 = 2$;

(2) $-20 \operatorname{div} 6 = -4$ and $-20 \operatorname{mod} 6 = 4$;

(3) $20 \operatorname{div} (-6) = -3$ and $20 \operatorname{mod} (-6) = 2$; and

(4) $0 \operatorname{div} 7 = 0$ and $0 \operatorname{mod} 7 = 0$.

The quotient and remainder upon the division of a by b can be obtained from the floor function. If $b > 0$, then $q = \left\lfloor \frac{a}{b} \right\rfloor$, and solving for r yields $r = a - b \cdot \left\lfloor \frac{a}{b} \right\rfloor$. If $b < 0$, perform the division with $|b|$ and negate the resulting quotient; in symbols, $q = -\left\lfloor -\frac{a}{b} \right\rfloor$ for $b < 0$. In particular, notice that $(-a) \operatorname{div} b$ and $a \operatorname{div} (-b)$ need not have the same value.

Example 3. If $a = 5k + 2$, then when a is divided by 5, k is the quotient and 2 is the remainder. However, if $a = 5k + (-2)$, then we should not conclude that the quotient is k and the remainder is -2, because the remainder must satisfy $0 \leq r < 5$. Rather, we must rewrite to obtain a nonnegative remainder. We find that $a = 5(k - 1) + 3$, so the quotient is $k - 1$ and the remainder is 3.

Example 4. Every integer n can be written in one and only one of the four forms, $4k$, $4k + 1$, $4k + 2$, or $4k + 3$, where k is an integer. This fact follows from the Division Theorem—just divide the dividend n by $b = 4$ and note that only four remainders are possible.

The next two examples demonstrate less obvious applications of Theorem 2.6. In each case, we use the Division Theorem to rewrite the general variable n in terms of the given divisor and appropriate remainder.

Example 5. Suppose that $n \operatorname{mod} 8 = 5$. What is the remainder of the division of $n^3 + 5n$ by 8?

Solution. By the Division Theorem, $n = 8k + 5$, for some integer k. Then

$$n^3 + 5n = (8k + 5)^3 + 5(8k + 5)$$
$$= 8^3 k^3 + 3(8^2 k^2)5 + 3(8k)5^2 + 5^3 + 5(8k) + 5^2$$
$$= 8(8^2 k^3 + 3(8k^2)5 + 3k5^2 + 5k) + 150$$
$$= 8(8^2 k^3 + 3(8k^2)5 + 3k5^2 + 5k + 18) + 6.$$

Thus (by the Division Theorem again), $n^3 + 5n = 8q + 6$, where $q = 8^2 k^3 + 3(8k^2)5 + 3k5^2 + 5k + 18$. Therefore, the remainder when $n^3 + 5n$ is divided by 8 is 6; i.e., $(n^3 + 5n) \operatorname{mod} 8 = 6$. $\square$

Example 6. Prove that $M = n(n + 1)(2n + 1)$ is divisible by 6 for all integers n.

Proof. For any n, by the Division Theorem, $n = 6k + r$, where k is an integer and r is an element of the set $\{0, 1, 2, 3, 4, 5\}$. Let us evaluate M for each possible value of r.

(1) If $r = 0$, then $M = 6k(6k + 1)(12k + 1)$.

(2) If $r = 1$, then

$$M = (6k + 1)(6k + 2)(12k + 3) = 6(6k + 1)(3k + 1)(4k + 1).$$

(3) If $r = 2$, then

$$M = (6k + 2)(6k + 3)(12k + 5) = 6(3k + 1)(2k + 1)(12k + 5).$$

(4) If $r = 3$, then

$$M = (6k + 3)(6k + 4)(12k + 7) = 6(2k + 1)(3k + 2)(12k + 7).$$

(5) If $r = 4$, then

$$M = (6k + 4)(6k + 5)(12k + 9) = 6(3k + 2)(6k + 5)(4k + 3).$$

(6) If $r = 5$, then

$$M = (6k + 5)(6k + 6)(12k + 11) = 6(6k + 5)(k + 1)(12k + 11).$$

As we see, in each of the cases $6|M$, and the proof is complete. $\qquad\square$

The previous example shows that by concentrating on the remainders one can reduce a problem of establishing a property of *infinitely* many integers to a problem of verifying the property for a *finite* number of cases. The importance of this idea is hard to overstate, and we will return to it when we study congruence relations in Chapter 7.

For our final example, we return to the problem given at the beginning of the chapter.

Example 7. Think of an integer between 41 and 59, inclusive. Subtract 25 from your number and write down the resulting 2-digit number. Now subtract 50 from your original number and square the result. Append this value to the one you wrote down earlier, forming a 4-digit number. Verify that this 4-digit number is the square of the original number, and prove that it happens in the general case as well.

First Thoughts. In problems like this, where we wish to find a general solution, it is imperative that we find a convenient way to represent the general number. In this case, if the first and second digits of our number are a and b, respectively, the form (ab) is not very helpful, because it cannot be manipulated algebraically. Before reading on, think for a moment how a (the tens digit) and b (the units digit) can be formed into an algebraic expression equaling the given 2-digit number. $\qquad\diamond$

Solution. If a and b are the tens and ones digits, respectively, then from the Division Theorem, our 2-digit number can be expressed as $10a + b$. Then, according to the given procedure, $10a + b - 25$ will be the first two digits of the square of the original number, while $(50 - (10a + b))^2$ will form the last two digits. Thus, if the method works, the square of our number will have $10a + b - 25$ hundreds and $(50 - (10a + b))^2$ ones. Algebraically, we have thus reduced the problem to showing that $(10a + b)^2$ and $100(10a + b - 25) + (50 - (10a + b))^2$ are equal. We leave it to the reader to use the algebraic rules of Property 2.1 to show this is the case. $\qquad\square$

Exercises and Problems.

(1) Compute the following floor function values.

   (a) $\lfloor \pi/2 \rfloor$

   (b) $\lfloor -\pi/2 \rfloor$

   (c) $\left\lfloor \frac{1+\sqrt{5}}{2} \right\rfloor$

   (d) $\left\lfloor -\sqrt{10} \right\rfloor$

   (e) $\lfloor 0 \rfloor$

(2) Suppose $n \in \mathbb{N}$. Prove that $\left\lfloor \frac{n}{2} \right\rfloor = \frac{n}{2}$ when n is even and $\left\lfloor \frac{n}{2} \right\rfloor = \frac{n-1}{2}$ when n is odd.

(3) The **ceiling function** of a real number x returns the smallest integer greater than or equal to x; it is denoted $\lceil x \rceil$. For instance, $\lceil 9.25 \rceil = 10$ and $\lceil -4.6 \rceil = -4$. Compute the following ceiling function values.

   (a) $\lceil \pi/2 \rceil$

   (b) $\lceil -\pi/2 \rceil$

   (c) $\left\lceil \frac{1+\sqrt{5}}{2} \right\rceil$

   (d) $\left\lceil -\sqrt{10} \right\rceil$

   (e) $\lceil 0 \rceil$

(4) Use Exercise 2 as a guide for determining $\left\lceil \frac{n}{2} \right\rceil$ when n is even and $\left\lceil \frac{n}{2} \right\rceil$ when n is odd.

(5) For each given value of a and b, find the appropriate values of q and r according to the Division Theorem.

   (a) $a = 30$ and $b = 7$.

   (b) $a = -30$ and $b = 7$

   (c) $a = -30$ and $b = -7$.

   (d) $a = 120$ and $b = 8$.

   (e) $a = -120$ and $b = 8$.

(6) Which of the following are true statements?

   (a) $3|52$.

   (b) $-4|52$.

   (c) $0|36$.

   (d) $4|0$.

   (e) $(a - b)|(a^n - b^n)$, where $a \neq b$ and n is a positive integer.

(7) Find a div b and a mod b for each problem below.

   (a) $a = 91, b = 7$.

   (b) $a = 344, b = 6$.

   (c) $a = -253, b = 11$.

   (d) $a = 162, b = -21$.

   (e) $a = 0, b = 10$.

(8) How can one denote that an integer k is even using mod notation? Similarly, how can one denote that an integer j is odd?

(9) (a) Is 0 even, odd, or neither? Explain your answer.

   (b) The hypotheses of the Division Theorem require that the divisor b is not 0. Which conclusion(s) of the Division Theorem fail to be true in the event that $b = 0$?

(10) Prove that the sum of two odd integers must be an even integer. Note the proof of part (1) of Theorem 2.4.

(11) Show that if $a|b$ and $b|a$, then $a = b$ or $a = -b$.

---

(12) Is the **converse**[1] of the statement in Problem 11 correct? Namely, if $a = b$ or $a = -b$, then must it follow that $a|b$ and $b|a$? Explain your answer.

(13) Construct the converse statements to Theorem 2.5 (1) and (7). Can you find a counterexample for each?

(14) For each of the following equations, determine the subset of real numbers for which it is true.

   (a) $\lfloor x \rfloor = -\lfloor -x \rfloor$.
   (b) $\lfloor x + 1 \rfloor = \lfloor x \rfloor + 1$.
   (c) $\lfloor x \rfloor = \lceil x \rceil$.
   (d) $\lfloor x \rfloor + 1 = \lceil x \rceil$.

(15) Prove that the only positive, common divisor of two consecutive integers is 1.

(16) Prove that the sum of any four consecutive integers is an even number.

(17) In each case below, determine whether or not there are integers x and y satisfying the given equation.

   (a) $16x + 10y = -22$.
   (b) $24x - 54y = 28{,}010$.

(18) Devise an efficient (as few strokes as possible) method for which you can use a calculator to determine the quotient and remainder when a is divided by b. For example, suppose $a = 23{,}920{,}534{,}206$ and $b = 172$. Ensure that your method works correctly when a or b is negative.

(19) Prove that the product of

   (a) three consecutive integers is always divisible by 3;
   (b) five consecutive integers is always divisible by 5.
   (c) Generalize statements (a) and (b). Can you prove your generalization?

(20) Suppose that $n \bmod 9 = 5$. What is $n(n^2 + 7n - 2) \bmod 9$?

(21) Prove that the difference of the squares of two consecutive odd integers is always divisible by 8.

(22) Show that a square of an integer cannot give the remainder 2 when divided by 3; i.e., $n^2 \neq 3k + 2$ for any integers n, k.

---

[1] See Section 4.1 for more discussion about the converse.

(23) Modify the technique given in Example 7 to work for integers other than those in the 41 to 59 range.

---

(24) Explain why the following fast method of squaring integers ending with digit 5 works. Let $N = (a5)$ where a is the number formed by all the digits of N but 5. Then N^2 can be obtained by multiplying a by $a + 1$ and attaching 25 at the end of the product.

For example: $35^2 = 1{,}225$ can be computed by multiplying $a = 3$ by $a + 1 = 4$ and attaching 25 to 12. Similarly, $235^2 = 55{,}225$ can be found by computing the product $23 \cdot 24 = 552$ and attaching 25.

(25) Prove that in a right triangle with integer side lengths, the length of at least one leg must be divisible by 3.

(26) Under what conditions will the sum of n consecutive integers be divisible by n? (For example, the number $16 + 17 + 18 + 19 + 20 + 21 + 22 = 133$ is divisible by 7.) Can you prove your answer?

(27) Prove that in a right triangle with integer side lengths, the length of at least one side must be divisible by 5.

(28) Prove that at least one of the last two digits of a square of an integer is even.

2.3 Matrices

Since their conception in the 1840s by Arthur Cayley and James Sylvester, matrices have grown to become one of the most important tools used in all areas of mathematics and the quantitative sciences. Algebra students use matrices to store and manipulate the coefficients from systems of equations. Matrices serve as the primary means of representing transformations from one geometric space to another. Computer programmers rely on matrices, referring to them as *two-dimensional arrays*. Matrices and matrix operations create crucial algebraic systems for many areas of abstract algebra. In discrete mathematics, one of the primary uses of matrices is in representing relations between elements of sets in a manner that is efficient for both storage and computation. This application of matrices will appear in Chapters 5 and 9. In this section, we define operations on matrices and compare their properties with those of real numbers given in Section 2.1.

A **matrix** is defined to be a rectangular array of numbers, meaning an arrangement of numbers in rectangular form with no empty positions. The plural form of matrix is **matrices**. The size of a matrix is determined by the number of horizontal **rows** and vertical **columns**. A matrix with m rows and n columns is said to have **size** $m \times n$, read "m by n", for positive integers m and n. When $m = n$, it is called a **square matrix**.

Each of the mn positions in an $m \times n$ matrix is assigned an "address" in a manner reminiscent of Cartesian coordinates. If the number appears at the intersection of Row i (counting from the top down) and Column j (counting from left to right), where $1 \leq i \leq m$ and $1 \leq j \leq n$, it is called the (i, j)-**entry** of the matrix. For instance, the 2×4 matrix

$$\begin{bmatrix} 1 & -4 & 0 & -2 \\ 0 & 5 & 7 & -1 \end{bmatrix}$$

has -4 as its $(1, 2)$-entry and 7 as its $(2, 3)$-entry.

Matrices are typically named with uppercase Latin letters, A, B, M, etc., while the entries are denoted with the corresponding lowercase letter. For instance, the (i, j)-entry of matrix A is denoted a_{ij}, giving a generic $m \times n$ matrix the form

$$A = \begin{bmatrix} a_{11} & a_{12} & \cdots & a_{1n} \\ a_{21} & a_{22} & \cdots & a_{2n} \\ \vdots & \vdots & \ddots & \vdots \\ a_{m1} & a_{m2} & \cdots & a_{mn} \end{bmatrix}.$$

We sometimes write $A = [a_{ij}]$ when a direct description of the entries of matrix A is needed. The **diagonal entries** of $A = [a_{ij}]$ are those for which $i = j$, namely a_{11}, $a_{22}, \ldots$, and particularly for square matrices, they are said to form the **main diagonal** of A. A square matrix for which the (i, j)-entry is 0 when $i \neq j$ is called a **diagonal matrix**.

Certain arithmetic operations can be defined on matrices of the same size as a straightforward extension of familiar operations on real numbers. If $A = [a_{ij}]$ and $B = [b_{ij}]$ are $m \times n$ matrices, the **sum** of A and B is the $m \times n$ matrix with (i, j)-entry equaling $a_{ij} + b_{ij}$; in symbols, $A + B = [a_{ij} + b_{ij}]$. Subtraction of matrices of the same size is also accomplished *componentwise*, namely $A - B = [a_{ij} - b_{ij}]$. A **scalar multiple** of a matrix A is obtained by multiplying each entry of A by a real number[2] k; that is, $kA = [ka_{ij}]$.

Example 8. Compute $B - 2C$ for $B = \begin{bmatrix} 1 & 5 \\ -2 & 6 \end{bmatrix}$ and $C = \begin{bmatrix} 3 & 0 \\ 4 & -2 \end{bmatrix}$.

Solution. As B and C are 2×2 matrices, $2C$ is 2×2, and the difference of B and $2C$ is defined as

$$\begin{aligned} B - 2C &= \begin{bmatrix} 1 & 5 \\ -2 & 6 \end{bmatrix} - 2\begin{bmatrix} 3 & 0 \\ 4 & -2 \end{bmatrix} \\ &= \begin{bmatrix} 1 & 5 \\ -2 & 6 \end{bmatrix} - \begin{bmatrix} 6 & 0 \\ 8 & -4 \end{bmatrix} \\ &= \begin{bmatrix} 1-6 & 5-0 \\ -2-8 & 6-(-4) \end{bmatrix} = \begin{bmatrix} -5 & 5 \\ -10 & 10 \end{bmatrix}. \end{aligned}$$
☐

Matrix multiplication is defined in a very different manner from the component-wise simplicity of matrix addition. Its definition arises from Cayley's use of matrix multiplication for the composition of two transformations (functions) in geometric settings. We first define the **inner product** of the ith row of A with the jth column of B as

$$\begin{bmatrix} a_{i1} & a_{i2} & \cdots & a_{ip} \end{bmatrix} \cdot \begin{bmatrix} b_{1j} \\ b_{2j} \\ \vdots \\ b_{pj} \end{bmatrix} = a_{i1}b_{1j} + a_{i2}b_{2j} + \cdots + a_{ip}b_{pj}.$$

Notice that this requires that the number of columns of A equals the number of rows of B. The **product** of matrices A and B is defined to be the matrix AB whose (i, j)-entry is the inner product of the ith row of A with the jth column of B. In contrast to matrix addition, which requires that matrices have the same size, the sizes required for matrix

---

[2]In the language of matrices and vectors, real numbers are often called **scalars**.

multiplication can be characterized as follows: if A is an $m \times p$ matrix and B is a $p \times n$ matrix, then the product AB is an $m \times n$ matrix.[3]

Example 9. Compute the product AB for matrices

$$A = \begin{bmatrix} 4 & -3 \\ 1 & 2 \\ 0 & -1 \end{bmatrix} \quad \text{and} \quad B = \begin{bmatrix} 1 & 5 \\ -2 & 6 \end{bmatrix}.$$

Solution. First note that the number of columns of A equals the number of rows of B (namely, two), hence the product AB is defined, and its size is 3×2. Using the definition of matrix multiplication,

$$AB = \begin{bmatrix} 4 & -3 \\ 1 & 2 \\ 0 & -1 \end{bmatrix} \begin{bmatrix} 1 & 5 \\ -2 & 6 \end{bmatrix}$$

$$= \begin{bmatrix} 4(1) + (-3)(-2) & 4(5) + (-3)(6) \\ 1(1) + 2(-2) & 1(5) + 2(6) \\ 0(1) + (-1)(-2) & 0(5) + (-1)(6) \end{bmatrix}$$

$$= \begin{bmatrix} 10 & 2 \\ -3 & 17 \\ 2 & -6 \end{bmatrix}. \qquad \square$$

Notice that the product BA is not defined for matrices A and B in Example 9, since the number of columns of B (two) does not equal the number of rows of A (three). This is our first indication that matrix multiplication is not commutative: BA need not equal AB. Of course, multiplication of square matrices of the same size is always defined, but this does not resolve the issue of commutativity. Let

$$B = \begin{bmatrix} 1 & 5 \\ -2 & 6 \end{bmatrix} \quad \text{and} \quad C = \begin{bmatrix} 3 & 0 \\ 4 & -2 \end{bmatrix}.$$

We invite the reader to gain practice multiplying matrices to confirm that

$$BC = \begin{bmatrix} 23 & -10 \\ 18 & -12 \end{bmatrix} \quad \text{and} \quad CB = \begin{bmatrix} 3 & 15 \\ 0 & 8 \end{bmatrix}.$$

The $n \times n$ matrix with 1's on the main diagonal and 0's elsewhere is called the $n \times n$ **identity matrix**, denoted I_n. Given the advantageous arrangement of 0's and 1's, appropriately sized identity matrices fulfill the role of a multiplicative identity (see Property 2.1). As an example, let A be defined as in Example 9; then

$$AI_2 = \begin{bmatrix} 4 & -3 \\ 1 & 2 \\ 0 & -1 \end{bmatrix} \begin{bmatrix} 1 & 0 \\ 0 & 1 \end{bmatrix} = \begin{bmatrix} 4(1) + -3(0) & 4(0) + -3(1) \\ 1(1) + 2(0) & 1(0) + 2(1) \\ 0(1) + (-1)(0) & 0(0) + (-1)(1) \end{bmatrix} = A$$

and similarly $I_3 A = A$. In general, for $m \times n$ matrix A, $I_m A = A$ and $AI_n = A$.

While matrix division is not an acceptable operation, the topic of multiplicative inverses of square matrices is crucial. Given a matrix A, if there exists a matrix B for which $AB = I_n$ for some n, then A is said to be **invertible** and B is called the **inverse**

---

[3]Geometrically, this indicates that AB maps n-dimensional space to m-dimensional space. It is the composition of B, mapping n-dimensional space to p-dimensional space, with A which maps p-dimensional space to m-dimensional space.

of A. It can be proven that A and B must be square matrices of size $n \times n$, $BA = I_n$, and the inverse of A is uniquely determined. The inverse of A is denoted A^{-1}. A procedure for computing the inverse of an $n \times n$ matrix may be found in standard texts in college algebra, pre-calculus, and linear algebra and is omitted here. We will describe the simple process for computing the inverse of any invertible 2×2 matrix and use this to demonstrate that many matrices are not invertible.

If

$$A = \begin{bmatrix} a & b \\ c & d \end{bmatrix}$$

with $ad - bc \neq 0$, the inverse of A is

$$A^{-1} = \frac{1}{ad - bc} \begin{bmatrix} d & -b \\ -c & a \end{bmatrix}.$$

The reader should verify that $AA^{-1} = I_2$ and $A^{-1}A = I_2$, with the suggestion that the scalar multiple $\frac{1}{ad-bc}$ be held in reserve until the matrix multiplication is complete. There are infinitely many 2×2 matrices for which $ad - bc$ is 0; this is the collection of all 2×2 matrices which are not invertible. This is in substantial contrast to the real numbers for which there is only one element that has no multiplicative inverse.

Besides the issues of commutativity of multiplication and matrix inverses, the properties of real numbers given in Property 2.1 have corresponding properties in matrix arithmetic. Considering both scalar multiplication and matrix multiplication, the list is more extensive for matrices. For ease of presentation, these properties will be described for the set of $n \times n$ matrices, for any fixed natural number n; we denote this set by M_n. Analogous properties also hold for nonsquare matrices with sizes appropriate for the operations.

Property 2.7. *Suppose A, B, and C are matrices in M_n and r and s are real numbers. Let **0** denote the matrix in M_n with all zero entries.*

(1) *Closure: The sum, difference, and product of two matrices in M_n are also in M_n.*

(2) *Commutative law: $A + B = B + A$.*

(3) *Associative laws: $(A + B) + C = A + (B + C)$, $(AB)C = A(BC)$, $r(sA) = (rs)A$, and $r(AB) = (rA)B = A(rB)$.*

(4) *Distributive laws: $A(B + C) = AB + AC$, $(A + B)C = AC + BC$, $r(A + B) = rA + rB$, and $(r + s)A = rA + sA$.*

(5) *Additive and Multiplicative Identities: $\mathbf{0} + A = A$ and $AI_n = A = I_nA$.*

(6) *Additive Inverse: There is an additive inverse in M_n, denoted $-A$, such that $A + (-A) = \mathbf{0}$.*

(7) *Multiplicative Inverse: If A is an invertible matrix, then $AA^{-1} = I_n = A^{-1}A$.*

First Thoughts. The notation $-A$ in part (6) simply indicates that each entry of A is negated. It is equal to the scalar multiplication $(-1)A$, as a result of using Theorem 2.2(8) of Section 2.1 for each entry. In the same way, each property of matrix arithmetic not only appears to be similar to a law for real numbers found in Property 2.1, but its truth (or "proof") is based on the corresponding law for real numbers, which is applied in each entry of the matrix.   ◇

Proof. Students will be asked to verify many of these claims for the special case of 2×2 matrices in the Exercises and Problems. Part (2) is shown here as a model to emulate.

Consider two arbitrary 2×2 matrices

$$A = \begin{bmatrix} a_{11} & a_{12} \\ a_{21} & a_{22} \end{bmatrix} \quad \text{and} \quad B = \begin{bmatrix} b_{11} & b_{12} \\ b_{21} & b_{22} \end{bmatrix}.$$

Then

$$\begin{aligned} A + B &= \begin{bmatrix} a_{11} & a_{12} \\ a_{21} & a_{22} \end{bmatrix} + \begin{bmatrix} b_{11} & b_{12} \\ b_{21} & b_{22} \end{bmatrix} \\ &= \begin{bmatrix} a_{11} + b_{11} & a_{12} + b_{12} \\ a_{21} + b_{21} & a_{22} + b_{22} \end{bmatrix} \\ &= \begin{bmatrix} b_{11} + a_{11} & b_{12} + a_{12} \\ b_{21} + a_{21} & b_{22} + a_{22} \end{bmatrix} \\ &= \begin{bmatrix} b_{11} & b_{12} \\ b_{21} & b_{22} \end{bmatrix} + \begin{bmatrix} a_{11} & a_{12} \\ a_{21} & a_{22} \end{bmatrix} = B + A. \end{aligned}$$

Take notice of the use of commutativity of addition of real numbers in each entry. $\square$

Two important properties of inverses follow directly from the discussion above. The order of the factors in the expanded form of $(AB)^{-1}$ may seem surprising at first, but the lack of commutativity of matrix multiplication dictates this order.

Property 2.8. *Suppose A and B are invertible matrices in M_n.*

(1) $\left(A^{-1} \right)^{-1} = A$.

(2) $(AB)^{-1} = B^{-1} A^{-1}$.

Proof. The proof of (1) will be provided as a pattern for proving the other results about inverses, which will be requested in the Exercises and Problems. The notation $\left(A^{-1} \right)^{-1}$ refers to the inverse of the matrix A^{-1}, which is the unique matrix X for which $A^{-1} X = I_n$. From Property 2.7(7), $X = A$ satisfies this matrix equation. By uniqueness, $\left(A^{-1} \right)^{-1} = A$. $\square$

For an $m \times n$ matrix $A = a_{ij}$, the **transpose** of A, written A^T, is the $n \times m$ matrix with (i, j)-entry equaling the (j, i)-entry of A; in symbols, $A^T = [a_{ji}]$. Thus, A and A^T have rows and columns interchanged. In the event that $A^T = A$, A is said to be a **symmetric matrix**. A symmetric matrix is necessarily a square matrix with symmetry about the main diagonal. In particular, $A = [a_{ij}]$ is an $n \times n$ symmetric matrix if and only if $a_{ji} = a_{ij}$ for $1 \le i, j \le n$.

Property 2.9. *Suppose A, B, and C are matrices in M_n, with C invertible, and suppose r is a real number.*

(1) $\left(A^T \right)^T = A$.

(2) $(A + B)^T = A^T + B^T$.

(3) $(AB)^T = B^T A^T$.

(4) $(rA)^T = rA^T$.

(5) $(C^T)^{-1} = (C^{-1})^T$.

Parts (1)–(4) of Property 2.9 are true for all nonsquare matrices for which the operations are defined. The sizes of the matrices in part (3) are noteworthy. If A is $m \times p$ and B is $p \times n$, giving AB size $m \times n$, then B^T is $n \times p$ and A^T is $p \times m$, making $B^T A^T$ size $n \times m$ as expected.

Exercises and Problems.

(1) Perform the matrix operations for the matrices defined as follows:

$$A = \begin{bmatrix} 6 & -10 \\ 2 & -3 \\ 4 & 0 \end{bmatrix},$$

$$B = \begin{bmatrix} 4 & 2 & 1 & 6 \\ -10 & 3 & 9 & 1 \\ 4 & -8 & 6 & -2 \end{bmatrix},$$

$$C = \begin{bmatrix} 0 & -8 & 5 \\ -3 & 4 & -6 \end{bmatrix},$$

$$D = \begin{bmatrix} 5 & 10 & 5 \\ -7 & 9 & -10 \\ 0 & 4 & -2 \end{bmatrix},$$

$$E = \begin{bmatrix} -2 & 4 & 8 \\ -7 & 6 & -1 \\ 4 & 5 & 3 \end{bmatrix}.$$

    (a) $D + E$

    (b) $4E - 3D$

    (c) DE

    (d) ED

    (e) DB

    (f) C^T

    (g) $A + C^T$

    (h) $B^T E$

(2) Positive powers of square matrices have the natural meaning. For natural number k, A^k is the product of k factors of matrix A. Refer to the matrices in Exercise 1.

    (a) Compute E^2.

    (b) Explain why A^2 is not defined.

(3) For A, B, C, D, and E, given in Exercise 1, determine which of the following operations are defined and perform those operations.

    (a) $A + D$

    (b) CB

    (c) BA

    (d) $B^T A$

    (e) $A^T C^T$

    (f) C^2

    (g) D^2

---

(4) Verify each property using generic 2×2 matrices in the manner modeled in the proof of Property 2.7(2).

    (a) Closure of matrix addition: if A and B are matrices in M_2, then $A + B$ is in M_2.

    (b) Closure of matrix multiplication: if A and B are in M_2, then AB is in M_2.

    (c) Associativity of matrix addition: $(A + B) + C = A + (B + C)$.

    (d) Left distributivity: $A(B + C) = AB + AC$.

    (e) Additive identity: $\mathbf{0} + A = A$.

    (f) Multiplicative identity: $AI_2 = A$.

    (g) Additive inverse: $A + (-A) = \mathbf{0}$.

    (h) Multiplicative inverse: $AA^{-1} = I_2$.

(5) Verify each property using generic 2×2 matrices in the manner modeled in the proof of Property 2.7(2).

    (a) $\left(A^T\right)^T = A$.

    (b) $(A + B)^T = A^T + B^T$.

    (c) $(AB)^T = B^T A^T$.

    (d) $\left(C^T\right)^{-1} = \left(C^{-1}\right)^T$.

(6) Verify each property by showing that the purported inverse satisfies the necessary equation as demonstrated in the proof of Property 2.8(1).

    (a) $(AB)^{-1} = B^{-1}A^{-1}$.

    (b) $\left(C^T\right)^{-1} = \left(C^{-1}\right)^T$.

(7) Let S denote the set of all diagonal 2×2 matrices. Note that a generic element of S has the form $D = \begin{bmatrix} d_1 & 0 \\ 0 & d_2 \end{bmatrix}$. For each of the following properties, determine whether the property is true for S. For each property that holds in S, demonstrate that it is true using generic matrices from S.

    (a) Closure of matrix addition

    (b) Closure of matrix multiplication

    (c) Additive identity in S

    (d) Multiplicative identity in S

    (e) Commutativity of multiplication

(8) Let $D = \begin{bmatrix} d_1 & 0 \\ 0 & d_2 \end{bmatrix}$ be a diagonal matrix.

    (a) Under what conditions will D be invertible?

    (b) If D is invertible, determine the form of D^{-1}.

    (c) If D is invertible, must D^{-1} be diagonal?

(9) Suppose k is a natural number and $D = \begin{bmatrix} d_1 & 0 \\ 0 & d_2 \end{bmatrix}$ is a diagonal matrix.

    (a) Compute D^2.

    (b) Determine a concise formula for D^k.

(10) Suppose A is a square matrix. Using the definition of symmetric matrix and the properties in this section, explain why $A + A^T$ must be a symmetric matrix.

(11) Prove that a 2×2 matrix A is not invertible if and only if one column is a multiple of the other. Hint: Use the condition that $A = \begin{bmatrix} a & b \\ c & d \end{bmatrix}$ is invertible if and only if $ad - bc \neq 0$.

The local university has four colleges: Professions, Sciences, Arts, and Humanities. Students are enrolled in classes within the four colleges in the following way.

- 800 students are taking classes in the Professions (P).
- 1,120 students are taking classes in the Sciences (S).
- 910 students are taking classes in the Arts (A).
- 1,230 students are taking classes in the Humanities (H).
- 450 students are taking classes in P and S.
- 250 students are taking classes in P and A.
- 300 students are taking classes in P and H.
- 470 students are taking classes in S and A.
- 520 students are taking classes in S and H.
- 610 students are taking classes in A and H.
- 110 students are taking classes in P, S, and A.
- 190 students are taking classes in P, S, and H.
- 120 students are taking classes in P, A, and H.
- 250 students are taking classes in S, A, and H.
- 60 students are taking classes in P, S, A, and H.

How many students are enrolled in the university?

3

Logic and Sets

When dealing with people, remember you are not dealing with creatures of logic....
—Dale Carnegie (1888–1955)

In mathematics, "logic" is used to refer to a particular type of formal reasoning. While logic is relevant to a wide range of pursuits, being able to use logical reasoning is essential for being able to understand and write mathematical proofs.

3.1 Propositions

The most basic building block of logic is the proposition. A **proposition**, or **statement**, is a declarative sentence that is either true or false, but not both. The requirement that a proposition be a declarative sentence means that a question or command is not a proposition. In addition, a sentence whose truth value is ambiguous—perhaps because it expresses personal opinion or contains terms that are not well-defined—is not a proposition. Thus, every proposition can be assigned an associated **truth value**, either *true* (T) or *false* (F).

Example 10. For each of the following sentences, determine whether it qualifies as a proposition. For each that is a proposition, determine its truth value.

(1) Earth is a planet.

(2) Is it raining?

(3) There exists a negative integer x such that $x^2 = 3x$.

(4) Buy tickets to the concert.

(5) $12|6$.

(6) $5 > 2$.

(7) $x + y = 5$.

(8) Today is September 1.

(9) Excellent!

(10) Georg is smarter than Leonard.

Solution. (1) This is a proposition; its truth value is T.

(2) This is a question rather than a declarative sentence and, thus, it is not a proposition.

(3) This is a proposition; its truth value is F.

(4) This is a command rather than a declarative sentence; therefore, it is not a proposition.

(5) This is a false proposition, because 12 is not a divisor of 6.

(6) This is a declarative sentence—written using mathematical symbols rather than words—and therefore, it is a proposition. The truth value is T.

(7) Because no specifications are given for the variables x and y, the truth value cannot be determined. Once x and y are identified, this statement will become either true or false, but as written, this is not a proposition. We will see later that a sentence with variables, for which the truth value changes depending on the input, is called a predicate.

(8) The word "today" may make the reader uncertain as to whether this declarative sentence has a truth value; the word "today" could be viewed as a variable, much like x and y in (7). However, if we make the reasonable assumption that "today" refers to the day on which the statement is made, then it is clear that the sentence is always either true or false, but not both. Therefore, it is a proposition.

(9) This is an exclamation rather than a declarative sentence; therefore, it is not a proposition.

(10) Even under the assumption that Georg and Leonard are uniquely determined, this is not a proposition, because the phrase "smarter than" is ambiguous, making it impossible to assign a truth value to the sentence. The statement

> *Georg has a higher IQ than Leonard*

is similarly problematic because there are multiple Intelligence Quotient tests, none of which are error-free. The statement

> *Georg got a higher score than Leonard on the Stanford-Binet Intelligence Scales test*

can be accepted as a declarative sentence that is either true or false, but not both, and therefore as a proposition. □

Although the definition of a proposition is easy to understand, sometimes it is not easy to decide whether a proposition is true or false. Furthermore, there are even propositions for which the truth value is unknown.

Example 11. Consider the following mathematical propositions.

(1) *Among all triangles of a given perimeter, the equilateral triangle has the greatest area.* This is true but not easy to prove with simple tools.

(2) *Given any collection of sixteen integers, there are two integers whose difference is divisible by 15.*

This is true, and an easy proof exists.

(3) *There exist three positive integers, x, y, and z, such that*

$$x^5 + y^5 = z^5.$$

This is false but difficult to prove.

(4) *There are infinitely many prime numbers that are one more than the square of an integer.*

It is easy to find examples of this statement: $5 = 2^2 + 1$, $17 = 4^2 + 1$, $37 = 6^2 + 1$, etc. However, the truth of the claim that infinitely many such prime numbers exist is unknown. Therefore, although we may be comfortable classifying the statement as a proposition—it is a declarative statement and we think it must be either true or false[1]—the truth value is unknown.

(5) *Every even integer greater than or equal to 4 can be expressed as the sum of two prime numbers.*

Again, it is easy to verify the statement for specific even integers: $4 = 2+2$, $6 = 3+3$, $8 = 3 + 5$, $10 = 5 + 5 = 3 + 7$, and so on. If a single even integer could be found for which the statement were *not* true, then we would know that the truth value of the proposition is F—but no such even integer has been found. But, neither has anyone been able to provide a proof that the statement is true for *every* even integer, which is required in order to establish that the truth value of the proposition is T. This proposition is known as *Goldbach's conjecture* (first stated in 1742), and until the truth value is known, it will remain a **conjecture**—that is, a statement that is believed to be true but has not been proven.

Propositions of the type given in Example 10 are **simple propositions** or **simple statements**. Simple propositions are typically represented with a single letter of the alphabet, such as p or q, with a colon immediately following to denote that the proposition is being defined. Using this notation, we could write

p : *Earth is a planet.*

More complex propositions, generally called **compound propositions**, can be constructed from simple propositions using logical operators, which we will discuss in Section 3.3.

3.2 Sets

Set theory is foundational to modern mathematics, and indeed most, if not all, of mathematics can be framed in the language of sets. Interestingly, though, the term **set** itself is a primary notion in mathematics and we do not define it. However, one may view a

---

[1]Mathematicians have learned that some statements are **undecidable** (i.e., impossible to determine whether true or false) inside a given mathematical system! Perhaps you have heard the history of Euclid's "fifth postulate" ... we will not discuss these issues in this text. However, we note that no one knows whether the given statement (*There are infinitely many prime numbers that are one more than the square of an integer*) is undecidable within usual arithmetic. If it is impossible to determine, then some mathematical philosophers would probably not call it a proposition.

set as a collection or group of objects (two other informal notions!). Objects that form a set are called **elements** of the set.

Sets are abstract in the sense that one need not be able to actually see the elements or collect them together in order for the set to exist. Examples of sets include the 50 states of the United States, the 26 letters of the English alphabet, all people with January birthdays, all even integers, and all prime numbers. Though sets theoretically may be formed using elements of different nature—such as the set consisting of the sun, Euclid's brain, Newton's dog, and the number π—in mathematics sets usually contain elements of the same type: a set of numbers, a set of people, a set of points, etc.

We begin by giving some standard notation and definitions. Sets are typically denoted by capital letters and their elements by lowercase letters. If a is an element of a set A, we write $a \in A$ and say "a is an element of A" or "a is in A" or "a belongs to A" or "A contains a". If a is not an element of A, we write $a \notin A$.

If a set contains only a few elements, it is usually represented by a list of its elements inside braces. For example, let $A = \{2, 5, 6, 10, 4\}$. Then $1 \notin A$ while $10 \in A$. The makeup of a set is not changed if one of its elements is listed more than once, and the order in which the elements are listed in the set does not matter. For example, $\{2, 6, 5, 10, 4\}$, $\{2, 5, 6, 6, 6, 10, 4, 4\}$, and $\{2, 5, 6, 2, 2, 10, 4, 5\}$ all represent the same set as $\{2, 4, 5, 6, 10\}$.

A set is **finite** if it contains finitely many elements. Otherwise, a set is called **infinite**. The number of (distinct) elements of a finite set A, also called the **cardinality** of A, is denoted by $|A|$. As examples, if $A = \{2, 5, 6, 10, 4\}$ and $B = \{3, 5, 3\}$, then $|A| = 5$ and $|B| = 2$.

We review here the symbols we will use to denote certain sets of numbers.

- $\mathbb{N}$ is the set of natural numbers (positive integers).
- $[n] = \{1, 2, 3, \ldots, n\}$.
- $\mathbb{N}_0$ is the set of whole numbers (nonnegative integers).
- $\mathbb{Z}$ is the set of integers.
- $\mathbb{Q}$ is the set of rational numbers.
- $\mathbb{R}$ is the set of real numbers.
- $\mathbb{R}^+$ is the set of positive real numbers.
- $\mathbb{C}$ is the set of **complex** numbers, those numbers which can be represented by the symbols $a + bi$, where $a, b \in \mathbb{R}$ and $i^2 = -1$.

Infinite or large finite sets are often represented with **set-builder notation**, in which the elements of the set are identified by a given "rule". In Chapter 4, we will discuss this kind of rule, more properly called a "predicate", in greater detail. For example, consider the set $A = \{n \in \mathbb{N} : n \text{ is even}\}$. The first expression, $n \in \mathbb{N}$, indicates that A contains only positive integers; the symbol : can be read "such that"; the phrase "n is even" specifies the rule for determining which positive integers are in A. Thus, $A = \{2, 4, 6, 8, \ldots\}$.

Set-builder notation can be used to define intervals of real numbers. Here are a few examples:

$$(a, b) = \{ x \in \mathbb{R} : a < x < b \},$$
$$(a, b] = \{ x \in \mathbb{R} : a < x \le b \},$$
$$[a, b] = \{ x \in \mathbb{R} : a \le x \le b \},$$
$$(-\infty, b) = \{ x \in \mathbb{R} : x < b \},$$
$$[a, \infty) = \{ x \in \mathbb{R} : a \le x \}.$$

Although it may seem strange, sets can contain other sets as elements. The set

$$S = \{1, 2, 3, 4, \{2, 5\}, \{1, \{2, 5\}\}\}$$

contains six elements (four integers and two sets), so $|S| = 6$. The set

$$T = \{\{\{1, 2\}, \{1, 3\}, \{2, 3\}\}, \{\{A, B, C\}, \{D, E, F\}, \{G\}\}\}$$

contains just two elements, each of which is a set containing three sets as its elements.

A set with no elements is called the **empty set** (or **null set**) and is denoted by $\varnothing$. Clearly, $|\varnothing| = 0$ and $\varnothing$ is the only set with zero elements. The sets $\{\}, \{ x \in \mathbb{Z} : 0 < x < 1 \}, \{ x \in \mathbb{Q} : x^2 = 2 \}$, and $\{ x \in \mathbb{R} : x^2 + 1 = 0 \}$ are all equal to $\varnothing$.

A set A is a **subset** of a set B, denoted $A \subseteq B$, if every element of A is an element of B. Note that by definition, for any set A, $A \subseteq A$. We write $A \not\subseteq B$ if A is not a subset of B.

We call two sets A and B equal, denoted $A = B$, if they contain exactly the same elements. For a working definition that is useful in proving such equality, we say A and B are **equal** if $A \subseteq B$ and $B \subseteq A$. If sets A and B are not equal, we write $A \ne B$.

We say that A is a **proper subset** of B and write $A \subset B$ if $A \subseteq B$ and $A \ne B$. Clearly, $A \subset B$ implies $A \subseteq B$. We write $A \not\subset B$ if A is not a proper subset of B.

Notice the obvious similarity between the symbols $\subset$ and $\subseteq$ for denoting when one set is a subset of another set and the symbols $<$ and $\le$ for denoting when one number is less than, or less than or equal to, another number. Of course, the equal symbol $(=)$ is used in both situations, with the distinction clear by context.

Example 12. After proper identification of numbers (like $5 = \frac{5}{1}$, $8/3 = 2.666\ldots$, $3 + 0i = 3$, etc.) we observe that

$$\mathbb{N} \subset \mathbb{Z} \subset \mathbb{Q} \subset \mathbb{R} \subset \mathbb{C}.$$

Example 13. Let $S = \{a, b, c, \{b, d\}, \{a, b, c\}\}$. Then $|S| = 5$, since S contains five elements, and each of the following assertions is true. (Think carefully about the meanings of $\in$, $\subset$, and $\subseteq$ in each expression.)

(1) $a \in S$, $b \in S$, $\{a, b, c\} \in S$, and $\{b, d\} \in S$.

(2) $d \notin S$ and $\{a\} \notin S$.

(3) $\varnothing \subset S$, $\{a\} \subseteq S$, $\{a\} \subset S$, $\{a, b\} \subset S$, $\{a, \{b, d\}\} \subset S$, and $S \subseteq S$.

(4) $a \not\subseteq S$, $\{b, d\} \not\subseteq S$, $\{a, \{b, c\}\} \not\subset S$, and $S \not\subset S$.

Exercises and Problems.

(1) Determine which of the following sentences are propositions.

      (a) Rainbows are pretty.

      (b) The sum of 7 and 3 is 15.

      (c) There was some place on Earth's surface where the temperature was at least 100° Fahrenheit on January 1, 2000.

      (d) Who is on first base?

      (e) Who is on first base. (Think of the Abbott and Costello routine.)

      (f) There are infinitely many prime numbers.

      (g) $2|6$.

      (h) You can lead a horse to water, but you can't make him drink.

(2) Determine which of the following sentences are propositions.

      (a) Abe Lincoln is the tallest of all U.S. presidents.

      (b) Discrete mathematics is a difficult subject.

      (c) He who hesitates is lost.

      (d) More males than females have been born since the beginning of time.

      (e) The equation $12x^2 - 5x - 2$ has two integer solutions.

      (f) Rich people tend to be greedy.

      (g) There are finitely many prime numbers.

      (h) This statement is false.

(3) List all elements of each set.

      (a) $\{x : x \text{ is a planet in our solar system}\}$

      (b) $\{n \in \mathbb{N} : 6|n \text{ and } n|72\}$

      (c) $\{x \in \mathbb{R} : 2x^2 - x - 7 = 0\}$

      (d) $\{n \in \mathbb{N} : n \text{ is a perfect square and a perfect cube and } n < 1000\}$.

(4) Do you think that the empty set, $\varnothing$, is a subset of every set? Explain your reasoning.

(5) Let $A = \{u, v, w, x, y, z\}$. Identify the true statements below. Each part contains at least one true statement.

      (a) $v \in A$; $\{v\} \in A$; $v \subseteq A$; $\{v\} \subseteq A$.

      (b) $t \in A$; $\{t, u\} \subset A$; $\{u, v\} \subset A$; $\{u, v, w\} \subseteq \{x, y, z\}$.

      (c) $\varnothing \in A$; $\varnothing \subset A$; $\{u, v, w, x, y, z\} \subset A$; $A \subseteq \{u, v, w, x, y, z\}$.

      (d) $w \in \varnothing$; $w \in \{w\}$; $\varnothing \subseteq \{w\}$; $\varnothing \subseteq \varnothing$.

(6) Let $B = \{\{1, 2\}, \{1, 3\}, \{2, 3, 4\}\}$. Identify the true statements below. Each part contains at least one true statement.

      (a) $\{1, 2\} \in B$; $\{1, 2\} \subset B$; $\{\{1, 2\}\} \subseteq B$; $\{\{1\}, \{2\}\} \subseteq B$.

      (b) $\{1, 2, 3\} \in B$; $\{1, 2, 3\} \subset B$; $\{\{1, 2, 3\}\} \subseteq B$; $\{\{1, 2\}, \{1, 3\}\} \subseteq B$.

      (c) $1 \in B$; $\{1\} \in B$; $\{1\} \subseteq B$; $\{1\} \subseteq \{1, 2\}$.

      (d) $|B| = 3$; $|B| = 4$; $|B| = 6$; $|B| = 7$.

(7) Let $A = \{1, 2, 3, \{1\}, \{1, 3\}\}$. Identify the true statements below and modify the false statements in a reasonable way to make them true.

    (a) $1 \in A$; $1 \subset A$; $\{1\} \in A$; $\{1\} \subset A$.

    (b) $2 \in A$; $2 \subset A$; $\{2\} \in A$; $\{2\} \subset A$.

    (c) $\{1, 2\} \in A$; $\{1, 2\} \subset A$.

    (d) $\{1, 3\} \in A$; $\{1, 3\} \subset A$.

    (e) $\{1, 2, 3\} \in A$; $\{1, 2, 3\} \subset A$.

    (f) $|A| = 5$.

(8) Use set-builder notation to describe the following sets:

    (a) the set of positive integers that are divisible by 10,

    (b) the set of all integers that are multiples of 6,

    (c) the set of squares of nonzero integers,

    (d) the set of all sets with cardinality four.

---

(9) Determine the elements in each of the following sets:

    (a) the set A consisting of the final digits of integers of the form 3^n (for example, since $3^8 = 6{,}561$, $1 \in A$),

    (b) the set B consisting of the final digits of integers of the form n^2,

    (c) the set C consisting of the final digits of integers of the form $n^2 + m^3$,

    (d) the set S consisting of remainders when an integer of the form n^2 is divided by 3,

    (e) the set T consisting of remainders when an integer of the form $n^2 + m^2$ is divided by 3.

(10) Determine which of the following statements are true in the case of three arbitrary sets P, Q, and R. Modify the false statements to make them true by replacing one symbol, if possible.

    (a) If $P \in Q$ and if $Q \subseteq R$, then $P \in R$.

    (b) If $P \in Q$ and if $Q \subseteq R$, then $P \subseteq R$.

    (c) If $P \subseteq Q$ and $Q \in R$, then $P \in R$.

    (d) If $P \subseteq Q$ and $Q \in R$, then $P \subseteq R$.

---

(11) Use set-builder notation to describe the following sets:

    (a) the set of all points (x, y) in the Cartesian plane lying inside the unit circle,

    (b) the set of all integers that are both perfect squares and perfect cubes,

    (c) the set of all integers that can be written as the sum of three perfect cubes,

    (d) the set of all 3-tuples of integers whose values form the side lengths of a right triangle.

(12) Let $A = \{a \in \mathbb{R} : \text{for all } x \in \mathbb{R}, (x^2 + 6x \geq a)\}$. Describe A using interval notation.

(13) Let $B = \{b \in \mathbb{R} : \text{there exists } x \in \mathbb{R}, (2x^2 - bx + 3 = 0)\}$. Describe B using interval notation.

3.3 Logical Operators and Truth Tables

He loves me, he loves me not. He loves me, he loves me not.

—A game of French origin

Simple propositions can be combined using logical operators to create compound propositions. Likewise, set operations can be used to create new sets from existing ones. In this section, we define the negation, conjunction, and disjunction logical operators, which closely correspond to the words "not", "and", and "or", respectively, in the English language, and the implication operator, which corresponds to the usage of "if ..., then" statements in English. In Section 3.4, we will examine comparable ways of combining sets.

3.3.1 Negation, Conjunction, Disjunction. The symbol $\neg$ is used in front of a proposition to denote the **negation** of the proposition. When symbols alone are used, the negation statement $\neg p$ is read "not p". When the statements are written out, the negation is usually written using typical sentence structures. For example, if p : *Earth is a planet*, then the negation is

$\neg p$: *It is not the case that Earth is a planet.*

Alternatively,

$\neg p$: *Earth is not a planet.*

We define the truth value of the negation of a proposition p to be the opposite of the truth value of p. That is, if p has a truth value of T, then $\neg p$ has a truth value of F; if p has a truth value of F, then $\neg p$ has a truth value of T.

If p and q are both propositions, then the symbol $\wedge$ is used to create a compound proposition, $p \wedge q$, read "p and q" and called a **conjunction**. We define the truth value of the proposition $p \wedge q$ to be T when both p and q have a truth value T; otherwise, we define the truth value of $p \wedge q$ to be F.

If p and q are both propositions, then the symbol $\vee$ is used to create a compound proposition, $p \vee q$, read "p or q" and called a **disjunction**. We define the truth value of the proposition $p \vee q$ to be T when at least one of p and q has a truth value T; if both p and q are false, we define the truth value of $p \vee q$ to be F.

The above rules for determining the truth value of a compound proposition based on the truth value of the component simple statements can be succinctly summarized in a **truth table**. See Table 3.1, where the first two columns provide all (four) possible pairs of truth values for the simple statements p and q.[2] The third, fourth, and fifth columns show the corresponding truth values for $\neg p$, $p \wedge q$, and $p \vee q$, respectively. We emphasize that the assignment of truth values for each of $\neg p$, $p \wedge q$, and $p \vee q$ constitutes a definition, or an axiom; they are not actually proven.

---

[2]It is critical that these two columns contain all four possible pairs of truth values for p and q. The order in which the pairs are shown in Table 3.1 is standard, although in theory, the order in which these pairs are given does not matter.

Table 3.1. Negation, Conjunction, Disjunction

p	q	$\neg p$	$p \wedge q$	$p \vee q$
T	T	F	T	T
T	F	F	F	T
F	T	T	F	T
F	F	T	F	F

Example 14. Let p, q, r, and s represent the following simple propositions.

p : *The temperature outside is below 32 degrees Fahrenheit.*

q : *It is sunny.*

r : *It is snowing.*

s : *Kurt is wearing snow boots.*

Write a sentence that corresponds to each compound proposition written symbolically below. Determine the truth value of each of the compound propositions, assuming that each of the original simple propositions has a truth value of T.

(1) $q \vee r$

(2) $q \wedge s$

(3) $p \wedge \neg r$

(4) $q \vee \neg s$

(5) $p \wedge \neg (r \vee s)$

Solution. In many cases, there is more than one way to write a correct sentence that corresponds to the given symbolic expression.

(1) *It is sunny or it is snowing.*

   If both q and r have a truth value of T, then $q \vee r$ also has a truth value of T.

(2) *It is sunny and Kurt is wearing snow boots*

   or even

   It is sunny but Kurt is wearing snow boots.[3]

   The statement $q \wedge s$ has a truth value of T.

(3) *The temperature outside is below 32 degrees but it is not snowing.*

   If the truth value of r is T, then the truth value of $\neg r$ is F. In order for a compound statement joined by $\wedge$ to have a truth value of T, both of the statements being connected must be true. Since $\neg r$ has a truth value of F, $p \wedge \neg r$ also has a truth value of F.

(4) *It is sunny or Kurt is not wearing snow boots.*

   Although the truth value of $\neg s$ is F, since the truth value of q is T, the truth value of $q \vee \neg s$ is T. (A compound statement using the "or" connector to join two or more statements is true if at least one of the component statements is true.)

---

[3]In common English, of course, the word "but" is used to suggest that what follows may not be expected, but it is worth noting that in mathematical statements the words "but" and "and" are equivalent.

Further Thoughts. Note that there is a subtle but important difference between the use of the word "or" in the typical English sentence and the logical use of the word "or". In common English usage, the word "or" often is used to indicate that *exactly one* of the components is true; consider, for instance, a menu that tells diners that "soup or salad is provided with each entree". This is an example of an **exclusive or**. The logical "or" used in mathematical discourse is an **inclusive or**, meaning that it is acceptable for both of the components to be true.   ◇

(5) Translating this expression into an English sentence is less clear than the previous examples. Here is one attempt: *The temperature outside is below 32 degrees, but it is not the case that it is snowing or Kurt is wearing snow boots.*

This is not entirely satisfactory, though, because the wording is awkward and the meaning is a bit unclear. Do any of the following accurately (and correctly) communicate the meaning?

a: The temperature outside is below 32 degrees, but it is not snowing or Kurt is not wearing snow boots.

b: The temperature outside is below 32 degrees, but it is not snowing and Kurt is not wearing snow boots.

c: The temperature outside is below 32 degrees, but it is not snowing or Kurt is wearing snow boots.

Though it may not be obvious at this point, we will see later that only *b* is another correct "translation" of $p \wedge \neg(r \vee s)$.   □

3.3.2 Logical Operators: Implication and Double Implication.

Two other connectives used to join simple statements are the implication and the double implication. If p and q are both propositions, then the symbol $\Rightarrow$ is used to create a compound proposition, $p \Rightarrow q$, called an **implication**, or a **conditional**. In the proposition $p \Rightarrow q$, the statement corresponding to p is called the **hypothesis**, or the **antecedent**, and the statement corresponding to q is called the **conclusion**, or the **consequent**. The implication $p \Rightarrow q$ is read in any of the following ways.

- p implies q.
- If p, then q.
- q if p.
- p is sufficient for q.
- q is necessary for p.

Let us consider a specific example in order to shed light on the way truth values are assigned to an implication.

Example 15. Let p and q be the following simple statements.

$$p: \textit{You ace your Discrete Math final exam.}$$
$$q: \textit{You pass your Discrete Math class.}$$

Suppose in an end-of-semester meeting with your Discrete Math professor, she promises you, "If you ace the final exam, then you will pass your Discrete Math class." This promise can be represented symbolically as $p \Rightarrow q$. What scenario(s) would need to unfold in order for you to be able to charge your professor had lied to you in making this statement? To answer this question (i.e., to find the truth values of the statement $p \Rightarrow q$), consider the following four possibilities.

Case 1: You do ace the final exam and you receive a passing grade in Discrete Math.

In this case, both p and q have a truth value of T, and you conclude $p \Rightarrow q$ also has a truth value of T. Your professor kept her promise; i.e., her statement was a true one.

Case 2: You do ace the final exam but, to your dismay, your final course grade in Discrete Math is an "F".

In this case, p has a truth value of T but q has a truth value of F. You conclude that your professor did not keep her promise, so her statement was false. Thus, $p \Rightarrow q$ has a truth value of F.

Case 3: You do not ace the final exam but you receive a passing grade in discrete math anyway.

In this case, p has a truth value of F while q has a truth value of T. Although you may initially think that this is a violation of what your professor told you, it is not. In fact, your professor did not explicitly address the outcome of a false p; that is, she didn't tell you what kind of grade you'd receive in Discrete Math if you did not ace the final exam. Thus, in essence, the statement $p \Rightarrow q$ is *vacuously true*, so we assign to $p \Rightarrow q$ a truth value of T. While in some cases, a ridiculous consequent will be vacuously true due to a false antecedent, here it is reasonable to assume that even if you had not aced the exam, you may still have received a passing grade in the class.

Case 4: You do not ace the final exam and you do not receive a passing grade in Discrete Math.

In this case, both p and q have a truth value of F. As noted in Case 3, your professor did not specify what your course grade would be if you did not ace the final exam, so we assign a truth value of T to $p \Rightarrow q$. $\square$

The truth values of the implication operator, as discussed in Example 15, and its negation are given in Table 3.2.

Table 3.2. Implication

p	q	$p \Rightarrow q$
T	T	T
T	F	F
F	T	T
F	F	T

The final logical connector to be discussed is the **double implication**, also called a **biconditional**. If p and q are both propositions, then the symbol $\iff$ is used to create a compound proposition, $p \iff q$, which can be read in any of the following ways.

- p if and only if q (often abbreviated as p iff q).
- p is necessary and sufficient for q.
- If p, then q, and conversely.

The symbolic representation $p \iff q$ is shorthand for $(p \Rightarrow q) \wedge (q \Rightarrow p)$. Therefore, the truth value of $p \iff q$ can be determined using the defined truth values for the implication and the conjunction, as shown in Table 3.3.

Table 3.3. Double Implication

p	q	$p \Rightarrow q$	$q \Rightarrow p$	$p \iff q$
T	T	T	T	T
T	F	F	T	F
F	T	T	F	F
F	F	T	T	T

In summary, $p \iff q$ is true in the cases when p and q share the same truth value (both T or both F) and false otherwise. In other words, $p \iff q$ is true if and only if p and q have the same truth value.

Exercises and Problems.

(1) Create a truth table for each statement.

   (a) $p \wedge (\neg p)$
   (b) $p \vee (p \wedge r)$

(2) Write the negation of each of the following sentences.

   (a) The integer n satisfies $n > 3$.
   (b) The integer m satisfies $4 \leq m \leq 10$.
   (c) The quadratic equation $f(x) = 0$ has exactly two solutions.
   (d) Integers m and n satisfy $m \geq 3$ and $n < 10$.
   (e) There was a day last week when the temperature was at least forty degrees.

(3) How do the final columns compare in the truth tables for the statements $p \Rightarrow q$ and $p \wedge (\neg q)$.

(4) Given propositions p and q, create the truth table (p, q, and final column) that corresponds to the "exclusive or" described in Example 14.

(5) Let p, q, and r be the following propositions about a fixed positive integer m:

   p : *m is divisible by 2*,
   q : *m is divisible by 5*,
   r : *m is divisible by 10*.

Write English statements corresponding to each of the following statements. In each case, determine whether the statement is true for all positive integers m.

   (a) $q \Rightarrow r$.
   (b) $\neg q \Rightarrow \neg r$.
   (c) $(p \wedge q) \Rightarrow r$.
   (d) $q \Rightarrow (\neg r)$.
   (e) $r \iff (p \wedge q)$.
   (f) $q \Rightarrow (p \vee r)$.
   (g) $\neg(p \vee q)$.

(6) Let p, q, and r be the following propositions about a fixed positive integer m.

> p : m is a perfect square (square of an integer),
> q : m is a perfect cube,
> r : m is a perfect sixth power.

Write English statements corresponding to each of the following statements. In each case, determine whether the statement is true for all positive integers m.

(a) $p \Rightarrow q$.

(b) $p \Rightarrow r$.

(c) $r \Rightarrow p$.

(d) $\neg(p \wedge q)$.

(e) $(p \wedge q) \Rightarrow r$.

(f) $(\neg q) \Rightarrow (\neg r)$.

(g) $r \iff (p \wedge q)$.

---

(7) Let p and q and r be the following propositions.

> p : $m/3$ is an integer,
> q : $n/2$ is an integer,
> r : $(nm)/6$ is an integer.

Write the following statements using p, q, r and the logical operator symbols $\vee$, $\wedge$, $\neg$, $\Rightarrow$, and $\iff$.

(a) "Neither $m/3$ nor $n/2$ is an integer."

(b) "If $m/3$ is an integer and $n/2$ is an integer, then $(nm)/6$ is an integer."

(c) "If $(nm)/6$ is an integer, then $m/3$ is an integer and $n/2$ is an integer." This is known as the **converse** of the statement in (a). Is the statement true or false? Give your reasons.

(d) "If $(nm)/6$ is not an integer, then it is not true that both $m/3$ and $n/2$ are integers." This is called the **contrapositive** of the statement in (a). Is the statement true or false? Give your reasons.

(8) In a playoff soccer match (in a hypothetical tournament), a team that has scored more goals than its opponent during regulation play is declared the winner. If the score is tied after regulation, an overtime period is played, and a team ahead at the end of the overtime period is declared the winner. If the score is still tied after the overtime period, a penalty shootout ensues until a winner is determined. Each team shoots five penalty kicks, and if one team converts more of its penalty kicks than does the other team, it is the winner. If the teams are tied after the penalty shootout, then the teams begin rounds of penalty kicks with one kick per team. The winning team is the first team to convert a penalty kick in a round in which the other team fails to convert its kick. Which of the following statements is true? (Note: We are assuming ties are not permitted.)

(a) In order to win a soccer match, it is sufficient to score more goals than the other team in regulation.

(b) In order to win a soccer match, it is necessary to score more goals than the other team during the overtime period.

(c) A team will win a soccer match if it converts more penalty kicks during a penalty shootout.

(d) In order to win a soccer match, if there is a penalty shootout, it is necessary and sufficient to convert more penalty kicks during the shootout.

(e) A team will win a soccer match only if there is a penalty-kick round, one kick per team, in which it converts a penalty kick and the other team fails to convert its kick.

(f) A team will win a soccer match if and only if it scores more total times than the other team (including all goals in regulation, overtime, and on penalty kicks).

(9) Find a compound statement A made out of propositions p, q, and r with the property that A is True if p, q, and r are all True, A is True if p, q, and r are all False, and A is False in all other cases.

(10) Using logical operators and the following propositions, write a logical statement that defines when it is permissible to drive in a High Occupancy Vehicle (HOV) lane. In other words, construct a symbolic statement using q, r, s, t, and u that is logically equivalent to p. (The requirement is that on a nonholiday weekday, between 6:00 a.m. and 9:00 a.m. and between 3:00 p.m. and 6:00 p.m., a car must have at least two occupants.)

> p : *You are permitted to drive in an HOV lane.*
> q : *The time is between 6:00 a.m. and 9:00 a.m.*
> r : *The time is between 3:00 p.m. and 6:00 p.m.*
> s : *The car has at least two occupants.*
> t : *It is Saturday or Sunday.*
> u : *It is a holiday.*

3.4 Operations on Sets

In much the same way that simple statements can be used as building blocks to create compound propositions, sets can be combined to create new sets. When performing operations on sets, we will assume that there is a **universal set**, Ω, which contains all of the elements from which the sets are created.

3.4.1 Set Operations. For $A \subseteq \Omega$, the **complement** of A (relative to Ω), denoted $\overline{A}$, is defined to be the set of all elements in Ω that are not in A. Using set-builder notation, we can write

$$\overline{A} = \{x \in \Omega : x \notin A\}.$$

The **union** of sets A and B, denoted $A \cup B$, is defined to be the set of all elements that appear in A or B or both. So

$$A \cup B = \{x : (x \in A) \vee (x \in B)\}.$$

The **intersection** of sets A and B, denoted $A \cap B$, is defined to be the set of all elements that appear in both A and B. So

$$A \cap B = \{x : (x \in A) \wedge (x \in B)\}.$$

Sets A and B are said to be **disjoint** if $A \cap B = \emptyset$.

Example 16. Given the universal set $\Omega = \{x \in \mathbb{N} : 1 \leq x \leq 10\}$, let $A = \{2, 3, 5, 7, 9\}$, $B = \{1, 6\}$, and $C = \{5, 6, 9\}$, verify the following.

(1) $A \cup B = \{1, 2, 3, 5, 6, 7, 9\}$.

(2) $C \cap B = \{6\}$.

(3) Sets A and B are disjoint.

(4) $\overline{C} = \{1, 2, 3, 4, 7, 8, 10\}$.

(5) $(A \cup B) \cap C = \{5, 6, 9\}$.

(6) $A \cup (B \cap C) = \{2, 3, 5, 6, 7, 9\}$.

(7) $(A \cup B) \cap (A \cup C) = \{2, 3, 5, 6, 7, 9\}$.   □

The complement, union, and intersection set operations have clear parallels to the logical operators negation, conjunction, and disjunction. In fact, combining sets in this way can be represented with a **membership table** that is essentially identical to the truth table defining the logical operators. For example, see Table 3.4. The first nonheader row is the only row that contains a value of T in the "$x \in A \cap B$" column, since that is the row depicting the situation in which the statements $x \in A$ and $x \in B$ are both true. Notice that each header is a proposition; otherwise, it would not make sense to use the values of T and F in the table.

Table 3.4. Complement, Union, and Intersection

$x \in A$	$x \in B$	$x \in \overline{A}$	$x \in A \cup B$	$x \in A \cap B$
T	T	F	T	T
T	F	F	T	F
F	T	T	T	F
F	F	T	F	F

The **set difference** $A \setminus B$ of two sets A and B, also denoted $A - B$, is defined to be the set of all elements that appear in A but not in B. So

$$A \setminus B = \{x \in \Omega : (x \in A) \wedge (x \notin B)\}.$$

Note that if A and B are disjoint, then $A \setminus B = A$; in particular, $\overline{A}$ may be expressed as $\Omega \setminus A$.

The **Cartesian product** of nonempty sets A and B, denoted $A \times B$, consists of all ordered pairs from A and B, respectively. Therefore,

$$A \times B = \{(a, b) : (a \in A) \wedge (b \in B)\}.$$

Note the similarity here to the coordinatization of the plane one encounters in a first algebra course, where points correspond to pairs of real numbers, (x, y). By convention, we denote $A \times A$ by A^2, and this is why the real Cartesian plane is often denoted as $\mathbb{R}^2$.

Finally, the **power set** of A, denoted $\mathcal{P}(A)$ or sometimes as 2^A, is the set of all subsets of A. That is, $\mathcal{P}(A) = \{B : B \subseteq A\}$.

Example 17. Given the universal set $\Omega = \{x \in \mathbb{N} : 1 \leq x \leq 10\}$, let $A = \{2, 3, 5, 7, 9\}$, $B = \{1, 6\}$, and $C = \{5, 6, 9\}$. Verify the following.

(1) $A \setminus C = \{2, 3, 7\}$.
(2) $B \times C = \{(1, 5), (1, 6), (1, 9), (6, 5), (6, 6), (6, 9)\}$.
(3) $\mathcal{P}(C) = \{\varnothing, \{5\}, \{6\}, \{9\}, \{5, 6\}, \{5, 9\}, \{6, 9\}, \{5, 6, 9\}\}$.
(4) $(A \setminus C) \cap (B \setminus C) = \varnothing$.   □

The use of the word "product" and the $\times$ sign in the Cartesian product $A \times B$ likely comes from the fact that for finite sets A and B, $|A \times B| = |A| \cdot |B|$. This is illustrated in Example 17(2), where $|\{1, 6\} \times \{5, 6, 9\}| = (2)(3) = 6$. The general case follows from interpreting multiplication as repeated addition: if $|A| = n$ and $|B| = m$, then for each of the n choices of an element a in A, there are m choices of an element b in B; this yields $m + m + \cdots + m = nm$ possible pairs (a, b).

The name and notation for the power set also has a connection to the number of elements in the set: if A is finite, $|\mathcal{P}(A)| = 2^{|A|}$. This is seen above, where the power set of the three-element set C has cardinality $2^3 = 8$. The general case will be proven in Theorem 8.5.

In summation notation, the index i is often a member of a set of consecutive, positive integers. Therefore, the sum

$$\sum_{i=1}^{5} (4i + 1) = 5 + 9 + \cdots + 21 = 65$$

could also be expressed as $\displaystyle\sum_{i \in [5]} (4i + 1)$.

If $B = \{2, 3, 5\}$, then

$$\sum_{i \in B} (4i + 1) = 9 + 13 + 21 = 43.$$

Just as uppercase sigma (Σ) is used for repeated sums, uppercase pi (Π) is used for repeated products. For instance,

$$\prod_{i=2}^{5} (2i) = (4)(6)(8)(10) = 1{,}920 \quad \text{and} \quad \prod_{i \in B} (2i) = (4)(6)(10) = 240.$$

When set operations are repeated with sets $A_1, A_2, \ldots, A_n$, for $n > 2$, a similar pattern of notation is used:

$$A_1 \cup A_2 \cup \cdots \cup A_n = \bigcup_{i=1}^{n} A_i = \{x : x \in A_i \text{ for some } i, 1 \leq i \leq n\},$$

$$A_1 \cap A_2 \cap \cdots \cap A_n = \bigcap_{i=1}^{n} A_i = \{x : x \in A_i \text{ for each } i, 1 \leq i \leq n\},$$

$$A_1 \times A_2 \times \cdots \times A_n = \{(a_1, a_2, \ldots, a_n) : a_i \in A_i \text{ for each } i, 1 \leq i \leq n\}.$$

Here, the index can also belong to an arbitrary set. For instance, if A_i contains all multiples of i in $[100]$, for $i \in \mathbb{N}$, then

$$\bigcap_{i \in B} A_i = \{2, 4, \ldots, 100\} \cap \{3, 6, \ldots, 99\} \cap \{5, 10, \ldots, 100\} = \{30, 60, 90\}.$$

3.4.2 Venn Diagrams. Venn diagrams provide a useful way to depict sets and their operations. In these diagrams, sets are often represented as circles or other geometric shapes, sometimes inside of a larger rectangle that represents the universal set. Venn diagrams should be drawn in a general fashion so that all given sets are shown to overlap, unless it is known that certain ones do not intersect. However, in a Venn diagram, representing two sets as overlapping circles does not imply that the sets necessarily have a nonempty intersection.

In Figure 3.1, the diagram in (1) represents $A \cup B$; the one in (2), $A \cap B$; and the one in (3), $\overline{A}$. Venn diagrams provide a visual representation of a given relationship, and while such a representation can be useful in providing guidance, diagrams are generally not sufficient when a proof is needed.

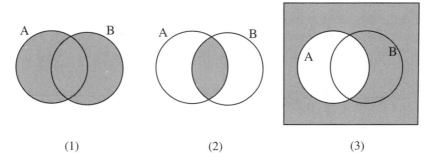

(1)   (2)   (3)

Figure 3.1. $A \cup B$, $A \cap B$, and $\overline{A}$

Example 18. Figure 3.2 depicts three Venn diagrams with various shaded components. For each diagram, provide a proper description of the shaded area, using set operations. (Several answers are possible.)

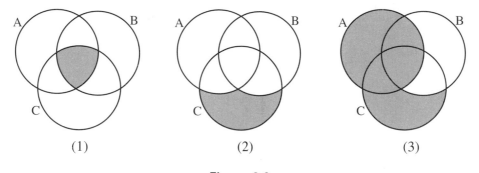

(1)   (2)   (3)

Figure 3.2

Solution. (1) The shaded area corresponds to the common elements of A, B, and C. The intersection of these three sets can be denoted $(A \cap B) \cap C$.

(2) The shaded area corresponds to elements that are in C but that are not in A or B. This set can be denoted as $C \setminus (A \cup B)$.

(3) The shaded area consists of all of A, in addition to elements of C that are not in B. This set is $A \cup (C \setminus B)$.   □

From the Venn Diagram depiction of $A \cup B$, we arrive at a useful counting principle for determining the number of elements in the union of sets. The general case for n sets (Inclusion-Exclusion Principle) is treated in Section 11.1, but we state the $n = 2$ case here.

Proposition 3.1. *For finite sets A and B, $|A \cup B| = |A| + |B| - |A \cap B|$.*

Exercises and Problems.

(1) If $|A| = 12$, $|B| = 7$, and $|A \cap B| = 4$, then how many elements are in $A \cup B$?

(2) If $|A| = 8$, $|B| = 13$, and $|A \cup B| = 16$, then how many elements are in $A \cap B$?

(3) At the local high school, 30 boys play basketball, 33 play soccer, and 34 play baseball. Furthermore, ten play both basketball and soccer, eleven play both soccer and baseball, and thirteen play both basketball and baseball. If six boys play all three sports, how many boys play exactly one of the three given sports? How many boys play at least one of the three given sports? Use a Venn diagram to solve the problem.

(4) Use a Venn diagram to solve the opening hook problem to this chapter, in which you are asked to determine the number of students enrolled in the university.

(5) Let $A = \{1, 4, 5, 6\}$, $B = \{2, 5, 6, 9\}$, and $C = \{4, 5, 8\}$. Find the following sets.

   (a) $A \cap B$

   (b) $A \cup B$

   (c) $A \setminus B$

   (d) $B \setminus (A \cap B)$

   (e) $A \setminus (B \cap C)$

   (f) $(A \cup B \cup C) \setminus (A \cap B \cap C)$

   (g) Why doesn't it make sense here to ask for the elements of $\overline{A}$?

(6) Create Venn diagrams for each of the following sets (for general sets A, B, and C), shading the given region.

   (a) $(A \cap B) \cup C$

   (b) $A \cap (B \cup C)$

   (c) $(A \cup B) \cap C \setminus (A \cup (B \cap C))$

   (d) $\overline{(A \setminus B) \cup C}$

   (e) $(A \cup B) \cap (A \cup C)$

(7) Figure 3.3 contains Venn diagrams with various shaded components. Provide a proper description of the shaded area in each diagram, using set operations.

(8) Consider the sets $A = \{0, 1, 2\}$, $B = \{1, 2\}$, and $C = \{0, 1\}$.

   (a) List all elements in the set $A \times B$.

   (b) List all elements in the set $C \times A$.

   (c) Which of the following are elements of the set $A \times B \times C$:

$$\emptyset, \ \{1\}, \ (0, 2, 1), \ (2, 2), \ \{(1, 1, 1)\}, \ (1, 0, 1)?$$

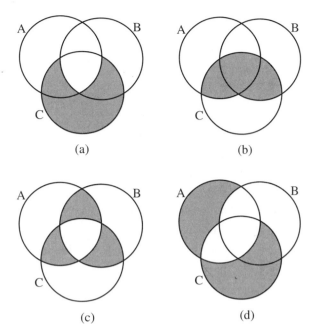

Figure 3.3. Diagrams for Exercise 7

(9) Let D_n denote the set of positive divisors of the integer n, and let $A = \{12, 20, 36\}$. Find each of the following.

(a) $\displaystyle\sum_{n \in A} (2n - 10)$

(b) $\displaystyle\prod_{n \in A} (n - 10)$

(c) $\displaystyle\bigcap_{n \in A} D_n$

(d) $\displaystyle\bigcup_{n \in A} D_n$

(10) Suppose that the set X is a subset of Y. Rewrite each of the following expressions in as simple a form as possible.

(a) $X \cup Y$

(b) $X \cap Y$

(c) $X \setminus Y$

(11) Let A, B, and C denote the following intervals of real numbers: $A = [1, 10]$, $B = (10, 20]$, $C = (5, 20)$. Then $A \setminus C$ can be expressed as $\{x \in \mathbb{R} : 1 \leq x \leq 5\}$ or $[1, 5]$. Write the following sets in similar forms, if possible.

(a) $A \setminus B$

(b) $A \cap B$

(c) $B \setminus C$

(d) $A \cap C$

(e) $A \cup B$

(f) $C \setminus B$

(12) Find the power set of each of the following sets.

   (a) $A = \{\clubsuit, \spadesuit\}$.
   (b) $B = \{a, b, c\}$.
   (c) $C = \{w, x, y, z\}$
   (d) $D = \{x, \{x\}\}$.

(13) Let A and B be arbitrary sets. For each set described in (a) through (d),

   (i) create a Venn diagram,
   (ii) create a membership table, and
   (iii) find a simple logical statement using $p : x \in A$ and $q : x \in B$ whose truth table is the same as the membership table in (ii).

   (a) $A \setminus B$
   (b) $B \setminus A$
   (c) $\overline{A \setminus B}$
   (d) $\overline{B \setminus A}$

---

(14) Let O be a point of a plane α. For each positive real number r, let $C(O, r) = \{P \in \alpha : OP = r\}$. What geometrical figure is defined by $C(O, r)$? Describe $\bigcup_{r \in \mathbb{R}, r > 0} C(O, r)$ and $\bigcup_{K \in \alpha} C(K, 3)$.

(15) Given $|A| = m$, $|B| = n$, and $|C| = p$, determine the lower and upper bounds for each of the following set sizes in terms of m, n, and p.

   (a) $|A \cap B \cap C|$
   (b) $|A \cup B \cup C|$
   (c) $|(A \cup B) \setminus C|$
   (d) $|A \times B \times C|$

(16) Does $A \setminus B = C$ imply $A = B \cup C$? If so, illustrate the property with a Venn diagram; if not, provide a counterexample.

(17) Does $A = B \cup C$ imply $A \setminus B = C$? If so, illustrate the property with a Venn diagram; if not, provide a counterexample.

(18) Let A be a set, with $A \neq \emptyset$. Does $A \times B = A \times C$ imply $B = C$? Justify your answer.

(19) Recall the notation $[n] = \{1, 2, 3, \ldots, n\}$. For fixed positive integer n, find $\bigcup_{i \in [n]} [i]$ and $\bigcap_{i \in [n]} [i]$.

(20) The **symmetric difference** of sets A and B, denoted $A \triangle B$, is defined to be the set of elements which are members of exactly one of A or B. For example

$$\{2, 4, 5, 6, 8\} \triangle \{1, 4, 5, 7, 8\} = \{1, 2, 6, 7\}.$$

   (a) Draw a Venn diagram depicting the symmetric difference of arbitrary sets A and B.
   (b) Find $\{2, 3, 6, 8\} \triangle \{1, 4, 6, 7, 8\}$.
   (c) With $\Omega = \{1, 2, 3, \ldots, 99, 100\}$, let A be the elements of Ω that are divisible by 4 and let B be the elements of Ω that are divisible by 6. Find $|A \triangle B|$.

(d) If $M \triangle P = N \triangle P$, is it necessary that $M = N$? Explain your answer.

(21) Does $(A \setminus B) \cup (B \setminus A) = (A \cup B) \setminus (A \cap B)$ for arbitrary sets A and B. If so, illustrate the property with a Venn diagram; if not, provide a counterexample. (See Problem 20.)

---

(22) Let $A = \{1, 2, 3, \ldots, 30\}$. For each integer $i \geq 2$, define

$$X_i = \{ik \; : \; k \in \mathbb{N} \text{ and } k \geq 2\}.$$

Describe the set $A \setminus \bigcup_{i=2}^{30} X_i$.

(23) (a) What is the union of the sets of points of all squares (with their interiors) that are inscribed in a given circle of radius r?

(b) What is the intersection of the sets of points of all squares (with their interiors) inscribed in a given circle?

(c) What is the union of the sets of points of all squares (with their interiors) circumscribed around a given circle?

(d) What is the intersection of the sets of points of all squares (with their interiors) circumscribed around a given circle?

(24) Solve Problem 23 where all squares with interiors are replaced by squares without their interiors.

3.5 Truth Values of Compound Propositions

What is the value in representing language symbolically? Why do we need these formal rules? One reason is that formal logic moves us away from the imprecision of most spoken and written languages, where determining whether a given statement is true or false or meaningless can be difficult. For example, how would you interpret the following statement?

"The cannibals had the missionary for dinner."

Of course, one can usually rewrite an English statement to remove ambiguities like the one above. What complicates the matter is that in spoken English we use words differently than they are used in formal mathematical settings. For example, suppose a waiter asks a customer whether she wants coffee or tea, and the customer says "Yes." Which of the following statements is accurate?

(1) *She enjoys coffee and tea equally and does not wish to choose.*

(2) *She would like both beverages with her meal.*

(3) *She prefers only one beverage, presumably the last option listed.*

In English, one will usually determine the intended meaning by context, whereas in formal logic, rules must be strictly followed. Using the rules of logic from Section 3.3, which of the following statements are true?

(1) *5 > 1 or Lincoln was the U.S. President in 1800.*

(2) *If cats are birds, then Lincoln became President of the U.S. in 1861.*

(3) *If Einstein was a physicist, then some men are mortal.*

Although one's intuitive reaction may be to dismiss the above statements as nonsensical, formal logic tells us that all are true:

(1) 5 > 1 *or Lincoln was the U.S. President in 1800.*

This can be written as $p \vee q$ where p : 5 > 1 has a truth value of T and q : *Lincoln was the U.S. President in 1800* has a truth value of F. Since $p \vee q$ has a truth value of T if at least one of p and q has a truth value of T, the given statement is true.

(2) *If cats are birds, then Lincoln became President of the U.S. in 1861.*

This can be written as $p \Rightarrow q$ where p : *Cats are birds* has a truth value of F and q : *Lincoln became President of the U.S. in 1861* has a truth value of T. By Table 3.2, $p \Rightarrow q$ has a truth value of T when the truth value of p is F (regardless of the truth value of q).

(3) *If Einstein was a physicist, then some men are mortal.*

Again, this can be written as $p \Rightarrow q$, this time with both p : *Einstein was a physicist* and q : *Some men are mortal* having truth values of T. (The word "some" means "at least one".) By Table 3.2, $p \Rightarrow q$ has a truth value of T when both p and q are true.

Although determining the veracity of such sentences is one use of formal logic, within mathematics, providing a framework for proof-writing is more important. To move toward the goal of using logic to create structure for mathematical proofs, we must first consider how to determine the truth value of more complicated compound propositions than what we have seen thus far.

In algebra, we frequently form complicated expressions out of simple ones. To compute the value of
$$c = ab^2 - (a - 3b)^3$$
for $a = 2$ and $b = 3$, one first computes b^2, ab^2, and $3b$, then $a - 3b$ and $(a - 3b)^3$; finally, one computes c.

A similar approach can be utilized in evaluating complicated symbolic statements. Consider, for example, the following proposition:
$$C : (\neg p \vee q) \Rightarrow (q \wedge p).$$
For a particular pair of truth values for p and q, we determine the truth value of C incrementally, using an order of operations that is similar to the one for numerical computations: first parentheses, then $\neg$, then $\wedge$, then $\vee$, and finally $\Rightarrow$.

To illustrate, we determine the truth value of the components of C, with an input of T for p and F for q, by using the order of operations in step-by-step fashion:
$$C = (\neg T \vee F) \Rightarrow (F \wedge T)$$
$$= (F \vee F) \Rightarrow (F \wedge T)$$
$$= F \Rightarrow F$$
$$= T.$$

Of course, this determines the truth value of C for just one of the four possible pairs of values for p and q. Using the proper order of operations, together with the rules (axioms) provided in Tables 3.1, 3.2, and 3.3 for the basic logical connectors, one can create a truth table as an efficient way to summarize the truth values for any specified proposition. In a truth table, the header row corresponds to simple or compound propositions while subsequent rows give corresponding truth values for the underlying propositions. Table 3.5 contains the truth table for C. Notice that p and C are logically equivalent; this might not be apparent without a truth table.

Table 3.5. Truth Table for C: $(\neg p \vee q) \Rightarrow (q \wedge p)$

p	q	$\neg p$	$\neg p \vee q$	$q \wedge p$	C
T	T	F	T	T	T
T	F	F	F	F	T
F	T	T	T	F	F
F	F	T	T	F	F

By now, the reader may have realized that a proposition can be thought of as a function with domain consisting of ordered sets of truth values that correspond to the truth values of each of the component simple statements. (See Chapter 5 for definitions of function-related terminology.) For example, if we think of $C : (\neg p \vee q) \Rightarrow (q \wedge p)$ as a function, the domain consists of all four possible ordered pairs of truth values for p and q; namely, the domain of C is $\{(T,T),(T,F),(F,T),(F,F)\}$. The range of any proposition will be $\{T,F\}$. Thus, the truth table shown above tells us that $C(T,T) = F$, $C(T,F) = T$, and so on.

Each additional simple proposition from which the compound proposition is constructed will necessitate a doubling of the number of rows in the truth table, in order to accommodate all possible combinations of truth values for the component simple propositions.

Example 19. Give truth tables for the following propositions.

(1) $(p \vee q) \Longleftrightarrow (\neg q \wedge \neg p)$.

(2) $(p \wedge \neg p) \Rightarrow q$.

(3) $\neg(q \Rightarrow p) \Longleftrightarrow \neg p$.

(4) $(p \Rightarrow q) \Rightarrow r$.

Solution. We provide truth tables below for each of the given propositions. The first set of (two or three) columns in each table gives the ordered sets of input values, i.e., the domain values of the proposition. The middle columns show intermediary steps, and the final column gives the output truth value corresponding to a particular input.

(1)

p	q	$p \vee q$	$\neg q$	$\neg p$	$\neg q \wedge \neg p$	$(p \vee q) \Longleftrightarrow (\neg q \wedge \neg p)$
T	T	T	F	F	F	F
T	F	T	T	F	F	F
F	T	T	F	T	F	F
F	F	F	T	T	T	F

(2)

p	q	$\neg p$	$p \wedge \neg p$	$(p \wedge \neg p) \Rightarrow q$
T	T	F	F	T
T	F	F	F	T
F	T	T	F	T
F	F	T	F	T

(3)

p	q	$q \Rightarrow p$	$\neg(q \Rightarrow p)$	$\neg p$	$\neg(q \Rightarrow p) \iff \neg p$
T	T	T	F	F	T
T	F	T	F	F	T
F	T	F	T	T	T
F	F	T	F	T	F

(4)

p	q	r	$p \Rightarrow q$	$(p \Rightarrow q) \Rightarrow r$
T	T	T	T	T
T	T	F	T	F
T	F	T	F	T
T	F	F	F	T
F	T	T	T	T
F	T	F	T	F
F	F	T	T	T
F	F	F	T	F

□

Some compound propositions, such as the one in Example 19(2), have a truth value of T for all combinations of T and F assigned to the simple propositions; that is, the final column in the truth table for the compound proposition consists entirely of T's. A proposition with this property is called a **tautology**. Other examples of tautologies are $p \vee \neg p$, $(p \wedge q) \Rightarrow p$, and $(p \vee q) \iff (q \vee p)$; the reader is invited to construct truth tables to verify that these are tautologies.

Other propositions, such as the one in Example 19(1), always have a truth value of F. Such a proposition is called a **contradiction** or an **absurdity**. Other examples of contradictions are $p \wedge \neg p$ and the negation of any of the tautologies listed in the previous paragraph.

3.6 Set Identities

You may have noticed that there was more than one way to describe the sets in Example 18, which indicates that identities are commonplace in set theory as well. We list some of the more common ones here.

Theorem 3.2. *Let A, B, and C be subsets of a universal set Ω. Then the following laws hold.*

(1) *Commutative Laws*

    (a) $A \cup B = B \cup A$ *and*

    (b) $A \cap B = B \cap A$.

(2) *Associative Laws*

    (a) $A \cap (B \cap C) = (A \cap B) \cap C$ *and*

    (b) $A \cup (B \cup C) = (A \cup B) \cup C$.

(3) *Distributive Laws:*

    (a) $A \cap (B \cup C) = (A \cap B) \cup (A \cap C)$ *and*

    (b) $A \cup (B \cap C) = (A \cup B) \cap (A \cup C)$.

(4) *Complement Laws*

    (a) $\overline{\overline{A}} = A$ *(the complement of $\overline{A}$ is A);*

    (b) $A \setminus B = A \cap \overline{B}$*;*

    (c) $A \cap \overline{A} = \varnothing$*; and*

    (d) $A \cup \overline{A} = \Omega$.

(5) *De Morgan's Laws for Sets*

    (a) $\overline{A \cap B} = \overline{A} \cup \overline{B}$ *and*

    (b) $\overline{A \cup B} = \overline{A} \cap \overline{B}$.

First Thoughts. Some of the statements may seem obvious based on Venn diagrams, but this is a good opportunity to practice proving set equality in a rigorous fashion. Notice that the following proofs are written in complete sentences; we will say more about that in Chapter 4 when we discuss the general format of proofs.

    Recall that sets D and E are equal if and only if $D \subseteq E$ and $E \subseteq D$. To show $D \subseteq E$, one typically picks an arbitrary element of D and shows that it is also an element of E. We will demonstrate this technique with the proofs of parts (1) and (3); the proofs of parts (2) and (5) make use of membership tables.   ◇

(1) The proofs of parts (a) and (b) are similar, so we will only prove the first part, that $A \cap B = B \cap A$. Let x be an arbitrary element of $A \cap B$. Then, by definition, $x \in A$ and $x \in B$. But having $x \in B$ and $x \in A$ is the exact requirement for x to be an element of $B \cap A$, so $x \in B \cap A$. This proves that $A \cap B \subseteq B \cap A$.

    Now let x be an arbitrary element of $B \cap A$. Then, by definition, $x \in B$ and $x \in A$. This implies that $x \in A \cap B$, which proves that $B \cap A \subseteq A \cap B$.

    We have shown that $A \cap B \subseteq B \cap A$ and $B \cap A \subseteq A \cap B$, which proves that $A \cap B = B \cap A$.

(2) We prove part (b) by using a membership table and leave the proof of (a) as an exercise. For the two sets in question to be equal, an arbitrary element x that is in one of them must also be in the other. Since each set is described in terms of sets A, B, and C, we will consider all eight possible scenarios of x lying in A, B, or C. Ultimately, the fact that the columns for $x \in (A \cup (B \cup C))$ and $x \in ((A \cup B) \cup C)$ are identical proves that the two sets are equal.

$x \in \dots$	A	B	C	$B \cup C$	$A \cup (B \cup C)$	$A \cup B$	$(A \cup B) \cup C$
	T	T	T	T	T	T	T
	T	T	F	T	T	T	T
	T	F	T	T	T	T	T
	T	F	F	F	T	T	T
	F	T	T	T	T	T	T
	F	T	F	T	T	T	T
	F	F	T	T	T	F	T
	F	F	F	F	F	F	F

(3) We again prove (b) and leave (a) as an exercise. Consider an arbitrary element x in $A \cup (B \cap C)$. By definition of union, $x \in A$ or $x \in B \cap C$. If $x \in A$, then $x \in A \cup B$ and $x \in A \cup C$ by definition of union. So, $x \in (A \cup B) \cap (A \cup C)$ by definition of intersection. On the other hand, if $x \in B \cap C$, then $x \in B$ and $x \in C$, by definition of intersection. This again means that $x \in A \cup B$ and $x \in A \cup C$, by definition of union. So, $x \in (A \cup B) \cap (A \cup C)$, which proves that $x \in A \cup (B \cap C) \subseteq (A \cup B) \cap (A \cup C)$. Now suppose $x \in (A \cup B) \cap (A \cup C)$. Then it is the case that x is an element of A or B (or both), and it is also the case that x is an element of A or C (or both). We consider two cases.

First, if $x \in A$, then $x \in A \cup (B \cap C)$ by definition of union, as we needed to show. On the other hand, if $x \notin A$, then x is necessarily in B (since $x \in A \cup B$) and x is necessarily in C (similarly), so $x \in B \cap C$. In both cases, $x \in A \cup (B \cap C)$, which proves that $(A \cup B) \cap (A \cup C) \subseteq A \cap (B \cup C)$. This proves that $(A \cup B) \cap (A \cup C) = A \cap (B \cup C)$.

(4) Proof left to the reader.

(5) We prove part (b) only, again using a membership table.

$x \in \dots$	A	B	$A \cup B$	$\overline{A \cup B}$	$\overline{A}$	$\overline{B}$	$\overline{A} \cap \overline{B}$
	T	T	T	F	F	F	F
	T	F	T	F	F	T	F
	F	T	T	F	T	F	F
	F	F	F	T	T	T	T

Since the columns headed by $x \in \overline{A \cup B}$ and $x \in \overline{A} \cap \overline{B}$ are identical, we have proven that the two sets in question are equal. $\square$

Exercises and Problems.

(1) Suppose that the truth values of the statements p and r are T and the truth values of the statements q and s are F. Find the truth values of the following compound statements.

   (a) $\neg(\neg(\neg p \vee q) \vee r) \vee s$.
   (b) $(p \Rightarrow q) \Rightarrow (r \Rightarrow s)$.
   (c) $(r \vee q) \Rightarrow \neg(p \wedge s)$.
   (d) $((r \vee s) \wedge (p \Rightarrow r)) \Rightarrow \neg(s \Longleftrightarrow p)$.

(2) Give truth tables for the following statements. Determine which are tautologies, which are contradictions, and which are neither.

(a) $(p \wedge q) \wedge \neg (p \vee q)$.

(b) $(p \wedge q) \vee ((\neg p) \Rightarrow q)$.

(c) $[(p \Rightarrow q) \wedge (q \Rightarrow r)] \Rightarrow \neg (p \Rightarrow r)$.

(d) $(\neg p) \Rightarrow (p \Rightarrow q)$.

(e) $((\neg p) \wedge q) \wedge (p \vee (\neg q))$.

(3) Let p and q be compound statements. What can be said about the truth value of the statement $(p \Rightarrow q)$ in each of the following cases?

(a) p is a contradiction.

(b) p is a tautology.

(c) q is a tautology.

(4) Determine how many rows a truth table of a compound statement would have if it used 2, 3, or 4 distinct simple statements, respectively. As examples, determine the number of rows necessary for the truth tables for these compound statements: $\neg p \wedge \neg q$, $p \vee q \vee \neg r$, $(p \vee q) \Rightarrow (\neg (r \wedge t))$. (You do not have to construct the complete table to answer this question.) Do you see any pattern in your answers? How many rows are required for the truth table of a compound statement that is constructed out of n simple statements?

(5) Prove that the set of all squares is a subset of the set of all quadrilaterals $ABCD$ for which $AC = BD$ (diagonals have equal length).

(6) Let $A = \{n \in \mathbb{Z} : n = 5q + 4 \text{ for some } q \in \mathbb{Z}\}$ and let $B = \{m \in \mathbb{Z} : m = 10k - 1 \text{ for some } k \in \mathbb{Z}\}$. Prove that $B \subseteq A$.

(7) Let $A = \{n \in \mathbb{Z} : n^2 + n \text{ is odd }\}$ and $B = \{m \in \mathbb{Z} : m^2 - 2m = 3\}$. Prove $A \subseteq B$.

(8) Let $M(n)$ be the set of all $n \times n$ matrices with real entries. Let $S = \{A = [a_{ij}] \in M(4) : a_{ij} = 0 \text{ for } i \neq j\}$. Let $T = \{A \in M(4) : A^T = A\}$. Prove $S \subseteq T$.

---

(9) Prove that $(A \setminus B) \cup (B \setminus A) = (A \cup B) \setminus (A \cap B)$ for any sets A and B.

(10) Prove that $\mathcal{P}(A) \cap \mathcal{P}(B)$ is equal to $\mathcal{P}(A \cap B)$ for any sets A and B. What can we say if the intersection operation is replaced by the union operation?

(11) Prove part (2)(b) of Theorem 3.2 using a membership table.

(12) Prove part (3)(a) of Theorem 3.2.

(13) Prove part (4) of Theorem 3.2.

(14) Prove part (5)(a) of Theorem 3.2 by using the definition of set equality. (Show that each of the given sets is a subset of the other.)

(15) Prove or disprove the following assertions involving arbitrary sets A, B, C, and D.

(a) $(A \setminus B) \setminus C = A \setminus (B \cup C)$.

(b) $(A \setminus B) \setminus C = (A \setminus B) \setminus (B \setminus C)$.

(c) $A \setminus (B \setminus C) = (A \setminus B) \cup (A \setminus \overline{C})$.

(d) $(A \subseteq C) \wedge (B \subseteq D) \Rightarrow A \times B \subseteq C \times D$.

(16) The truth table of a compound proposition (i.e., a logical formula) containing the four propositions p, q, r, and s will have $2^4 = 16$ rows.

    (a) Provide a proposition (use some or all of the standard operations of negation, disjunction, conjunction, and implication) whose truth table results in a T in the final column of the truth table in the first row and F's in the final column of the fifteen other rows.

    (b) Provide a proposition that will result in a T in the final column of the truth table in the sixth row and F's in the final column of the fifteen other rows.

    (c) Provide a proposition that will result in a T in the final column of the truth table in the first and sixth rows and F's in the final column of the fourteen other rows.

    (d) Is there a proposition for which the final column of the truth table contains 3 T's and 13 F's? If so, give an example.

    (e) Is there a proposition for which the final column of the truth table contains 5 T's and 11 F's? If so, give an example.

3.7 Infinite Sets and Paradoxes

3.7.1 Infinite Sets. With a finite set M, we define the **cardinality** of M to be the number of elements in M, denoted $|M|$, and we say sets A and B are **equivalent** if $|A| = |B|$. While $\mathbb{Z}$ (integers), $\mathbb{Q}$ (rationals), and $\mathbb{R}$ (reals), are examples of infinite sets of numbers, there are (infinitely) many other examples. Does it make sense to talk about two infinite sets being equivalent? What would this even mean?

It is possible to show that finite sets A and B are equivalent without knowing the size of either set. For example, in a classroom, think about how one might determine that the set of chairs is equivalent to the set of students in the room. Obviously, if each student is sitting in exactly one chair and each chair holds exactly one student, then we need not count the number of either to determine they are equal in number. More generally, if there is a one-to-one correspondence between the elements of A and the elements of B, then $|A| = |B|$. Formally, a one-to-one correspondence can be established by defining a one-to-one and onto function f from set A to set B. Such a function need not be defined algebraically, depending on the sets involved; in the previous example, the obvious choice for f assigns each student to the chair in which he or she is sitting! (The reader who needs a refresher on one-to-one and onto functions should skip ahead and read about them in Chapter 5, where they are discussed thoroughly.)

With finite sets A and B, it is apparent that if $A \subset B$, then A and B are not equivalent. With infinite sets, this is not so clear. How does one compare one infinite quantity to another? For example, let $\mathbb{E}^+$ and $\mathbb{N}$ denote the sets of positive even integers and positive integers, respectively. Could these sets be equivalent, even though $\mathbb{E}^+$ is a proper subset of $\mathbb{N}$? In the next example, we show that indeed they are.

Example 20. Prove that the sets of positive integers and positive even integers are equivalent.

Solution. Define a function f from $\mathbb{N}$ to $\mathbb{E}^+$ via $f(n) = 2n$. This function is easily shown to be one-to-one and onto, which establishes a one-to-one correspondence between the elements of $\mathbb{E}^+$ and the elements of $\mathbb{N}$. Thus $\mathbb{E}^+$ and $\mathbb{N}$ are equivalent sets. $\square$

While at first glance it may be startling for the set of integers to be equivalent to one of its proper subsets, the concept is quite reasonable. After all, cutting an infinite set "in half" (or doubling it) should still leave one with an infinite set. But what if the "ratio" between the set sizes is not so obvious? For example, what about $\mathbb{Z}$ and $\mathbb{Q}$? Clearly, $\mathbb{Z} \subset \mathbb{Q}$, but there is no obvious two-to-one, three-to-one, or even n-to-one correspondence between the elements of $\mathbb{Q}$ and $\mathbb{Z}$, let alone a one-to-one correspondence! However, consider an infinite table of positive rational numbers that begins as follows. In row n, the sum of the numerator and denominator of each number is $n + 1$:

$$\frac{1}{1}$$
$$\frac{1}{2}, \frac{2}{1}$$
$$\frac{1}{3}, \frac{2}{2}, \frac{3}{1}$$
$$\frac{1}{4}, \frac{2}{3}, \frac{3}{2}, \frac{4}{1}$$
$$\frac{1}{5}, \frac{2}{4}, \frac{3}{3}, \frac{4}{2}, \frac{5}{1}$$

Our goal is to turn the numbers in the above table into an ordering of the set of positive rational numbers. A moment's thought should convince you that each rational number will eventually appear in the table, so all we need to do is proceed downwards, moving left to right and eliminating any number that has already appeared previously in a reduced form. For example, $\frac{2}{2}$ (in the third line) equals $\frac{1}{1}$ (in the first line), so we eliminate it. Here are the first five lines of the revised table, after following this procedure:

$$\frac{1}{1}$$
$$\frac{1}{2}, \frac{2}{1}$$
$$\frac{1}{3}, \frac{3}{1}$$
$$\frac{1}{4}, \frac{2}{3}, \frac{3}{2}, \frac{4}{1}$$
$$\frac{1}{5}, \frac{5}{1}$$

This process can be continued indefinitely, and the corresponding sequence (list) of rational numbers is uniquely determined:

$$1, \frac{1}{2}, 2, \frac{1}{3}, 3, \frac{1}{4}, \frac{2}{3}, \frac{3}{2}, 4, \frac{1}{5}, 5, \ldots .$$

Think about what we have just done. In creating a sequence[4] of ALL positive rational numbers, we have just put them in one-to-one correspondence with $\mathbb{N}$, the positive integers! Note that we did not explicitly define an algebraic function from $\mathbb{N}$ to $\mathbb{Q}^+$, as we did in Example 20, but that is not necessary; our construction demonstrates the

---

[4]Obviously we cannot write down every element of an infinite list, but we did provide an algorithm for listing as many as desired, and it should be clear that any given element of $\mathbb{Q}^+$ will eventually appear in the list.

existence of a one-to-one correspondence, and that is sufficient to show that $\mathbb{Q}^+$ and $\mathbb{N}$ are equivalent.

The hard work is done, and from here it is not difficult to show that $\mathbb{N}$ and $\mathbb{Q}$ are also equivalent—given a list of the positive rational numbers, simply insert each negative rational immediately after its positive counterpart.

Any set that can be put in one-to-one correspondence with a subset of $\mathbb{N}$ is said to be **countable**; more specifically, such a set is either **countably finite** or **countably infinite**, depending on whether it is finite or not. Thus, although $\mathbb{N} \subset \mathbb{Z} \subset \mathbb{Q}$, all three sets are countably infinite. Moreover, any set whose elements can be algorithmically put into an ordered list is countable. The question now begging to be asked is whether $\mathbb{R}$, the set of real numbers, is countable.

We address this question by first showing that a subset of $\mathbb{R}$, the open interval $(0, 1)$, is not equivalent to $\mathbb{N}$. If these two sets were equivalent, then there would be a function f from $\mathbb{N}$ onto $(0, 1)$. This means that every number in $(0, 1)$ would appear in the sequence $f(1), f(2), f(3), \ldots$. We demonstrate that this is not possible by constructing a number α in $(0, 1)$ that is not in that sequence. To wit, define the value of the digit appearing k places to the right of the decimal point in α to be different than the digit k places to the right of the decimal point in $f(k)$. By this definition, $\alpha \neq f(k)$ for any k, so α cannot be in the list!

For example, suppose the enumerated list of numbers in $(0, 1)$ begins as follows:

$$f(1) = 0.\underline{3}4252363474578426\ldots,$$
$$f(2) = 0.3\underline{2}639021640769065\ldots,$$
$$f(3) = 0.62\underline{9}01570256304216\ldots.$$

In constructing α, we simply choose the first digit to not be 3, the second digit to not be 2, and the third digit to not be 9, and so on. One systematic way to do this is as follows: if $f(k)$ has a 3 in the kth digit, then define α to have a 7 there, and if not, then define α to have a 3 there. In the example above, α would start $0.733\ldots$, and we see that $\alpha \neq f(1)$, $\alpha \neq f(2)$, and $\alpha \neq f(3)$. Continuing, α will differ from $f(k)$ for every $k \in \mathbb{N}$. This proves that the natural numbers cannot be put in one-to-one correspondence with the numbers in the open interval $(0, 1)$, so $(0, 1)$ is uncountable.

Example 21. Are the intervals $(0, 1)$ and $(0, 1]$ equivalent?

First Thoughts. The two sets seem quite close in size, because $(0, 1] = (0, 1) \cup \{1\}$. However, since the sets are uncountable, we cannot simply line up the elements of one set and "shift" them by just one number to get the other set. Rather, we will seek a function that maps each number in $(0, 1]$ to itself, except for a countable sequence of them. Consider how this might be done before reading on.   ◇

Solution. Define $f(x)$ on the interval $(0, 1]$ as follows. If $x = 2^{-k}$ for some $k \in \mathbb{N}$, then $f(x) = 2^{-(k+1)}$. So, $f(1) = 1/2$, $f(1/2) = 1/4$, $f(1/4) = 1/8$, etc. For all other values of x, let $f(x) = x$. Clearly, f is a one-to-one and onto function from $(0, 1]$ to $(0, 1)$, so this establishes the equivalence of the two sets.   □

Example 22. Prove that the set $(0, 1)$ is equivalent to $\mathbb{R}^+$.

Solution. The function $f(x) = \frac{1}{x} - 1$ is one of many one-to-one functions on $(0,1)$ whose range is the set of positive real numbers. Thus, there is a one-to-one correspondence between the two sets, and they are equivalent. $\square$

Example 23. Prove that $\mathbb{R}^+$ and $\mathbb{R}$ are equivalent.

Solution. The function $\ln(x)$, defined on $\mathbb{R}^+$, provides a one-to-one correspondence between the two sets, which proves they are equivalent. $\square$

Based on the previous examples and Exercises 6 and 7 at the end of this section, we conclude that $\mathbb{N}$, $\mathbb{Z}$, and $\mathbb{Q}$ are equivalent to each other, as are the sets $\mathbb{R}$, $\mathbb{R}^+$, (a,b), $(a,b]$, $[a,b)$, and $[a,b]$ for any real numbers $a < b$. The cardinality of the countable sets $\mathbb{N}$ (and $\mathbb{Z}$ and $\mathbb{Q}$) is denoted $\aleph_0$ (pronounced "aleph-naught"); the cardinality of $\mathbb{R}$ is denoted $\aleph_1$ or $\mathfrak{c}$. The cardinals $\aleph_0$ and $\aleph_1$ thus denote two different levels of infinity.

A very natural question, formulated by Georg Cantor in 1878, is whether there is another level of infinity between $\aleph_0$ and $\aleph_1$. In other words, is there a set B for which $\mathbb{Z} \subset B \subset \mathbb{R}$ such that B is equivalent to neither $\mathbb{Z}$ nor $\mathbb{R}$? The *continuum hypothesis* says the answer is "no", but the question has been shown to be undecidable (i.e., neither the hypothesis nor its negation causes a contradiction within the usual axioms of set theory). The hypothesis continues to garner significant interest in mathematics.

We have noted that the power set of a finite set A contains $2^{|A|}$ elements. Thus, since $n < 2^n$ for all $n \in N$, a finite set is not equivalent to its power set. The following theorem says this is true for infinite sets as well.

Theorem 3.3. *There is not a one-to-one correspondence between the elements of a set A and the elements of $\mathcal{P}(A)$.*

Proof. Suppose f is a function from a set A to $\mathcal{P}(A)$. Thus, for each $a \in A$, $f(a)$ is a subset of A. Our plan is to construct a set $M \subseteq A$ for which $M \neq f(a)$ for any $a \in A$. To this end, define M as follows:

$$M = \{a \in A : a \notin f(a)\}.^5$$

To prove the claim, suppose to the contrary that $f(a') = M$ for some $a' \in A$. By definition, each element m of M has the property that $m \notin f(m)$. If $a' \in M$, then $a' \notin f(a') = M$, which is a contradiction. If $a' \notin M$, then $a' \in f(a') = M$, which is also a contradiction. We conclude there is not a function from a set onto its power set. $\square$

We can use the previous theorem to build a chain of sets with increasing cardinalities:

$$\mathbb{Z}, \mathcal{P}(\mathbb{Z}), \mathcal{P}(\mathcal{P}(\mathbb{Z})), \mathcal{P}(\mathcal{P}(\mathcal{P}(\mathbb{Z}))), \dots .$$

Since $\mathcal{P}(\mathbb{Z})$ and $\mathbb{R}$ are equivalent (see Problem 9), the infinite cardinals of these sets may be denoted $\aleph_0, \mathfrak{c} = \aleph_1, \aleph_2, \aleph_3, \dots$, respectively.

---

[5] For example, suppose $A = \{1,2,3,4\}$ and $f(1) = \{1,4\}$, $f(2) = \{1,3,4\}$, $f(3) = \{2,3\}$, and $f(4) = \{3\}$. Then $M = \{2,4\}$.

3.7.2 Paradoxes. The nonintuitive nature of many of the previous results may be simultaneously bewildering and intriguing. Indeed, these results led to paradoxes such as those described below, which were deeply unsettling to many mathematicians in the early days of set theory. In the early 1900s, the paradox framed by Bertrand Russell (1872–1970) raised questions about the very notion of a set and ultimately led to greater attentiveness to how sets are defined.

Suppose the barber in a small town is given a directive to shave those and only those men who do not shave themselves. It seems we have constructed two sets in a reasonable fashion, one set consisting of men who shave themselves and a second set consisting of men who do not shave themselves (and who are thus shaved by the barber). The paradox arises when we consider which set the barber, himself, belongs to. If the barber is in the first set, then he shaves himself, so he is not shaved by the barber, which is false. On the other hand, if the barber is in the second set, then he does not shave himself, so he must be shaved by the barber, which means he shaves himself, which is false. The directive cannot be followed by someone who is trying to simultaneously function as both the barber and a man in the town.

Consider a reframing of the barber paradox involving the set of all books. Some books may refer to themselves within the book while other books make no reference to themselves. Let S denote the subset of all books that reference themselves, and let B be a book that lists only those books not in S. Everything seems clear here, and, for example, we would not expect to find *Journey into Discrete Mathematics* among the pages of book B, due to this sentence.

However, consider the question: "Is $B \in S$?" In words, "Does the book B reference itself?" Suddenly we are at an impasse. For if B is in S, then that must mean that B refers to itself somewhere in the book. But that would mean that B is not in S!

Russell's Paradox generalizes the previous two scenarios and goes as follows. Let

$$S = \{A \,:\, A \text{ is a set and } A \notin A\}.$$

In words, S is the set of all sets that do not contain themselves as members. The paradoxical question is, "Is S an element of itself?" One cannot answer this question as either "yes" or "no" without obtaining a contradiction.

Example 24. Most computer programs have a finite run time, but others may get stuck in an infinite loop and will not terminate. The **halting problem** is the determination of whether a given program will terminate or run forever. Suppose that a program F were written that could perform this determination. This program F would receive a program x as input and then print either the statement "Program x will terminate" or the statement "Program x will not terminate."

Now consider another program P that will receive as input a program y and will do one of two things, depending on what happens when program F is given y as input. If F prints "Program y will terminate," then program P will loop forever. If F prints "Program y will not terminate," then P will terminate and print "Thanks for playing." We now arrive at the perhaps not-unexpected question: "What happens when P is given P as input?"

Solution. If P terminates and prints "Thanks for playing" upon receiving itself as input, then we conclude that F must have printed "Program P will not terminate." On the other hand, if P loops forever, then it must be that F had printed the statement "Program P will terminate." Both of these statements are contradictions. It seems such a program F must not exist. $\square$

The question as to whether a program F could be written to solve the halting problem was posed and answered by Alan Turing (1912–1954). It does not deal with sets per se, but the parallels to the previous examples are evident. In both cases we run into difficulties because of self-referential statements.

Two methods have been proposed for avoiding paradoxes that arise from self-referential statements. One, called **type theory**, was devised by Russell and Alfred North Whitehead (1861–1947). In this theory, "types" are created and arranged hierarchically; elements are then classified by type and it is only permissible to create elements of a given type from elements classified at a lower level in the hierarchy. Thus, while it is possible for an element to be a set, the elements in that set cannot lie in the same level of the hierarchy as the set itself.

The second method involves using some version of **axiomatic set theory**, in which precise rules are given as to how sets can be constructed. The most commonly used one is Zermelo-Fraenkel set theory, which combines the initial ideas of Ernst Zermelo (1871–1953), a modification by Abraham Fraenkel (1891–1965), and the Axiom of Choice. This version of set theory does not assume there is always a set of elements meeting any given property. Rather, given any set and a well-defined property, there will be some subset of that set whose elements satisfy the property. Under these rules, the set of all sets cannot be constructed, and Russell's Paradox does not occur.

Exercises and Problems.

(1) Show that the sets of natural numbers and whole numbers are equivalent.

(2) Use Example 20 as a guide to show that the following pairs of sets are equivalent.

  (a) The set of even integers and the set of integers.

  (b) The set of odd integers and the set of integers.

(3) Show that $\mathbb{N}$ and $\mathbb{Z}$ are equivalent by enumerating the elements of $\mathbb{Z}$ in a sequence.

(4) Explain why the barber's paradox and the paradox of the self-referential books are both specific contextualizations of Russell's paradox.

---

(5) Classify each set as either countable or uncountable.

  (a) The set of all right triangles with integer side lengths.

  (b) The set of all points in the Cartesian plane with rational coordinates.

  (c) The set of all unit circles in the plane whose center has rational coordinates.

  (d) The set of all circles in the plane whose center has rational coordinates.

  (e) Any infinite set of nonoverlapping open intervals on the real number line.

  (f) The set of all lines in the plane going through the point $(0, 3)$.

  (g) The set of all lines in the plane containing two points that have rational coordinates.

(6) Show that the open interval (a, b) is equivalent to the open interval (c, d).

---

(7) Show that $(0, 1)$ and $[0, 1]$ are equivalent.

(8) Let A be a finite subset of $\mathbb{N}$. Show that $(0, 1) \cup A$ and $(0, 1)$ are equivalent.

(9) Show that the power set of $\mathbb{N}$ is equivalent to $[0, 1]$.

(10) Let S be the set of all positive integers that are not describable in fewer than fifteen English words. According to the Well-Ordering Principle (discussed in Section 6.2), every nonempty set of positive integers has a smallest element. Since there are a finite number of English words, there are a finite number of ways to construct a phrase using fifteen of them. Therefore S must be nonempty and necessarily has a smallest element, n. Explain the paradox that arises from the phrase, "the smallest positive integer that is not describable in fewer than fifteen English words".

Complete the following Sudoku puzzle. Each row, column, and 3 × 3 subgrid must contain each of the digits from 1 to 9.

2	7	1	5					8
5			9		2		4	
				7	1			
8	1			6				9
		5	7		1	6		
4			3	5			1	7
		4				8	2	
	3		4		5			1
7				2	9	4		3

4

Logic and Proof

"Contrariwise," continued Tweedledee, "if it was so, it might be; and if it were so, it would be; but as it isn't, it ain't. That's logic."

—from *Alice's Adventures in Wonderland*
Lewis Carroll (Charles Dodgson) (1832–1898)

Logical thought processes have been utilized by humanity since the beginning of our existence, even before we developed language and human rhetoric. While efforts to express logic using mathematical notation were made subsequently by various mathematicians, it was not until the mid-19th century with the systematic work of George Boole (1815–1864) and Augustus De Morgan (1806–1871) that formal logic began to be viewed as a rigorous framework for studying mathematics. In this chapter, we will see how to use the language and notation of logic developed in Chapter 3 to prove mathematical statements in a formal way.

4.1 Logical Equivalences

We say that propositions A and B are **logically equivalent** if the proposition $A \Longleftrightarrow B$ is a tautology; in this case, the statement $A \Longleftrightarrow B$ is called a **logical equivalence**. This notion is analogous to that of an *identity*, where two expressions have equal numerical values when evaluated for all possible values of the variables. We are all familiar with identities in algebra and trigonometry. For example,

$$a^2 - b^2 = (a - b)(a + b) \quad \text{for all } a, b \in \mathbb{R} \text{ and}$$

$$\cos^2 t + \sin^2 t = 1 \quad \text{for all } t \in \mathbb{R}.$$

When viewed as functions, logically equivalent propositions will have identical truth value outputs for each choice of truth value inputs.[1]

---

[1]Note that this necessitates that logically equivalent propositions operate on the exact same "component" statements; i.e., in the language of functions, they must have the same domains.

Identities have appeared in previous chapters, and many have been important enough to warrant specific names. For instance, the *associative law for addition* in Property 2.1(3) is an algebraic identity: $(a + b) + c = a + (b + c)$ for all real values of a, b, and c. Similarly, the *associative law for intersection* in Theorem 3.2(2)(a) is a set identity: $A \cap (B \cap C) = (A \cap B) \cap C$ for all subsets A, B, and C of a universal set. Based on our experience with logical truth tables and set membership tables in Chapter 3, a direct correspondence between the set identities in Theorem 3.2 and certain logical equivalences is evident. The same names are often used.

Theorem 4.1. *Suppose p, q, and r are logical propositions. Then the following laws hold.*

(1) *Commutative Laws*

   (a) $p \vee q \Longleftrightarrow q \vee p$ *and*

   (b) $p \wedge q \Longleftrightarrow q \wedge p$.

(2) *Associative Laws*

   (a) $((p \vee q) \vee r) \Longleftrightarrow (p \vee (q \vee r))$ *and*

   (b) $((p \wedge q) \wedge r) \Longleftrightarrow (p \wedge (q \wedge r))$.

(3) *Distributive Laws:*

   (a) $((p \vee q) \wedge r) \Longleftrightarrow (p \wedge r) \vee (q \wedge r)$ *and*

   (b) $((p \wedge q) \vee r) \Longleftrightarrow (p \vee r) \wedge (q \vee r)$.

(4) *Double Negation Law:* $\neg(\neg p) \Longleftrightarrow p$

(5) *De Morgan's Laws:*

   (a) $\neg(p \wedge q) \Longleftrightarrow (\neg p \vee \neg q)$ *and*

   (b) $\neg(p \vee q) \Longleftrightarrow (\neg p \wedge \neg q)$.

(6) *Idempotent Laws:*

   (a) $(p \vee p) \Longleftrightarrow p$ *and*

   (b) $(p \wedge p) \Longleftrightarrow p$.

Some of the above laws of logic may seem obvious or intuitive, while others may not. De Morgan's Laws may be new to the reader, but they are commonplace methods of reasoning. For example, if a friend asks if you are hot and tired after a long day of yard work, and you respond, "No," then your friend must conclude that you are either not hot *or* not tired (or neither). If a college requires a student to score at least 600 on the Math SAT or at least 1250 on the combined SAT in order to be exempt from taking Math 101 and a particular student is not exempt, we conclude that the student did not score at least 600 on the Math SAT *and* the student did not score at least 1250 on the combined SAT.

One method of proving a logical equivalence is by way of a truth table. The goal is to demonstrate that each compound expression is a tautology by showing that all truth values are T. We illustrate this in the following example.

Example 25. Use truth tables to verify the first of De Morgan's Laws given above.

Table 4.1. De Morgan's First Law

p	q	$p \vee q$	$\neg(p \vee q)$	$\neg p$	$\neg q$	$\neg p \wedge \neg q$	$\neg(p \vee q) \iff (\neg p \wedge \neg q)$
T	T	T	F	F	F	F	T
T	F	T	F	F	T	F	T
F	T	T	F	T	F	F	T
F	F	F	T	T	T	T	T

Solution. In Table 4.1 we construct a truth table that includes columns for both sides of the logical equivalence given in De Morgan's First Law, $\neg(p \vee q)$ and $(\neg p \wedge \neg q)$.

The fact that the columns for $\neg(p \vee q)$ and $(\neg p \wedge \neg q)$ are identical proves their logical equivalence. Alternatively, the logical equivalence can be verified by observing that the final column in the table confirms that $\neg(p \vee q)$ is true if and only if $(\neg p \wedge \neg q)$ is true; that is, $\neg(p \vee q) \iff (\neg p \wedge \neg q)$ is a tautology. $\square$

Logical equivalences often become useful in allowing the proof-writer to replace one statement with an equivalent one. Just as in algebra one can replace chosen variables on each side of an equation with an arbitrary expression and claim that the new equality is also valid, one can replace components of a logical equivalence to derive new logical equivalences. For example, substituting $p = a \vee b$ and $q = \neg b \Rightarrow (a \wedge b)$ into the tautology $(p \wedge q) \Rightarrow p$ yields another tautology:

$$((a \vee b) \wedge (\neg b \Rightarrow (a \wedge b))) \Rightarrow (a \vee b).$$

The logical equivalence of an implication with its **contrapositive** is particularly handy in proof-writing. This logical equivalence says

$$(p \Rightarrow q) \iff (\neg q \Rightarrow \neg p),$$

as verified in Table 4.2.

Table 4.2. Implication and Contrapositive Are Equivalent

p	q	$p \Rightarrow q$	$\neg q$	$\neg p$	$\neg q \Rightarrow \neg p$
T	T	T	F	F	T
T	F	F	T	F	F
F	T	T	F	T	T
F	F	T	T	T	T

This logical equivalence allows for any proposition that is an implication to be replaced with an equivalent implication (its contrapositive), which may be easier to prove.

Example 26. Form the contrapositive of each of the following propositions.

(1) *If Joel is drinking alcohol legally, then Joel is at least 21 years of age.*

(2) *It will not snow if the dew point is above 32 degrees Fahrenheit.*

(3) *If a student is married or a senior, then the student may live off campus.*

(4) *If your filing status is single or married filing jointly and if your taxable income is less than $100,000 and if you claim no dependents[2], then you are eligible to file your taxes with the 1040EZ form.*

Solution. We will first represent each proposition symbolically and then form the contrapositive (both symbolically and in words).

(1) Label the statements as

$$r: \ \textit{Joel is drinking alcohol legally} \text{ and}$$
$$s: \ \textit{Joel is at least 21 years of age.}$$

Symbolically, the statement *If Joel is drinking alcohol legally, then Joel is at least 21 years of age* can be written simply as

$$r \Rightarrow s.$$

The contrapositive is

$$\neg s \Rightarrow \neg r,$$

which corresponds to the statement *If Joel is less than 21 years of age, then he is not drinking alcohol legally.*

Of course, the negation of *r* could mean either that Joel is not drinking alcohol or that he is drinking alcohol illegally.

(2) This example is very much like the preceding one, except that it is slightly less obvious which of the simple statements is the antecedent and which is the consequent. Since the sentence could be rewritten as *If the dew point is above 32 degrees Fahrenheit, then it will not snow*, the antecedent is *p* : *The dew point is above 32 degrees Fahrenheit* and the consequent is *q*: *It will not snow.*

The contrapositive is

$$\neg q \Rightarrow \neg p,$$

which corresponds to the statement *If it snows, then the dew point is 32 degrees or lower.*

(3) Label the statements as:

a : *A student is married.*

b : *A student is a senior.*

c : *The student may live off campus.*

Symbolically, the statement *If a student is married or is a senior, then the student may live off campus* can be written as

$$(a \vee b) \Rightarrow c.$$

The contrapositive is

$$\neg c \Rightarrow \neg(a \vee b);$$

by De Morgan's Law, this is equivalent to

$$\neg c \Rightarrow (\neg a \wedge \neg b).$$

With the given assignments of *a*, *b*, and *c*, this corresponds to the statement *If a student may not live off campus, then the student is neither married nor a senior.*

---

[2]In fact, this is only a partial list of the conditions that must be met in order to use the 1040EZ form!

(4) Consider the following statements.

   s : *Your filing status is single.*

   j : *Your filing status is married filing jointly.*

   t : *Your taxable income is less than \$100,000.*

   d : *You claim no dependents.*

   f : *You are eligible to file your taxes with the 1040EZ form.*

Symbolically, the statement *If your filing status is single or married filing jointly and if your taxable income is less than \$100,000 and if you claim no dependents, then you are eligible to file your taxes with the 1040EZ form* can be written as

$$((s \vee j) \wedge t \wedge d) \Rightarrow f.$$

The contrapositive is

$$\neg f \Rightarrow \neg((s \vee j) \wedge t \wedge d).$$

Using De Morgan's Laws again, this is equivalent to

$$\neg f \Rightarrow (\neg(s \vee j) \vee \neg t \vee \neg d),$$

which in turn is equivalent to

$$\neg f \Rightarrow ((\neg s \wedge \neg j) \vee \neg t \vee \neg d).$$

In words, this can be written as

If you are not eligible to file your taxes with the 1040EZ form, then you are neither single nor married filing jointly or your taxable income is at least \$100,000 or you do claim dependents. □

There are two other statements that can be formed from an implication. The **converse** of the statement $p \Rightarrow q$ is

$$q \Rightarrow p.$$

The **inverse** of the statement $p \Rightarrow q$ is

$$\neg p \Rightarrow \neg q.$$

Although the contrapositive of an implication is logically equivalent to the implication, this is not true for the converse or the inverse, as can be illustrated by a simple example. Let p and q be the following simple propositions.

   p : *Quadrilateral Q is a rectangle.*

   q : *Quadrilateral Q has two pairs of parallel sides.*

The **implication** $(p \Rightarrow q)$ says: *If quadrilateral Q is a rectangle, then Q has two pairs of parallel sides.*

The **converse** $(q \Rightarrow p)$ says: *If quadrilateral Q has two pairs of parallel sides, then Q is a rectangle.*

The **inverse** $(\neg p \Rightarrow \neg q)$ says: *If quadrilateral Q is not a rectangle, then Q does not have two pairs of parallel sides.*

The **contrapositive** $(\neg q \Rightarrow \neg p)$ says: *If quadrilateral Q does not have two pairs of parallel sides, then Q is not a rectangle.*

Those who are familiar with basic geometric shapes will recognize that the converse and inverse statements above are false: a parallelogram is a quadrilateral with two pairs of parallel sides, which may not be a rectangle. Table 4.3 also confirms that the converse and inverse of an implication are not, in general, logically equivalent to the implication, since the truth values in both the fourth column (the converse) and the last column (the inverse) are different from those in the third column (the implication). Note, however, that the converse and inverse are logically equivalent to each other. (I.e., the converse is the contrapositive of the inverse, and vice versa.)

Table 4.3. Converse and Inverse Truth Values

p	q	$p \Rightarrow q$	$q \Rightarrow p$	$\neg p$	$\neg q$	$\neg p \Rightarrow \neg q$
T	T	T	T	F	F	T
T	F	F	T	F	T	T
F	T	T	F	T	F	F
F	F	T	T	T	T	T

Exercises and Problems.

(1) Write a statement that is equivalent to the negation of the converse of $a \Rightarrow b$ in symbolic form without using the $\Rightarrow$ symbol.

(2) State the converse, inverse, and contrapositive of each conditional statement.

   (a) If she won the lottery, then she is a millionaire.
   (b) If the flower pot is on the ground, then someone pushed it off of the stand.
   (c) When it rains, Susanna carries an umbrella.
   (d) If I don't get my way, then I scream and stomp my feet.
   (e) If my car is out of gas and the bus is not running, then I need to walk to work.
   (f) If two integers are odd, their sum is even.

(3) Translate each of these proverbs into if/then form, if they aren't already. Then write the converse and contrapositive of each one.

   (a) If it ain't broke, don't fix it.
   (b) If you want something done right, you have to do it yourself.
   (c) An apple a day keeps the doctor away.
   (d) He who hesitates is lost.
   (e) Spare the rod, spoil the child.
   (f) You can't make an omelet without breaking a few eggs.
   (g) A watched pot never boils.

(4) Write the negation of each sentence below.

   (a) The message was short and sweet.
   (b) The groom was tall, dark, and handsome.
   (c) For dessert, you may have either chocolate cake or ice cream.
   (d) The integer n is divisible by 5 or 6.

(5) Use a truth table to show that $p \Rightarrow q$ and $\neg p \vee q$ are logically equivalent.

(6) Prove De Morgan's Second Law

$$\neg(p \vee q) \iff \neg p \wedge \neg q$$

by creating a truth table.

(7) Prove that $\neg(p \iff q)$ is logically equivalent to $\neg(p \Rightarrow q) \vee \neg(q \Rightarrow p)$.

4.2 Predicates

Many statements, both within mathematics and otherwise, contain variables. Sentences such as the ones below are all examples of sentences that contain variables.

(1) *x is an integer and $5x = 20$.*

(2) $3x - x^2 > 0$.

(3) *x is Ernst's grade in Math 170.*

(4) *She is a member of the United States Senate.*

In the first three sentences, the variable is x; in the last, the variable is the pronoun "she". In each of these examples, inserting a particular value for the variable results in a proposition, but the truth value depends on the variable chosen. For instance, if $x = 4$, the statement *x is an integer and $5x = 20$* is true; for any other choice of x, the resulting statement is false.

Functional notation is used to denote a sentence that describes a property of a variable x. For example, we might write

$$P(x) : 3x - x^2 > 0.$$

Such a sentence is not a proposition because its truth value depends on the x that is selected from a particular set A; when a specific choice is made for x, the sentence becomes a proposition. For the above $P(x)$, we see that $P(0) : 3 \cdot 0 - 0^2 > 0$ is false while $P(2) : 3 \cdot 2 - 2^2 > 0$ is true. Such a sentence is called a **predicate** (or **propositional form** or **open sentence**) **on** A.

The earlier definitions of $\neg$, $\wedge$, $\vee$, $\Rightarrow$, and $\iff$ can be extended to predicates, thus allowing for the creation of more complicated predicates from simple ones.

Example 27. For each of the following compound predicates, find the set of real values of x that yield a true proposition.

(1) $(2x + 4 = 12) \vee (x - 7 = 23)$.

(2) $(x^2 > x) \wedge (x < 10)$.

(3) $(x^2 > x) \Rightarrow (x > 1)$.

(4) $(x^2 - x - 6 > 0) \wedge (|x| \leq 5)$.

Solution. (1) $\{4, 30\}$.

(2) $(-\infty, 0) \cup (1, 10)$.

(3) Since an implication is only false when the hypothesis is true and the conclusion is false, we only need to exclude the x values for which $x^2 > x$ and $x \leq 1$. This is the set of negative numbers, so the predicate is a true proposition for $x \geq 0$.

(4) $[-5, -2) \cup (3, 5]$. $\square$

A sentence that contains a variable is not necessarily a predicate. For example, the following sentences are all propositions.

(1) *For each real number x, $3x - x^2 > 0$.* (The truth value is F.)

Note that although $P(x) : 3x - x^2 > 0$ is a predicate, the assertion that the predicate is true for *all* real numbers x makes the entire sentence a proposition.

(2) *There exists a real number x such that $3x - x^2 > 0$.* (The truth value is T.)

(3) *There exists x in the set $\{A, B, C, D, F\}$ such that x is Ernst's grade in Math 170.* (It is reasonable to assume that the truth value is T.)

(4) *For each x in the set of people who are in the United States Congress, x is a member of the U.S. Senate.* (The truth value is F.)

The phrases "for all" (alternatively, *for any*, *for each*, or *for every*)[3] and "there exists" are called **quantifiers** and are denoted by the symbols $\forall$ and $\exists$, called the **universal** and **existential** quantifiers, respectively. The proposition "For each real x, $3x - x^2 > 0$" can be written symbolically as

$$\forall x \in \mathbb{R}, P(x).$$

If $A = \{a_1, a_2, \dots, a_n\}$ is a finite set, the proposition $\forall x \in A$, $P(x)$ is logically equivalent to

$$P(a_1) \wedge P(a_2) \wedge \cdots \wedge P(a_n).$$

The proposition $\exists x \in A$, $P(x)$ is logically equivalent to

$$P(a_1) \vee P(a_2) \vee \cdots \vee P(a_n).$$

When a set cannot reasonably be described with words or by listing all of its elements, quantifiers become useful. In such cases, a set is usually defined using set-builder notation, with a predicate $P(x)$ describing its elements. Here are some examples.

(1) $\{x : x \in \mathbb{Z} \text{ and } |x| \le 3\}$. There are other ways of describing the same set, including

$$\{x : (x \in \mathbb{Z}) \wedge (|x| \le 3)\},$$
$$\{y \in \mathbb{Z} : |y| \le 3\}, \text{ and}$$
$$\{-3, -2, -1, 0, 1, 2, 3\}.$$

Notice that the choice of variable used in the predicate has no effect on the set itself.

(2) $\{x : (x \in \mathbb{R}) \wedge (x^2 - 6x - 7 = 0)\}$

$$= \{x \in \mathbb{R} : x^2 - 6x - 7 = 0\}$$
$$= \{x \in \mathbb{R} : (x - 7)(x + 1) = 0\}$$
$$= \{-1, 7\}.$$

(3) $\{(x, y) \in \mathbb{R}^2 : x^2 + y^2 = 1\}$. This set is the unit circle in the coordinate plane.

---

[3]When just one variable is present, these terms have the same meaning, but we will see later that with two variables and two quantifiers, the meanings can be different when used in sentence form.

(4) The set of natural numbers which, when divided by 5, give remainder 1 is the same set as

$$\{1, 6, 11, 16, 21, ...\} \quad \text{and} \quad \{n \in \mathbb{N} : 5|(n-1)\}.$$

One more way to describe this set is as the set of natural numbers having 1 or 6 as the last digit in their decimal representation.

(5) $\{a : a$ is a letter of the English alphabet$\} = \{a, b, c, ..., x, y, z\}$.

Remark. Although the two sets are indeed equal, we should point out that the first representation is not ideal, because the letter a is used as the variable describing elements in the set, even though it is also an element of the set itself. While not wrong, such a representation is best avoided.   ⋄

Example 28. Determine whether each of the following is a predicate or a proposition. Identify the truth value of each of those that are propositions.

(1) $x > 0$.

(2) $\forall x \in \mathbb{Z}, x > 0$.

(3) $\exists x \in \mathbb{Z}, x > 0$.

(4) $\forall x \in \mathbb{Z}, x^2 > 0$.

(5) $y = x + 2$.

(6) $\exists x \in \mathbb{Z}, y = x + 2$.

(7) She has taken calculus.

(8) Every student currently enrolled at Acme University has taken calculus.

Solution. (1) This is a predicate.

(2) Because the sentence includes the phrase "for every x in the integers", this is a proposition. The truth value is clearly F since there are many nonpositive integers.

(3) This is a proposition. Because the statement only requires the existence of an integer that is greater than 0, the truth value is T.

(4) This is a proposition. Although the statement is true for all integers other than $x = 0$, since the proposition claims that $x^2 > 0$ for *all* integers, the truth value is F.

(5) This is a predicate with two variables; the truth value of the equation depends on the particular choices for x and y.

(6) This is a predicate. Although the domain for x has been specified, there are no restrictions on y.

(7) Unless we know who the pronoun "she" is referring to in the sentence, it functions as a variable, so the statement is a predicate.

(8) This is a proposition. The truth value is F if there is at least one student currently enrolled at Acme University who has *not* taken a calculus course, or T if it can be verified that *every* student currently enrolled at Acme University *has* taken a calculus course.   □

Any proposition can be negated, but care must be taken when negating propositions that include quantifiers. For example, consider the last proposition in Example 28:

Every student enrolled at Acme University has taken calculus.

To negate the statement, we must declare an exception to the given universal quantifier *every* by using an existential quantifier with regard to a student at Acme University. One such statement is *There is a student enrolled at Acme University who has not taken calculus.* The next example provides an explanation for how to form the negations of some of the other propositions in Example 28.

Example 29. Determine the meaning of each of the following propositions.

(1) $\neg(\forall x \in \mathbb{Z}, x > 0)$.

(2) $\neg(\exists x \in \mathbb{Z}, x > 0)$.

(3) $\neg(\forall x \in \mathbb{Z}, x^2 > 0)$.

Solution. As we translate each proposition, we also consider alternative symbolic representation.

(1) The simplistic way to translate this symbolic statement is: *It is not the case that every integer x is greater than 0.* However, using the phrase "it is not the case that ..." is neither satisfying nor enlightening. Without the $\neg$ symbol, the proposition made an assertion about *all* integers; to negate the original statement requires that there is *at least one* integer for which the given property is not true. That is, the negation claims that there exists at least one integer that is not greater than 0. Symbolically, this can be written

$$\exists x \in \mathbb{Z}, x \le 0.$$

Note that this is a proposition whose truth value is T.

(2) This time, we are negating a statement about existence of an integer. In negating such a statement, we are claiming that there is *no* integer for which the property is true. That is, the negation states: *For any integer x, x $\le$ 0.* Symbolically, this can be written as

$$\forall x \in \mathbb{Z}, x \le 0.$$

This proposition clearly has a truth value of F.

(3) Once again, we are negating a universal proposition. Doing so results in an existential proposition, namely

$$\exists x \in \mathbb{Z}, x^2 \le 0.$$

This, of course, is a true proposition, since there is an integer (namely 0) whose square is less than or equal to 0.   □

The quantified propositions examined in Example 29 illustrate how the universal and existential quantifiers are interchanged when forming negations:

$$\neg(\exists x\, P(x)) \iff \forall x, (\neg P(x)),$$
$$\neg(\forall x\, P(x)) \iff \exists x\, (\neg P(x)).$$

Notice that the two formulations above are actually extensions of De Morgan's Laws to any number of conjunctions. In the first equivalence, the statement on the left side asserts that there is no x for which $P(x)$ is true. Thus, $P(x)$ is false for all values of x, which is the meaning of the right-hand statement. In the case where the quantifier x belongs to the finite set $A = \{x_1, x_2, \ldots, x_n\}$, this extension of De Morgan's Law would be written as

$$\neg(P(x_1) \vee P(x_2) \vee \cdots \vee P(x_n)) \iff (\neg P(x_1) \wedge \neg P(x_2) \wedge \cdots \wedge \neg P(x_n)).$$

Similarly, the second equivalence takes the form

$$\neg(P(x_1) \wedge P(x_2) \wedge \cdots \wedge P(x_n)) \iff (\neg P(x_1) \vee \neg P(x_2) \vee \cdots \vee \neg P(x_n)).$$

4.3 Nested Quantifiers

Incorporating more than one variable using **nested quantifiers** can render predicates significantly more complicated. Even so, as long as quantifiers are applied to all variables, the predicate will be a proposition with a truth value of T or F.

Example 30. Let $P(x, y) : x^2 + y = 10$. Translate each of the following propositions and determine their truth values, assuming the domain of discourse is the set of real numbers (i.e., x and y are real numbers).

(1) $\forall x \exists y, \ P(x, y)$.

(2) $\exists x \forall y, \ P(x, y)$.

(3) $\exists x \exists y, \ P(x, y)$.

Solution. (1) This is translated as, *For every real number x, there exists a real number y such that $x^2 + y = 10$.* This could equivalently be written as, *For any real number x, there exists a real number y such that $y = 10 - x^2$,* which is clearly true.

(2) This time, the statement reads, *There exists a real number x such that for all real numbers y, $x^2 + y = 10$.* The proposition is false since for each real x, there is only one y value that makes the statement true, namely $y = 10 - x^2$.

(3) This proposition states, *There exists a real number x, for which there exists a real number y, such that $y = 10 - x^2$.* Since we easily can find a pair (x, y) satisfying $y = 10 - x^2$, the proposition has a truth value of T. ☐

The collection of quantifiers and symbols used with compound statements may disguise what is actually a fairly simple mathematical statement. That is not the purpose of the symbolism, of course! The symbolic versions of statements are precise and this lack of ambiguity is important for building mathematical theory. With experience, you will become comfortable translating between formal symbolic statements and mathematical statements written in prose.

Example 31. Convert each statement to a simple English sentence.

(1) $\forall x \in \mathbb{R}^+, \ \exists y \in \mathbb{R}^+, \ y < x$.

(2) $\forall x \in \mathbb{Q}, \ \exists a \in \mathbb{Z}, \ \exists b \in \mathbb{Z}, \ x = \dfrac{a}{b}$.

Solution. The first statement can be translated literally as, *For every positive real number x, there is a positive real number y such that y is less than x.* A much simpler interpretation is this: *There is no smallest positive real number.* The second statement has a literal translation of "For every rational number x, there is an integer a and an integer b such that $x = a/b$." Again, the simpler version isn't encumbered with variables: *Every rational number can be written as the quotient of two integers.* ☐

Example 32. Let b and g denote a boy and a girl, respectively, in a particular middle school class. Let $P(b, g)$ denote the predicate "b admires g". Interpret each of the following quantified propositions. Which of them mean the same thing?

(1) $\forall b\, \exists g,\, P(b, g)$.

(2) $\exists g\, \forall b,\, P(b, g)$.

(3) $\exists b\, \forall g,\, P(b, g)$.

(4) $\forall g\, \exists b,\, P(b, g)$.

(5) $\exists b\, \exists g,\, P(b, g)$.

(6) $\exists g\, \exists b,\, P(b, g)$.

(7) $\forall b\, \forall g,\, P(b, g)$.

(8) $\forall g\, \forall b,\, P(b, g)$.

Solution. (1) This can be translated as the sentence *Each boy in the class admires some girl in the class.*

(2) Although all of the quantifiers are the same as in (1), the meaning is different. This symbolic statement would be translated as *There is a girl in the class who is admired by every boy in the class.*

(3) This statement means *There is a boy in the class who admires every girl in the class.*

(4) This translates as *Each girl in the class is admired by some boy in the class.*
   The remaining solutions are left as an exercise.   □

There are several observations to make about Example 32. First, the propositions $\forall x\, \forall y,\, P(x, y)$ and $\forall y\, \forall x,\, P(x, y)$ have the same meaning; likewise, the propositions $\exists x\, \exists y,\, P(x, y)$ and $\exists y\, \exists x,\, P(x, y)$ are equivalent. On the other hand, the propositions $\forall x\, \exists y,\, P(x, y)$ and $\exists x\, \forall y,\, P(x, y)$ have different meanings, in general.

- In $\forall x\, \exists y,\, P(x, y)$, we claim that for any value of x that is selected from the domain, a value of y can be found such that $P(x, y)$ has truth value T. This value of y may depend on x and differ from one value of x to another.

- In $\exists x\, \forall y,\, P(x, y)$, we claim that there is a value of x in the domain such that, no matter which value of y is considered, $P(x, y)$ will have a truth value of T.

These subtleties also influence our choice of word when translating a symbolic statement into words. Do you see a difference between the statements *Each boy in the class admires some girl* and *All boys in the class admire some girl*? In the first statement, the choice of the girl depends on the boy; in the symbolic statement, g depends on b. On the other hand, depending on which word is emphasized when spoken, some may interpret the statement *All boys in the class admire some girl* to mean that all boys admire the same girl. In this case, g will not depend on b, and the proper symbolic form of the sentence is $\exists g\, \forall b\, P(b, g)$. For this reason, the universal quantifier in statements of the form $\forall x\, \exists b\, P(x, y)$ are often translated as "For each" or "For any" rather than "For all" or "For every", to minimize ambiguity.

The rules noted at the end of Section 4.2 for the negation of quantifiers also apply to propositions with nested quantifiers. In such cases, the negation must be "distributed" across the quantifiers. In the symbolic formulation, each universal quantifier is replaced with the existential quantifier, and vice versa. Consider the following example.

Example 33. Form the negation of each proposition in Example 32 and translate into a natural sentence in English.

Figure 4.1. DILBERT ©2001 Scott Adams. Used by permission of
ANDREWS MCMEEL SYNDICATION. All rights reserved.

Solution. We give solutions to the odd parts and leave the even parts to Exercise 10.

(1) Original: $\forall b\,\exists g,\,P(b,g)$.

The negation can be rewritten as $\exists b\,\forall g,\,\neg(P(b,g))$. This may be translated as *There is a boy who does not admire any girl in the class* or *There is a boy who admires none of the girls.*

(3) Original: $\exists b\,\forall g,\,P(b,g)$.

The negation is equivalent to $\forall b\,\exists g,\,\neg(P(b,g))$. A translation is *Each boy has some girl that he does not admire.*

(5) Original: $\exists b\,\exists g,\,P(b,g)$.

The negation is $\forall b\,\forall g,\,\neg(P(b,g))$, which is translated as *Each boy does not admire any of the girls* or as *None of the boys admire any of the girls.*

(7) Original: $\forall b\,\forall g,\,P(b,g)$. The symbolic representation of this negation is $\exists b\,\exists g,\,\neg(P(b,g))$, which may be translated as

There is a boy and a girl that he does not admire or maybe

There is a boy who does not admire one of the girls but **not**

There is a boy who does not admire a single girl.   □

Exercises and Problems.

(1) Classify each of the following as a predicate or a proposition. Identify the truth value of each of those that are propositions.

   (a) p is a prime number.

   (b) $\exists x \in \mathbb{Z},\, 4x^2 = 5$.

   (c) $\forall x \in \mathbb{R},\, e^x > 0$.

   (d) He is left-handed.

   (e) Everybody loves Raymond.

(2) Versions of statements i and ii below about real numbers appeared in Property 2.1 in Section 2.1.

   i. There exists a y such that for all x, $x + y = x$.

   ii. For each x, there exists y such that $x + y = 0$.

(a) Rewrite each one using nested quantifiers.

(b) Determine the value of y in each statement. (Note that only one of them depends on x.)

(3) Find the truth values of the following statements. Explain your reasoning.

(a) There is an integer x such that $xy \leq 0$ for all integers y.

(b) There is an integer x such that $x + y > 0$ for all integers y.

(c) For every integer x, there is an integer y such that $x + y = 0$.

(d) For every integer x, there is a real number y such that $xy = 1$.

(4) Let $A = \{1, 2, 3, \ldots, 10\}$. Find the truth values of the following statements and construct their negations.

(a) $\forall a \in A, a \mid 720$.

(b) $\exists a \in A, a^2 = 50$.

(c) $\forall a \in A, |a - 5|^2 \leq 25$.

(5) Find the truth values of the following statements. Then construct their negations.

(a) There exists an integer x such that $(x^2 - x - 1)(x + 1) = 0$.

(b) For all real numbers x and y, $x^2 + y^3 \geq 0$.

(c) There exist distinct integers x and y such that $x^2 = y^3$.

(d) There exist distinct integers x and y such that $x^4 = y^5$.

(6) Find the truth values of the following statements, where the universe of discourse is the real numbers.

(a) For all x, $(x^2 - x - 1)(x + 1) = 0$ implies $x^2 - x - 1 = 0$.

(b) For all x, $(x^2 + 1)(x^3 - x - 10) = 0$ implies $x^3 - x - 10 = 0$.

(c) For all x, $x^2(x + 4)^2 > 1$ implies $x(x + 4) > 1$.

(d) For all x, $\sqrt{5x + 1} < 9$ implies $x < 16$.

(7) Give solutions to parts (5)–(8) of Example 32.

(8) Let s be a student enrolled at Acme University in the fall semester, let c be a class taught at Acme University in the fall semester, let g be a grade, and let

$$Q(s, c, g) : \text{Student } s \text{ took course } c \text{ and received a grade of } g.$$

Provide a translation for each symbolic proposition.

(a) $\exists s, \; Q(s, \text{Math } 170, B)$.

(b) $\forall s \exists c \exists g, \; Q(s, c, g)$.

(c) $\forall c \exists s, \; Q(s, c, F)$.

(d) $\forall s \exists c, \; Q(s, c, A)$.

---

(9) Convert each English sentence to a symbolic statement or vice versa.

(a) Every nonzero real number has a multiplicative inverse.

(b) $\forall x \in \mathbb{R}, \exists y \in \mathbb{R}, x + y = 0$.

(c) $\forall m \in \mathbb{Z}, \forall n \in \mathbb{Z}, m + n \in \mathbb{Z}$.

(d) Natural numbers are not closed under subtraction.

(e) $\forall n \in \mathbb{N}$, n is prime $\Rightarrow (k|n \Rightarrow (k = 1) \vee (k = n))$.

(f) $\forall n \in \mathbb{Z}$, $\exists p \in \mathbb{N}$, (p is prime $) \wedge (p > n)$.

(10) Finish Example 33 by forming the negations and translating the propositions in (2), (4), (6), and (8) of Example 32.

(11) (a) Find examples of predicates $P(x)$ and $Q(x)$ over the reals (i.e., x is a real number) that show that these two compound propositions are not necessarily logically equivalent.

     (i) $\exists x \in \mathbb{R}$, $P(x)$ and $\exists x \in \mathbb{R}$, $Q(x)$.

     (ii) $\exists x \in \mathbb{R}$, $(P(x) \wedge Q(x))$.

  (b) Show that these two compound propositions are logically equivalent for all predicates $P(x)$ and $Q(x)$.

     (i) $\forall x \in \mathbb{R}$, $P(x)$ and $\forall x \in \mathbb{R}$, $Q(x)$.

     (ii) $\forall x \in \mathbb{R}$, $(P(x) \wedge Q(x))$.

(12) Let $P(x, y)$ be the predicate $x = 2y + 1$. Find the truth values of the following statements and explain your answer.

(a) $\exists x \in \mathbb{Z}$, $\exists y \in \mathbb{Z}$, $P(x, y)$.

(b) $\exists x \in \mathbb{Z}$, $\forall y \in \mathbb{Z}$, $P(x, y)$.

(c) $\forall x \in \mathbb{Z}$, $\exists y \in \mathbb{Z}$, $P(x, y)$.

(d) $\forall x \in \mathbb{Z}$, $\forall y \in \mathbb{Z}$, $P(x, y)$.

(13) Label each of the following statements as true or false. Prove your answers.

(a) For all real numbers b, $x^2 + bx + 3 = 0$ has a real solution, x.

(b) For all positive integers m and n, there are integers x and y such that $x^m = y^n$.

(c) $\forall x \in \mathbb{R}$, $\forall y \in \mathbb{R}$, $\dfrac{x^2 + y^2}{2} \geq xy$.

(d) $\exists x \in \mathbb{R}$, $\exists y \in \mathbb{R}$, $\dfrac{x}{y} + \dfrac{y}{x} < 2$.

(e) $\exists x \in \mathbb{R}$, $\exists y \in \mathbb{R}$, $(x > 0) \wedge (y > 0) \wedge \left(\dfrac{x}{y} + \dfrac{y}{x} < 2 \right)$.

(14) Find the truth values of the following statements.

(a) For all real numbers x, $x^2 - 7x \geq 12$.

(b) For all real numbers x and y, $x^2 - 3xy + y^2 \geq 0$.

(c) For all real numbers x and y, $x^2 + xy + y^2 \geq 0$.

(15) Determine whether each statement is true or false. Explain.

(a) $\exists m \in (0, 1]$, $\forall x \in (0, 1]$, $m \leq x$.

(b) $\forall x \in (0, 1]$, $\exists y \in (0, 1]$, $x < y$.

(c) $\forall x \in (0, 1]$, $\exists y \in (0, 1]$, $y < x$.

(d) $\exists M \in (0, 1]$, $\forall x \in (0, 1]$, $x \leq M$.

(16) Write the negation of each statement in Problem 15 in symbolic form.

(17) Determine whether each statement is true or false. Explain.

(a) $\forall x \in \mathbb{Z}$, $\forall y \in \mathbb{Z}$, $\exists z \in \mathbb{Z}$, $x < y \Rightarrow x < z < y$.

(b) $\forall x \in \mathbb{Q}$, $\forall y \in \mathbb{Q}$, $\exists z \in \mathbb{Q}$, $x < y \Rightarrow x < z < y$.

(18) Is the following statement true? Prove your answer.
$$\forall n \in \mathbb{N}, \exists x \in \mathbb{Z}, \exists y \in \mathbb{Z}, x^2 - y^2 = 2n + 1.$$

(19) Suppose a, b, and c are integers. Are the following statements true? Prove your answers.

    (a) $\forall a \exists b \exists c, b^2 + c^2 = a^2$.

    (b) $\forall a \neq 0, \exists b \neq 0, \exists c \neq 0, b^2 + c^2 = a^2$.

---

(20) Find propositions $P(x, y)$ and $Q(x, y)$ that demonstrate that
$$(\forall x \exists y, P(x, y)) \wedge (\forall x \exists y, Q(x, y))$$
is different than $\forall x \exists y, (P(x, y) \wedge Q(x, y))$.

(21) Let a, b, c be real numbers. Is the following statement correct? Prove your answer.
$$\forall a \forall b \forall c, (a + b + c = 0 \Rightarrow a^3 + b^3 + c^3 = 3abc).$$

(22) A real-valued function $f(x)$ is said to be **continuous** at a point x_0 if, for each $\epsilon \in \mathbb{R}$, $\epsilon > 0$, there is a real number $\delta > 0$ such that $|f(x) - f(x_0)| < \epsilon$ whenever $|x - x_0| < \delta$.

    (a) Write the above definition using quantifiers and symbols.

    (b) Use symbols to write the negation of the statement "f is continuous at x_0." Do not simply use the $\neg$ symbol in front of your solution to (a)!

    (c) Write a sentence definition of what it means for $f(x)$ to be discontinuous at a point x_0. You may use symbols and mathematical notation, but your basic structure should be in sentence form.

(23) A function g is **continuous on an interval** (a, b) if it is continuous at all points of the interval.

    (a) Use symbols to write the definition of g being continuous on (a, b).

    (b) Write a sentence definition of what it means for $f(x)$ to be discontinuous on an interval (a, b).

(24) A function h is **uniformly continuous on an interval** (a, b) if for each $\epsilon > 0$ there is a real number $\delta > 0$, such that for every $x_0 \in (a, b)$, $|h(x) - h(x_0)| < \epsilon$ whenever $|x - x_0| < \delta$.

    (a) Use symbols to write the definition of h being uniformly continuous on (a, b). Your statement should be similar to the one formed in Problem 23, but the meaning is quite different. These concepts will be taken up in a real analysis or advanced calculus course.

    (b) Write a sentence definition of what it means for $f(x)$ to not be uniformly continuous on an interval (a, b). Compare your statement to the one in Problem 23.

4.4 Rules of Inference

Experimentation, trial and error, or intuition may lead one to a **conjecture**—that is, a statement that one suspects is true. The validity of a mathematical conjecture, however, is not conclusively established until a proof is given. While intuition and experimentation may be convincing, they do not serve as a proof in mathematics.

A mathematical proof that consists of a sequence of statements that begins with **hypotheses** (or **premises**) and ends with a **conclusion** is also called a formal **argument**. The final conclusion of a proof establishes a **theorem**, a statement that has been proven to be true using other mathematical statements and accepted rules of reasoning.

Not every mathematical statement can be proven. There must be statements that are accepted but left unproven; these statements, the **axioms**, should specify the essential assumptions upon which the system will be built. Within a mathematical system, all truths must be derived from the axioms of the system. For example, the arithmetic properties in Property 2.1 of Section 2.1 and the rules for the logical connectors defined in the truth tables in Tables 3.1–3.3 in Section 3.3 serve as axioms. While these axioms do not comprise what would be considered a complete set of axioms, they will be sufficient for our purposes.

4.4.1 Valid Argument Forms. A mathematical argument is **valid** if the conclusion has a truth value of T any time the premises all have truth values of T, regardless of what the particular premises are. Thus, valid arguments can be characterized by their structure. The most basic forms of valid arguments are called **rules of inference**.

Example 34. Consider the following arguments. Which of them should be deemed valid?

(1) P_1 : Tonight Gottlob will study or he will watch a movie.

P_2 : If Gottlob does not have a test tomorrow, then he will not study tonight.

P_3 : Gottlob does not have a test tomorrow.

C : Therefore, Gottlob will watch a movie tonight.

(2) P_1 : Any great city has at least one college.

P_2 : Harrisonburg has four colleges.

C : Therefore, Harrisonburg is a great city.

(3) P_1 : All Discrete Math students know how to write a valid proof.

P_2 : Jacob is not a Discrete Math student.

C : Therefore, Jacob does not know how to write a valid proof.

(4) P_1 : The sum of any two even integers is an even integer.

P_2 : The sum of x and y is not an even integer.

C : Therefore, x and y are not both even integers.

(5) P_1 : The sum of any two odd integers is an even integer.

P_2 : The sum of x and y is an even integer.

C : Therefore, x and y are both odd integers.

First Thoughts. The reader may already have an intuitive sense of which of the above arguments are valid. However, determining with confidence the validity of a given argument is often best accomplished by translating each statement into symbolic form, so that the structure of the argument itself can be examined. This will be demonstrated for part (1); other parts will be seen later in the section or as exercises.   ◇

Solution. We first identify the following simple statements in (1).

p : *Gottlob will study tonight.*
q : *Gottlob will watch a movie tonight.*
r : *Gottlob has a test tomorrow.*

Now, the argument can be represented symbolically in the following form.

$$p \vee q$$
$$\neg r \Rightarrow \neg p$$
$$\underline{\neg r \qquad\qquad}$$
$$\therefore q$$

Each statement above the horizontal line is a premise in the argument; the final statement is the conclusion. Note that the conclusion is preceded by the symbol $\therefore$, which can be read as "therefore" or "thus".

Recall that we have defined an argument as being valid if the conclusion has a truth value of T any time the premises all have truth values of T. In this case, this means that the argument for the above is valid if

$$((p \vee q) \wedge (\neg r \Rightarrow \neg p) \wedge (\neg r)) \Rightarrow q$$

is a tautology. This is shown to be the case in the truth table in Table 4.4, where, in order to save space, we let $D : (p \vee q) \wedge (\neg r \Rightarrow \neg p) \wedge (\neg r)$.   □

Table 4.4. Truth Table for Example 34(1)

p	q	r	$P_1 : p \vee q$	$P_3 : \neg r$	$\neg p$	$P_2 : \neg r \Rightarrow \neg p$	D	$D \Rightarrow q$
T	T	T	T	F	F	T	F	T
T	T	F	T	T	F	F	F	T
T	F	T	T	F	F	T	F	T
T	F	F	T	T	F	F	F	T
F	T	T	T	F	T	T	F	T
F	T	F	T	T	T	T	T	T
F	F	T	F	F	T	T	F	T
F	F	F	F	T	T	T	F	T

While one can always use a truth table to determine the validity of an argument, it is beneficial to establish a collection of basic rules of inference forms, from which more complex arguments can be efficiently constructed. We present these basic forms in Table 4.5.

Example 35. Identify each of the following as corresponding to one of the argument forms given in Table 4.5.

Table 4.5. Rules of Inference

Rule of Inference	Tautology	Name
p $p \Rightarrow q$ $\therefore q$	$(p \wedge (p \Rightarrow q)) \Rightarrow q$	Modus ponens
$\neg q$ $p \Rightarrow q$ $\therefore \neg p$	$(\neg q \wedge (p \Rightarrow q)) \Rightarrow \neg p$	Modus tollens
p q $\therefore p$	$(p \wedge q) \Rightarrow p$	Conjunctive simplification
$p \vee q$ $\neg p$ $\therefore q$	$((p \vee q) \wedge \neg p) \Rightarrow q$	Disjunctive syllogism
$p \Rightarrow q$ $q \Rightarrow r$ $\therefore p \Rightarrow r$	$((p \Rightarrow q) \wedge (q \Rightarrow r)) \Rightarrow (p \Rightarrow r)$	Hypothetical syllogism

(1) If the moon is visible, then it is nighttime.
   The moon is visible.
   Therefore, it is nighttime.

(2) An integer that is divisible by 12 is also divisible by 6.
   An integer that is divisible by 6 is also divisible by 3.
   Therefore, an integer that is divisible by 12 is also divisible by 3.

(3) Abraham must take a history course or a literature course.
   Abraham will not take a literature course.
   Therefore, Abraham will take a history course.

(4) If the price of gasoline rises, then people will drive less.
   The number of miles driven has increased.
   Therefore, the price of gasoline has not risen.

Solution. In each case we begin the solution by assigning letters to represent the simple statements.

(1) Let p : *The moon is visible* and q : *It is nighttime* be the statements.
   The given argument can be represented symbolically as

$$p \Rightarrow q$$
$$p$$
$$\therefore q.$$

   This is an example of modus ponens.

Further Thoughts. The conclusion of this argument may not be true: it is possible for the moon to be visible during the daytime. However, the argument *form* is still valid. Remember that an argument form is valid if the conclusion will be true *when all premises are true*; if one or more premises are false, then the conclusion may indeed be false without invalidating the logical legitimacy of the argument.   ◇

(2) Assign p, q, and r to be the simple statements

   p : *An integer is divisible by 12*,

   q : *An integer is divisible by 6*, and

   r : *An integer is divisible by 3*.

   The given argument can be represented as

$$p \Rightarrow q$$
$$q \Rightarrow r$$
$$\overline{}$$
$$\therefore p \Rightarrow r.$$

   This is an example of hypothetical syllogism.

(3) Assign h and l to be the simple statements

   h : *Abraham takes a history course* and

   l : *Abraham takes a literature course*.

   The given argument can then be represented as

$$h \vee l$$
$$\neg l$$
$$\overline{}$$
$$\therefore h.$$

   This is an example of disjunctive syllogism.

(4) Assign g and d to be the simple statements

   g : *The price of gasoline goes up* and

   d : *People drive less*.

   The given argument can then be represented as

$$g \Rightarrow d$$
$$\neg d$$
$$\overline{}$$
$$\therefore \neg g.$$

   This is an example of modus tollens.   □

   The basic argument forms given in Table 4.5 can be combined to create more complex arguments.

Example 36. Consider again the argument given in Example 34(1).

P_1 : *Tonight Gottlob will study or he will watch a movie.*

P_2 : *If Gottlob does not have a test tomorrow, he will not study tonight.*

P_3 : *Gottlob does not have a test tomorrow.*

C : *Therefore, Gottlob will watch a movie tonight.*

The validity of this argument was verified previously by writing the given statements symbolically and using a truth table, with

$$p \vee q, \quad \neg r \Rightarrow \neg p, \quad \text{and} \quad \neg r$$

as premises and q as the conclusion. Now, verify the argument's validity by framing it as a sequence of arguments from Table 4.5.

Solution. The premises can be combined in any helpful way using any of the argument forms found in the table to reach intermediate conclusions. These intermediate conclusions can then also be used as building blocks in the basic argument forms. The goal is to ultimately reach a conclusion matching the one at the end of the original argument.

$$\begin{array}{ll} \neg r \Rightarrow \neg p & \text{[Premise]} \\ \neg r & \text{[Premise]} \\ \hline \therefore \neg p & \text{[Modus ponens]} \\ p \vee q & \text{[Premise]} \\ \hline \therefore q & \text{[Disjunctive syllogism]}. \end{array}$$

$\square$

Typically, mathematical axioms, conjectures, and theorems are given as quantified statements. Proving an existentially quantified statement requires simply the demonstration of one particular element in the domain that satisfies the given condition. On the other hand, proving a universally quantified statement requires that the given property must be shown true for *all* elements of a specified set.

Universal modus ponens is an argument form that provides the framework for deductive reasoning in mathematics. The structure of universal modus ponens is as follows:

$$\begin{array}{l} \forall x \in S, P(x) \Rightarrow Q(x). \\ \text{For a particular } s \in S, P(s). \\ \hline \therefore Q(s). \end{array}$$

Example 37. The following is an example of the use of universal modus ponens.

$$\begin{array}{l} \text{For all } n \in \mathbb{Z}, \text{ if } n \text{ is even, then } n^2 \text{ is even.} \\ \text{18 is an even integer.} \\ \hline \text{Therefore, } 18^2 \text{ is an even integer.} \end{array}$$

Universal modus tollens is an argument form that is frequently used in proofs by contradiction. The template for this argument form is

$$\begin{array}{l} \forall x \in S, P(x) \Rightarrow Q(x). \\ \text{For a particular } s \in S, \neg Q(s). \\ \hline \therefore \neg P(s). \end{array}$$

Example 38. The following is an example of the use of universal modus tollens.

$$\begin{array}{l} \text{For all } n \in \mathbb{Z}, \text{ if } 5n - 3 \text{ is even, then } n \text{ is odd.} \\ \text{24 is an even integer.} \\ \hline \text{Therefore, } 5(24) - 3 \text{ is an odd integer.} \end{array}$$

4.4.2 Logical Fallacies. Logical fallacies occur when an argument form is used that is not valid. Note again that the emphasis is on the *form* of the argument, not the truth of the individual statements that comprise the premises and conclusion. Of course, the conclusion of any argument may be false because one or more of the premises is false, even if a valid argument form is used.

The logical fallacy known as **affirming the conclusion** (or **affirming the consequent**) takes the form

$$p \Rightarrow q$$
$$\underline{q \qquad\qquad}$$
$$\therefore p.$$

If the implication is given as a universally quantified statement, the form will look like this instead:

$$\forall x \in S, P(x) \Rightarrow Q(x).$$
$$\underline{\text{For a particular } s \in S, Q(s).}$$
$$\therefore P(s).$$

The reader may recognize that at the heart of this fallacy is an assumption that the converse of an implication is logically equivalent to the implication. Part (2) of Example 34 demonstrates this type of fallacy.

A second common type of logical fallacy, known as **denying the hypothesis** (or **denying the antecedent**) has the form

$$p \Rightarrow q$$
$$\underline{\neg p \qquad\qquad}$$
$$\therefore \neg q.$$

Here is the form of this fallacy if the implication is given as a universally quantified statement:

$$\forall x \in S, P(x) \Rightarrow Q(x).$$
$$\underline{\text{For a particular } s \in S, \neg P(s).}$$
$$\therefore \neg Q(s).$$

The faulty logic underlying this fallacy is the assumption that the inverse of an implication is logically equivalent to the implication. Part (3) of Example 34 demonstrates denying the hypothesis.

Exercises and Problems.

If you can lead it to water and force it to drink, it isn't a horse.

—Unknown

(1) Consider the argument used in Example 34(4). If the argument is valid, name the form of reasoning used (from the choices in Table 4.5). If it is not valid, name the type of logical fallacy used.

(2) Consider the argument used in Example 34(5). If the argument is valid, name the form of reasoning used (from the choices in Table 4.5). If it is not valid, name the type of logical fallacy used.

(3) Identify each of the following as utilizing one or more valid rules of inference (name the rules used) or as demonstrating a logical fallacy.

(a) In order to receive college credit for a class, it is necessary to enroll in the class.
Béla has not received college credit for General Psychology.
Therefore, Béla has not enrolled in General Psychology.

(b) All environmentalists are vegetarians.
Deirdre is a vegetarian.
Therefore, Deirdre is an environmentalist.

(c) All geometers love mathematics.
Kenny is a geometer.
Therefore, Kenny loves mathematics.

(d) If Owen doesn't get enough sleep, then he is grumpy.
Owen is not grumpy.
Therefore, he got enough sleep.

(e) The authors of this book do not like questions using their names.
No names are mentioned in this question.
Therefore, the authors like this question.

(4) Identify each of the following as utilizing one or more valid rules of inference (name the rules used) or as demonstrating a logical fallacy.

(a) If it is a Friday during Lent, Ada will not be eating fish.
Ada is eating fish.
Therefore, it is either not Friday or not the Lenten season.

(b) If my alarm goes off, then I will catch the bus.
If I catch the bus, I will be on time to work.
If I am on time to work, then I will get a raise.
Therefore, if my alarm goes off, then I will get a raise.

(c) When I forget to carry my umbrella, it always rains.
Yesterday it rained.
Therefore, I forgot my umbrella yesterday.

(d) If a number is the product of distinct primes, then it is composite.
The number 7,331 is not the product of distinct primes.
Therefore, 7,331 is composite.

(e) Every nonzero rational number has a multiplicative inverse.
The number -1 is a rational number not equaling zero.
Therefore -1 has a multiplicative inverse.

(5) The logic game WFF 'N PROOF, first released in the 1960s, contains the following question (known as the **Tardy Bus Problem**).

The following statements are premises.

P_1 : If Charles takes the bus, then Charles misses his appointment, if the bus is late.

P_2 : Charles shouldn't go home, if (i) Charles misses his appointment, and (ii) Charles feels downcast.

P_3 : If Charles doesn't get the job, then (i) Charles feels downcast, and (ii) Charles should go home.

Using rules of inference, determine the validity of the following conclusions.

(a) If Charles takes the bus, then Charles does get the job, if the bus is late.
(b) Charles does get the job, if (i) Charles misses his appointment, and (ii) Charles should go home.
(c) If the bus is late, then (i) Charles doesn't take the bus, or Charles doesn't miss his appointment, if (ii) Charles doesn't get the job.
(d) Charles doesn't take the bus, if (i) the bus is late, and (ii) Charles doesn't get the job.
(e) If Charles doesn't miss his appointment, then (i) Charles shouldn't go home, and (ii) Charles doesn't get the job.
(f) Charles feels downcast, if (i) the bus is late, or (ii) Charles misses his appointment.
(g) If Charles does get the job, then (i) Charles doesn't feel downcast, or (ii) Charles shouldn't go home.
(h) If (i) Charles should go home, and Charles takes the bus, then (ii) Charles doesn't feel downcast, if the bus is late.

(6) Variations of the following "Murder Mystery" have been floating around the Internet for years. The original author is unknown. Assuming each of the following statements is true, what conclusion can be reached regarding the identity of the murderer?

(a) If Mr. Barton had an intruder in his apartment, then Mr. Barton shot the intruder.
(b) Either Mr. Thompson was a CPA who worked at an accounting firm or he broke into Mr. Barton's apartment at midnight.
(c) Mrs. Scott and Mr. Thompson were both employed at the local high school, and they were good friends.
(d) Mr. Barton's gun was found in his apartment, and the bullet in Mr. Thompson's calf matched those used by Mr. Barton's gun.
(e) If the person who shot Mr. Thompson also stabbed him, then Mr. Thompson's body was found in the parking garage.
(f) Mr. Thompson's body was found at 1:20 a.m.
(g) Mrs. Scott would not kill anyone with whom she works.
(h) Either Mr. Thompson's body was found in the park or his body was found before 1:00 a.m.
(i) A knife was found in the parking garage or Mr. Barton's gun was found in the elevator.
(j) Mrs. Scott waited in the lobby of the apartment building from 12:15 a.m. to 12:45 a.m. She was either waiting for her husband to get off work or she was the person who killed Mr. Thompson.
(k) If Mrs. Scott was friends with Mr. Thompson, then her husband was jealous of the friendship.
(l) If Mrs. Scott was waiting in the lobby for her husband to get off work, then her husband was the elevator operator.

4.5 Methods of Proof

The rules of inference presented in the previous section form the foundation for proof-writing. However, while it may be helpful to practice translating sentences into symbolic form and identifying rules of inference as they are utilized, this method is too stilted for use by experienced proof-writers. In fact, good proof-writing is something of an art! A well-written proof is most often presented in paragraph form, using complete sentences, with sufficient justification and explanation given for the reader to follow, but not so much as to be laborious. The level of detail provided should be relative to the level of mathematical maturity of the intended audience, chosen accordingly to be convincing to the reader. For the purposes of this introduction to proof-writing, we will err on the side of including more explanation than may be needed, and we will write our proofs as prose.

Most typically, theorem statements are of the form "If A, then B" or the form "A if and only if B". The proof of such a statement can be *direct* or *indirect*. We introduce both of these methods of proof, including both *proof by contradiction* and *proof by contrapositive* as types of indirect proof.[4] Here are some tips to be mindful of when writing proofs.

(1) While examples may be valuable for gaining confidence that a statement is true and gaining intuition as to why it is true, *examples are not sufficient to prove a general statement*.

(2) Use variables to represent a particular, but arbitrary, element of the domain of discourse. Be sure to use different variables to represent distinct elements of the domain.

(3) Provide justification for each step of the proof by citing a definition, an axiom, a previously proven theorem, or a rule of inference.

(4) All of the hypotheses of a theorem statement are assumed to be true; it is often helpful to identify and specifically state these assumptions at the beginning of a proof of the theorem.

The remainder of this chapter will introduce several specific types of proof, as explained above. For this introduction, the proof methods are neatly divided into subsections with accompanying examples. Naturally, in more realistic problem-solving situations, this handy categorization is not provided; indeed, there may be multiple ways to prove a given statement, although one proof may be preferred over others for its elegance or simplicity. There are few firm rules for determining which method will be the most fruitful when attempting to write a proof; as with many other things, there is no substitute for experience in the quest to become more adept at proof-writing.

The theorems we will prove in the examples that follow come from simple number theory. Recall that an **even integer** is an integer divisible by 2; i.e., n is even if $n = 2k$ for some integer k. Similarly, an **odd integer** is an integer n that can be expressed as $n = 2j + 1$ for some integer j. The set of integers is **closed** under the operations of addition and multiplication; i.e., for any integers m and n, $m + n \in \mathbb{Z}$ and $mn \in \mathbb{Z}$.

4.5.1 Direct Proof. To prove a statement of the form "if A, then B" directly, begin by assuming that A is true. Then, making use of axioms, definitions, previously proven

---

[4]*Proof by mathematical induction*, an important subcategory of direct proof, is given its own chapter (see Chapter 6) and will not be addressed here.

theorems, and rules of inference, proceed directly until B is reached as a conclusion. Thus, a direct proof employs modus ponens. Direct proofs are most easily employed when establishing the general form of the antecedent is straightforward.

Example 39 (seen previously as Theorem 2.4(3)). Prove that the sum of an even integer and an odd integer is odd.

First Thoughts. Although it may not appear so at first glance, this theorem statement is of the form "if A, then B". The statement could be rewritten as

For any integers x and y, if x is even and y is odd, then $x + y$ is odd.

Thus, A is the statement that x is even and y is odd, while B is the statement that $x+y$ is odd. The domain of discourse is the integers. For a direct proof, we begin by assuming that the entirety of A is true.   ⋄

Proof. Let x be an even integer and let y be an odd integer. By the definition of even integer, there is some $k \in \mathbb{Z}$ such that $x = 2k$, and by the definition of odd integer, there is some $j \in \mathbb{Z}$ such that $y = 2j + 1$. Substituting,

$$x + y = (2k) + (2j + 1)$$
$$= 2(k + j) + 1.$$

Closure of the integers ensures under addition that $k + j \in \mathbb{Z}$. Thus, by definition, $x + y$ is an odd integer.   □

Further Thoughts. In the above proof, note that we chose different letters, k and j, for use in the definitions of even and odd integers (for x and y, respectively). This is important. Using the same letter—say, $x = 2k$ and $y = 2k + 1$—would have implied that x and y are consecutive integers (i.e., they differ by 1). This is more restrictive than the premises of the theorem statement specify. *One should always retain the stated level of generality when proving a theorem.*   ⋄

Example 40. Prove that if n is an even integer, then $(-1)^n = 1$.

Proof. Let n be any even integer. By definition, $n = 2k$ for some integer k. Then utilizing rules of exponents,

$$(-1)^n = (-1)^{2k} = ((-1)^2)^k = (1)^k = 1.$$ □

4.5.2 Proof by Contradiction.
The technique known as **proof by contradiction** is one type of indirect proof. In a proof by contradiction, in order to prove a statement of the form "If A, then B", one assumes that both A and $\neg B$ are true. The goal is then to reach a contradiction, which allows one to conclude that A and $\neg B$ can never both be true. That is, whenever A is true, B must also be true. This method of proof is useful when assuming $\neg B$ allows you to easily utilize a definition or theorem.

Example 41. Prove that if the sum of a real number and itself equals itself, then the number must be 0.

First Thoughts. If we let $A(x) : x + x = x$ and $B(x) : x = 0$, then the statement to be proved can be written symbolically as

$$\forall x \in \mathbb{R}, \, A(x) \Rightarrow B(x).$$

To write a proof by contradiction, we will assume that $A(x)$ and $\neg B(x)$ are simultaneously true and hope to reach a contradiction. ◇

Proof. Let x be any real number. Assume that $x + x = x$, but $x \neq 0$. Then $2x = x$, and since $x \neq 0$, we can divide each side by x to obtain $2 = 1$, which is clearly a contradiction. We conclude that if $x + x = x$, then $x = 0$. □

Remark. Notice that our First Thoughts involved carefully identifying the statements $A(x)$ and $B(x)$ and determining how to structure a proof by contradiction. However, the proof itself contains no reference to $A(x)$ or $B(x)$ or the underlying proof structure! This is typical of the way that proofs are written: much of what goes into understanding why a theorem is true and how to write the proof is omitted from the final, "polished" proof. ◇

Further Thoughts. As an alternative, the statement given in this example could easily be proven directly: Let x be any real number. Assume that $x + x = x$; that is, $2x = x$. By subtracting x from each side, we obtain $x = 0$. ◇

4.5.3 Proof by Contrapositive.
Proof by contrapositive makes use of modus tollens, which relies on the equivalence of an implication with its contrapositive. The proof begins by assuming $\neg B$ is true. Referencing axioms, definitions, previously proven theorems, and rules of inference, the proof ultimately reaches the conclusion that $\neg A$ is true. In other words, this is a direct proof on the contrapositive of the original statement.

Example 42. Prove that if the sum of two integers is even, then either both integers are even or both integers are odd.

First Thoughts. By experimenting, it is easy to be convinced that the statement we wish to prove is true. However, as we have stressed, providing an example is not sufficient for proof of a universally quantified statement. Intuitively, it seems that the statement must be true since if an even integer and an odd integer were added together, there would be an "extra" 1 that would cause the sum to be odd.

Writing a direct proof of this statement is somewhat challenging because one must start with a sum and from the sum extract information about the addends, but the intuitive understanding of what happens points us toward a proof by contrapositive. What follows below may look quite daunting, so we should emphasize that mathematicians hardly ever write out steps like this in symbols. We do it this once to demonstrate the validity of the argument, but after much practice such thoughts become second nature to the experienced proof-writer.

Let $E(n) : n$ is even.

Then the statement that we want to prove can be written symbolically as

$$\forall x \forall y, \ E(x + y) \Rightarrow [(E(x) \wedge E(y)) \vee (\neg E(x) \wedge \neg E(y))].$$

Symbolically, the contrapositive of the above statement can be written as

$$\forall x \forall y, \ \neg[(E(x) \wedge E(y)) \vee (\neg E(x) \wedge \neg E(y))] \Rightarrow \neg E(x + y),$$

or equivalently (by applying De Morgan's Law to the antecedent),

$$\forall x \forall y, \ [\neg(E(x) \wedge E(y)) \wedge \neg(\neg E(x) \wedge \neg E(y))] \Rightarrow \neg E(x + y).$$

Applying De Morgan's Law once more, the above can be rewritten as

$$\forall x \forall y, \ [(\neg E(x) \vee \neg E(y)) \wedge (E(x) \vee E(y))] \Rightarrow \neg E(x + y).$$

At this point, pausing to think about what this symbolic statement is telling us may be helpful. The expression $(\neg E(x) \vee \neg E(y))$ tells us that at least one of x or y must be not even—that is, at least one of x or y must be odd—while the expression $(E(x) \vee E(y))$ requires that at least one of x or y must be even. Since these two expressions are joined by a conjunction $(\wedge)$, the antecedent requires exactly one of x or y to be odd (it does not matter which one) and the other to be even; the consequent asserts that the sum $x + y$ will be odd. In summary, the negation of "both integers are even or both integers are odd" is that one integer is even and the other one is odd. We thus start the proof by contrapositive with such a statement.   ◇

Proof. To prove the contrapositive, let x be an odd integer and let y be an even integer. From Example 39, $x + y$ is odd, so the contrapositive is true. Thus the given statement is true as well.   □

Example 43. Prove that the product of two integers is odd if and only if both of the integers are odd.

First Thoughts. This theorem statement is of the form "*A* if and only if *B*". Since a biconditional is the conjunction of two conditional statements, proving a biconditional really requires the proof of *two* conditional statements: "if *A*, then *B*" (denoted as $\Rightarrow$) and "if *B*, then *A*" (denoted as $\Leftarrow$). In this case, the statement "if *A*, then *B*" ($\Rightarrow$) is

if the product of two integers is odd, then both of the integers are odd,

while the statement "if *B*, then *A*" ($\Leftarrow$) is

if two integers are both odd, then their product is odd.

The two component proofs can be given in either order and a different type of proof may be used for each one.   ◇

Proof. ($\Leftarrow$) This statement appeared as part (4) of Theorem 2.4, but we did not prove it at that time. Let x and y be odd integers. By the definition of odd, there exist integers k and m such that $x = 2k + 1$ and $y = 2m + 1$. By substitution,

$$xy = (2k + 1)(2m + 1) = 4km + 2k + 2m + 1 = 2(2km + k + m) + 1.$$

By closure of the integers under addition, $2km + k + m$ is an integer. Therefore, by definition, xy is odd.

($\Rightarrow$) This statement appeared as part (5) of Theorem 2.4. Assume that x and y are integers but they are not both odd; that is, at least one of x or y is even. Due to the symmetry of x and y in this case, without loss of generality we may assume that x is even. Then, by the definition of even, $x = 2n$ for some integer n. It follows upon substitution that

$$xy = (2n)y = 2(ny).$$

By closure of the integers, ny is an integer, so by definition, xy is even. By contrapositive, if xy is odd, then x and y must both be odd.

Thus, we have proven that the product of two integers is odd if and only if both of the integers are odd.   □

4.5.4 Constructive Proofs, Counterexamples, and Vacuous Proofs.

While proofs of universally quantified statements are more commonly encountered, knowing how to prove an existentially quantified statement is essential. Recall that an existentially quantified statement simply makes a claim about the existence of a particular entity. If a single example of the desired object can be produced, the statement has been proven. Such a proof is often called a **constructive proof**.

Example 44. Prove that there exists an integer n such that $\dfrac{n^2 + n}{3n + 8} = 1$.

First Thoughts. In this case, algebra should reveal a candidate for n. We work backwards to find n, noting that each step is reversible, before presenting the actual proof.

$$\frac{n^2 + n}{3n + 8} = 1 \qquad\qquad \Longleftrightarrow$$
$$n^2 + n = 3n + 8 \qquad\qquad \Longleftrightarrow$$
$$n^2 - 2n - 8 = 0 \qquad\qquad \Longleftrightarrow$$
$$(n - 4)(n + 2) = 0.$$

Thus, we can let $n = 4$ or $n = -2$ to prove the statement. ◇

Proof. Let $n = 4$. Then

$$\frac{n^2 + n}{3n + 8} = \frac{4^2 + 4}{(3 \cdot 4) + 8} = \frac{20}{20} = 1.$$

This proves the statement. □

Example 45. Prove that the irrational numbers are not closed under multiplication.

First Thoughts. The statement *p: irrational numbers are closed under multiplication* is a universal statement. However, the given statement, which may be read as, *It is not the case that the irrational numbers are closed under multiplication*, is the negation of p. This means the given statement is logically equivalent to an existential statement. We can prove it false if we can produce two irrational numbers whose product is rational (i.e., has the form a/b for some integers a and b). ◇

Proof. Let $x = \sqrt{2}$ and $y = \sqrt{8}$. Then x and y are both irrational, but $xy = 4$ is rational. Thus the irrational numbers are not closed under multiplication. □

Occasionally, one is presented with a statement that may or may not be true and is asked to prove *or* disprove the given statement. In this case, experimentation may be required in order to decide whether to attempt a proof or a disproof. To disprove a universally quantified statement, providing a single **counterexample**—that is, a single example in which the given statement does *not* hold—is sufficient; thus disproof of a universally quantified statement is constructive. On the other hand, disproving an existentially quantified statement amounts to *proving* a quantified statement: one must show that the given statement does not hold for *any* elements of the domain of discourse.

Example 46. Prove or disprove that if n^2 is divisible by 9, then n is divisible by 9.

First Thoughts. Before going on, the reader should decide (by experimentation with particular integers or by reasoning) whether she thinks the statement is true or false. This will determine whether to strive to write a proof or a disproof.   ◇

Disproof. If $n = 3$, then n^2 is divisible by 9; however, n is not divisible by 9. Thus the statement is false.   □

In summary,

(1) a single example *cannot prove* a universally quantified statement (unless the domain of discourse contains only one element);

(2) a single counterexample *can disprove* a universally quantified statement;

(3) a single example *can prove* an existentially quantified statement;

(4) a single counterexample *cannot disprove* an existentially quantified statement (unless the domain of discourse contains only one element).

Finally, we consider the situation in which a statement of the form "if A, then B" is to be proven, but the statement A is *never* true. Since a conditional statement is true any time the antecedent is false (regardless of whether the consequent is true or false), we would regard such a statement as **vacuously true**. This notion is actually used in common English with statements "If A, then pigs will fly" or "If A, then I'm a monkey's uncle" or "If A, then hell will freeze over," where the speaker is voicing his opinion about the impossibility of A occurring.

Proving that A is always false is sufficient for proving statements of the form "if A, then B".

Example 47. For all $x \in \mathbb{R}$, if $x^2 < 0$, then $3x^2 + 5 = -7x$.

First Thoughts. Note that this statement is in the form "if $A(x)$, then $B(x)$", with statements $A(x) : x^2 < 0$ and $B(x) : 3x^2 + 5 = -7x$ over the domain of discourse $\mathbb{R}$.   ◇

Proof. For any $x \in \mathbb{R}$, $x^2 \geq 0$. Thus, since the antecedent ($x^2 < 0$) is always false, the implication is vacuously true.   □

Exercises and Problems.

> *For want of a nail the shoe was lost.*
> *For want of a shoe the horse was lost.*
> *For want of a horse the rider was lost.*
> *For want of a rider the message was lost.*
> *For want of a message the battle was lost.*
> *For want of a battle the kingdom was lost.*
> *And all for the want of a horseshoe nail.*
>
> —Proverb

(1) Prove that no integer is both even and odd.

(2) Given sets A, B, and C, prove that if $A \subseteq B$ and $B \subseteq C$, then $A \subseteq C$. (Note: A Venn diagram is not a proof! You should provide a proof in paragraph form, using the definition of subset repeatedly.)

(3) Prove that the product of an even integer and any other integer is an even integer.

(4) Prove or disprove each of the statements.

    (a) The product of any two consecutive integers is even.

    (b) The product of any integer and its square is an even integer.

    (c) The sum of two positive odd integers is never prime.

    (d) The sum of two even integers is always composite.

    (e) The sum of two prime numbers is sometimes a prime number.

    (f) The square of any even integer is divisible by 4.

    (g) Every positive integer can be written as the sum of two squares.

    (h) The sum of two rational numbers is a rational number.

    (i) The sum of two irrational numbers is an irrational number.

(5) Prove that if $a \bmod n = b \bmod n$, then $n|(a - b)$.

(6) Return to the Sudoku problem at the beginning of the chapter. Prove each of the following statements by making a sequence of logical statements.

    (a) The middle entry of the puzzle is a 9.

    (b) The entry in row 4, column 3 is a 7.

    (c) The entry in row 4, column 6 is a 4.

    (d) The entry in row 5, column 2 is a 2.

    (e) The entry in row 5, column 1 is a 3.

---

(7) Prove that if $a \bmod n = r$, then $a^2 \bmod n = r^2 \bmod n$.

(8) For sets A and B, prove that the following three statements are equivalent.

    (a) A and B are disjoint.

    (b) $A \setminus B = A$.

    (c) $B \setminus A = B$.

(9) Prove that for every integer x, if x is odd, then there exists an integer y such that $x^2 = 8y + 1$.

(10) In a popular word game, one is given a starting word and then must change one letter at a time, forming a valid word at each step, and end with another given word. For example, given the words DOG and CAT, a valid sequence of words is DOG-COG-COT-CAT. Can you find a valid sequence of words to move from the word SHIRT to the word PANTS? Prove that in any such sequence at least one of the words of the sequence must contain two vowels.

(11) Prove that for every integer x, $x - 3$ is odd if and only if $x + 12$ is even.

(12) Use proof by contradiction to prove that if x is rational and y is irrational, then $x + y$ is irrational.

(13) Prove that if $a < b$ and $b < c$, then $a < c$. Use the following definition of $<$ in your proof: given real numbers x and y, we say $x < y$ if there exists a positive number z such that $y = x + z$.

(14) Prove that if $3|(a^2 + b^2)$ for some integers a and b, then $3|a$ and $3|b$.

(15) There are two ways one might interpret the statement "If p, then q, if r." Write the statement in symbolic form in two different ways, using the implication symbol and parentheses, and prove they are logically equivalent.

(16) Recall that the **symmetric difference** of sets A and B, denoted $A \triangle B$, is defined to be the set of elements which are members of exactly one of A or B. Prove that $A \triangle B = (A \setminus B) \cup (B \setminus A) = (A \cup B) \setminus (A \cap B)$.

Suppose an open necklace contains a string of four red beads and four blue beads. The pattern of the necklace can be described by using a string of eight characters (four R's and four B's). There is no first bead, so two necklace patterns should be considered identical if one necklace may be obtained from the other by reversing the order of the beads and pushing a substring of beads around the necklace to join the other side. For example, the patterns *BBBRRBRR*, *RRBRRBBB*, and *BRRBBBRR* would all be considered identical. Create a list of all distinct patterns (one representative for each pattern).

5

Relations and Functions

Mathematicians do not study objects, but relations among objects; they are indifferent to the replacement of objects by others as long as relations do not change. Matter is not important, only form interests them.

—Henri Poincaré (1854–1912)

It would scarcely be an overstatement to say that most of mathematics, including the contents of this book, can be represented using a mathematical structure called a relation. Nonmathematical relationships—whether familial, business, or romantic—can also be described via relations.

They can also describe relationships, whether familial, business, or romantic. Viewed generally as a pairing of elements from two sets, binary relations can describe any personal or mathematical relationship that you can imagine. This includes tasks such as comparing, ordering, grouping, or labeling. Inasmuch as relations describe any assignment of one item to another, functions are merely a special category of relations. Special types of functions, such as permutations and sequences, arise in discrete mathematics and are covered in this text. In this chapter, we provide an introduction to relations, functions, and sequences.

5.1 Relations

Let A and B be sets, and recall that $A \times B$ is the set of all ordered pairs (a, b) where $a \in A$ and $b \in B$. A **binary relation** R **from** A **to** B is any subset of $A \times B$. We will usually refer to a binary relation as simply a **relation**. In essence, a relation is any pairing of (some or all of) the elements of one set, A, with (some or all of) the elements of another set, B; symbolically, $R \subseteq A \times B$. In the case where A and B are equal sets, we simply say that R is a **relation on** A. As above, the letter R is often used to represent a relation. However, we will also denote relations by mathematical symbols or lowercase letters, which can be Greek (such as α, β, or γ) or Latin (such as f, g, or h).

Example 48. Let $A = \{1, 5\}$ and $B = \{x, y, z\}$. The following are examples of relations from A to B:

$$\alpha = \{(1, x), (1, y), (5, z)\}, \qquad\qquad \beta = \{(1, x), (1, y)\},$$
$$\theta = \{(5, x), (5, y), (5, z)\}, \qquad\qquad f = \{(1, x), (5, z)\}.$$

When an ordered pair (a, b) is an element of a relation R, we say a **is related to** b and denote it in one of the following two ways:

$$(a, b) \in R \quad \text{or} \quad aRb.$$

In Example 48, we have $(1, y) \in \alpha$, $1 \beta x$, $5 \theta z$, and $(1, x) \in f$. Many familiar mathematical symbols (such as $=, \cong, \sim, \equiv, <, \leq$) are used to represent well-known relations. With a familiar relation such as $<$, although it would be technically correct to write $(3, 5) \in <$ to denote that $3 < 5$, this sentence might be the only time it has ever happened.

Collecting the first coordinates of all the pairs that form a relation ϕ, we get a subset of A called the **domain** of ϕ, denoted by $\mathrm{Dom}(\phi)$. In set notation, this definition is written as

$$\mathrm{Dom}(\phi) = \{a \in A : \exists b \in B, (a, b) \in \phi\}.$$

In Example 48, $\mathrm{Dom}(\alpha) = \{1, 5\} = A$, $\mathrm{Dom}(\beta) = \{1\}$, $\mathrm{Dom}(\theta) = \{5\}$, and $\mathrm{Dom}(f) = \{1, 5\} = A$.

Collecting the second coordinates of all the pairs that form a relation ϕ, we get a subset of B called the **range** of ϕ, denoted by $\mathrm{Range}(\phi)$. This definition can also be written as

$$\mathrm{Range}(\phi) = \{b \in B : \exists a \in A, (a, b) \in \phi\}.$$

In Example 48, $\mathrm{Range}(\alpha) = \{x, y, z\} = B$, $\mathrm{Range}(\beta) = \{x, y\}$, $\mathrm{Range}(\theta) = \{x, y, z\} = B$, and $\mathrm{Range}(f) = \{x, z\}$.

A relation can be represented by a **graph** of the ordered pairs, as shown in Figure 5.1, typically with a horizontal axis corresponding to the set of first coordinates (the elements of A) and the vertical axis corresponding to the set of second coordinates (the elements of B) in the pairs of the relation. This method of representing relations likely calls to mind a familiar graphical way of representing equations, such as $y = 2x$, in the Cartesian plane.

More commonly within discrete mathematics, a relation can be represented by its **directed graph** (also called a **digraph**), as shown in Figure 5.2. In a digraph, each (directed) edge corresponds to an ordered pair; the edge originates at the first coordinate and terminates at the second coordinate, with each element of A or B represented by a vertex of the digraph. The number of edges that originate at vertex v is the **out-degree** of v; the number of edges that terminate at vertex v is the **in-degree** of v. In the case when R is a relation on A (i.e., $A = B$), an element may be related to itself; the directed graph of such a relation would include an edge that begins and ends at the same vertex, called a **loop**.

The graph and digraph of a relation allow us to visualize the domain and range. When representing a relation ϕ by its graph, we see that $\mathrm{Dom}(\phi)$ and $\mathrm{Range}(\phi)$ are the projections of ϕ onto the sets A and B, respectively. In a digraph representation, the range and domain consist of the elements at the initial and terminal points of the arrows, respectively.

Often relations are defined by predicates, as in the next example.

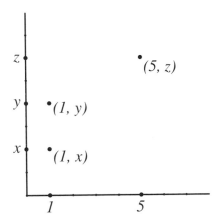

Figure 5.1. Graph of α

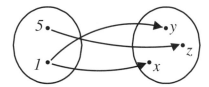

Figure 5.2. Digraph of α

Example 49. Consider the relation

$$\phi = \{(x, y) : x, y \in \mathbb{R}, \ (x^2 + y^2 - 25)(x^2 + (y - 8)^2 - 9) = 0\}.$$

There are infinitely many pairs of real numbers in ϕ, including the following:

$$A : (-5, 0), \quad B : (-3, -4), \quad C : (\sqrt{11}, \sqrt{14}), \quad D : (3, 8), \quad E : (0, 11).$$

The graph of this relation is the union of two tangent circles in the xy-plane, one of radius 5 with center at the origin and the other with radius 3, centered at $(0, 8)$. (See Figure 5.3.) The domain and range are the intervals of real numbers $[-5, 5]$ and $[-5, 11]$, respectively.

In everyday English, the word "relation" has many meanings, but often one can identify a mathematical formulation of the situation. For example, consider the relationship between U.S. airports and international airports: "There is a direct flight from one airport to the other." The mathematical formulation can be thought of as the relation from the set U of all United States airports to the set I of international airports, formed by ordered pairs (a, b), where there is a direct flight from airport a to airport b. Of course, the relation could be understood without regard to geographic location if it were defined on $W = U \cup I$, the set of all airports in the world. In this case, $(a, b) \in W \times W$ if there is a flight from airport a to airport b. While similar, these two relations are different because they are defined on different sets.

Here are several more examples. Notice that some of them are examples of information that could be stored in a database, so the application to data management is apparent.

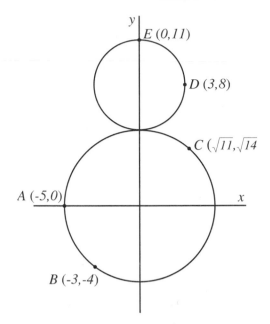

Figure 5.3.   Graph for Example 49

Example 50. Let A be the set of all students currently enrolled at Andersen High School.

(1) At most high schools, each student is given a personal ID number; let B be the set of all student ID numbers at Andersen High. We can define a relation γ by $(a, b) \in \gamma$ if b is the student ID number for student a. $\text{Dom}(\gamma) = A$ and $\text{Range}(\gamma)$ is the set of student ID numbers currently in use at Andersen High.

(2) Let C be the set of math classes taught at Andersen High School. Then a relation δ can be defined as $(c, a) \in \delta$ if student a is on the class roster for course c. So, for instance, (*Geometry Section 5, Brahmagupta*) $\in \delta$ means that Brahmagupta is a student in Section 5 of the geometry course. The domain is the set of math courses currently being taught at Andersen High and the range is the set of Andersen students who are currently taking a math class.

(3) An alternative relation, H, for the situation in (2) can be created by letting C be as before and considering $\mathcal{P}(A)$, the set of all possible subsets of the students enrolled at Andersen High. Now, define $(c, r) \in H$ if $r \in \mathcal{P}(A)$ is the set of students on the class roster for class c. In this case, $\text{Dom}(H)$ is the set of math courses currently being taught at Andersen High, and $\text{Range}(H)$ is the set of all class rosters, a subset of $\mathcal{P}(A)$.

Example 51. Let A be the set of all NFL football teams. Define a relation θ on A by $a\theta b$ if team a beat team b in a regular season game during a specified season. A digraph for θ on a subset of six teams in A is given in Figure 5.4. Notice that the graph is drawn on one copy of A only, since θ is a relation from A to A.

It happened that New Orleans beat Atlanta twice in the season in question, but there is just one arrow from New Orleans to Atlanta because a relation is a set and each ordered pair can occur only once. Since Green Bay and Chicago played each other

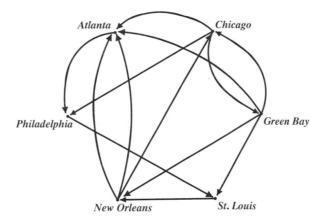

Figure 5.4. Digraph for Example 51

twice and each team beat the other once, there is a pair of oppositely directed arrows between Green Bay and Chicago.   □

In addition to digraphs, binary relations can also be represented using matrices. This is most advantageous in the world of computer programming, where two-dimensional arrays are used to store, examine, and manipulate relations. Suppose $A = \{a_1, a_2, \ldots, a_m\}$ and $B = \{b_1, b_2, \ldots, b_n\}$, noting that an implied order has been imposed on the elements of these sets, and suppose $R \subseteq A \times B$. The **matrix of relation** R, denoted M_R, is the $m \times n$ matrix whose (i, j)-entry is 1 if $(a_i, b_j) \in R$ and it is 0 if $(a_i, b_j) \notin R$, for $i = 1, 2, \ldots, m$ and $j = 1, 2, \ldots, n$.

For the relation $\alpha = \{(1, x), (1, y), (5, z)\}$ from $A = \{1, 5\}$ to $B = \{x, y, z\}$ found in Example 48, M_α is the following 2×3 matrix:

$$\begin{bmatrix} 1 & 1 & 0 \\ 0 & 0 & 1 \end{bmatrix}.$$

Notice that rows 1 and 2 correspond to elements 1 and 5, while columns 1, 2, and 3 correspond to elements x, y, and z, respectively. It should be clear how the three ordered pairs in R correspond to the three occurences of 1 in the matrix, as well as to the three edges in the digraph in Figure 5.2. If R is a relation on a set A with $|A| = n$, then M_R is a square matrix of size n. The following relation is an example of this.

Example 52. Let $A = \{1, 2, 3, 4, 5, 6, 7, 8, 9\}$ and define γ on A by $a\gamma b$ if $3|(a - b)$; that is, $\gamma = \{(a, b) \in A \times A : 3|(a - b)\}$. Represent γ with a directed graph and with a matrix.

First Thoughts. Note that $(7, 7) \in \gamma$ since $7 - 7 = 0$ and $3|0$. Likewise, $(a, a) \in \gamma$ for each $a \in A$. This property is portrayed in the digraph of γ by a looping arrow from each element to itself. (See the digraph in Figure 5.5.)

Since $7 - 4 = 3$ and $3|3$, $(7, 4) \in \gamma$; since $4 - 7 = -3$ and $3|(-3)$, $(4, 7) \in \gamma$ also. This property is shown in the digraph of γ by having arrows both from 7 to 4 and from 4 to 7. More generally, this relation has the property that if $(a, b) \in \gamma$ for some a and b, then $(b, a) \in \gamma$ also.   ◇

Solution. The digraph of R appears in Figure 5.5. The matrix of this relation is

$$M_R = \begin{bmatrix} 1 & 0 & 0 & 1 & 0 & 0 & 1 & 0 & 0 \\ 0 & 1 & 0 & 0 & 1 & 0 & 0 & 1 & 0 \\ 0 & 0 & 1 & 0 & 0 & 1 & 0 & 0 & 1 \\ 1 & 0 & 0 & 1 & 0 & 0 & 1 & 0 & 0 \\ 0 & 1 & 0 & 0 & 1 & 0 & 0 & 1 & 0 \\ 0 & 0 & 1 & 0 & 0 & 1 & 0 & 0 & 1 \\ 1 & 0 & 0 & 1 & 0 & 0 & 1 & 0 & 0 \\ 0 & 1 & 0 & 0 & 1 & 0 & 0 & 1 & 0 \\ 0 & 0 & 1 & 0 & 0 & 1 & 0 & 0 & 1 \end{bmatrix}.$$

□

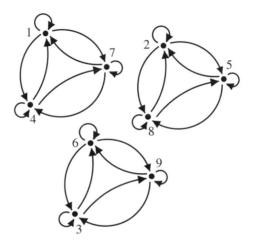

Figure 5.5.   Digraph for Example 52

Further Thoughts. A careful comparison of the digraph and the matrix of R is in order, as every feature of one will be observed in the other. For instance, each loop in the digraph is represented by a 1 on the diagonal of the matrix. The fact that for each edge in the digraph there is a second edge in the opposite direction is indicated by the symmetry of the matrix.

Finally, notice that in this example, if $(a, b) \in \gamma$ and $(b, c) \in \gamma$ for some a, b, and c, then $(a, c) \in \gamma$. This property can be confirmed by checking all cases for small sets like A; alternatively, it can be proven in a general fashion, as we will do in Chapter 7.   ◇

5.2 Properties of Relations on a Set

The relation in Example 52 exhibits several significant properties (reflexive, symmetric, transitive). We define them below, along with other notable properties of relations.

A binary relation, ϕ, on A is called

- **reflexive** if for all a in A, (a, a) is in ϕ;
- **irreflexive** if for all a in A, (a, a) is not in ϕ;
- **symmetric** if for all a and b in A, when (a, b) is in ϕ, then (b, a) is in ϕ also;

- **asymmetric** if for all a and b in A, if (a, b) is in ϕ, then (b, a) is not in ϕ;
- **antisymmetric** if for all a and b in A, when (a, b) and (b, a) are in ϕ, then $a = b$;
- **transitive** if for all a, b, and c in A, when (a, b) and (b, c) are in ϕ, then (a, c) is in ϕ also;
- an **equivalence relation** if ϕ is reflexive, symmetric, and transitive;
- a **partial ordering** on A if ϕ is reflexive, antisymmetric, and transitive.[1] A set together with a partial ordering is called a **partially ordered set** or a **poset**.

Example 53. Determine which of the above properties hold on each of the following binary relations. Which of them are equivalence relations? Which are partial orderings?

(1) Let A be all citizens of the U.S. Define a relation ϕ_1 on A by $a \, \phi_1 \, b$ if a and b have the same biological mother.

(2) Define a relation ϕ_2 on $\mathbb{N}$ by $a \, \phi_2 \, b$ if a is a divisor of b. (The symbol $|$ is commonly used for this relation, in which case $a|b$ means that a is a divisor of b.)

(3) Define a relation ϕ_3 on $\mathbb{R}$ by $\phi_3 = \{(a, b) : a \leq b\}$.

(4) Define a relation ϕ_4 on $\mathbb{R}$ by $\phi_4 = \{(a, b) : a < b\}$.

(5) Define a relation ϕ_5 on $\mathbb{Z}$ by $a \, \phi_5 \, b$ if $a - b$ is a multiple of 7. This is very similar to the relation given in Example 52. Another way of stating this relation is that $a \, \phi_5 \, b$ if a mod $7 = b$ mod 7, based on the result in Problem 5 of Section 4.5.

Solution. (1) ϕ_1 is reflexive: Any citizen of the U.S. has the same biological mother as him/herself, so $(a, a) \in \phi_1$ for any $a \in A$.

ϕ_1 is symmetric: If a and b have the same biological mother, then both $(a, b) \in \phi_1$ and $(b, a) \in \phi_1$ are true.

ϕ_1 is transitive: If $(a, b) \in \phi_1$, then a and b have the same mother; if $(b, c) \in \phi_1$, then b and c have the same mother. Hence, a, b, and c all have the same mother, so $(a, c) \in \phi_1$.

Thus, ϕ_1 is an equivalence relation.

(2) ϕ_2 is reflexive: If a is any positive integer, then $a = a \cdot 1$, so by definition a divides a; thus, $(a, a) \in \phi_2$.

ϕ_2 is antisymmetric: For $a, b \in \mathbb{N}$, suppose that (i) $(a, b) \in \phi_2$ (a is a divisor of b), in which case $b = am$ for some positive integer m, and (ii) $(b, a) \in \phi_2$, in which case $a = bk$ for some positive integer. By substitution, this means that $a = amk$. Therefore, $mk = 1$, so $m = k = 1$ and $a = b$.

Transitivity follows from Theorem 2.5(2).

Since ϕ_2 is reflexive, antisymmetric, and transitive, ϕ_2 is a partial ordering.

(3) ϕ_3 is reflexive: For any real number a, $(a, a) \in \phi_3$, since $a \leq a$.

ϕ_3 is antisymmetric: For real numbers a and b, if $a \leq b$ and $b \leq a$, then $a = b$.

ϕ_3 is transitive: For real numbers a, b, and c, if $a \leq b$ and $b \leq c$, then $a \leq c$.

Since ϕ_3 is reflexive, antisymmetric, and transitive, ϕ_3 is a partial ordering.

---

[1]The relation is called a "partial" ordering because there may be a pair of elements of A that are not related to each other; elements of A that are not related to each other are called **incomparable**.

(4) ϕ_4 is irreflexive: For any real number a, $a \not< a$, so $(a, a) \notin \phi_4$.

ϕ_4 is asymmetric: If $a < b$, then it cannot be the case that $b < a$.

ϕ_4 is (vacuously) antisymmetric: Since ϕ_4 is asymmetric, $(a, b) \in \phi_4$ and $(b, a) \in \phi_4$ will never both occur.

ϕ_4 transitive: For real numbers a, b, and c, if $a < b$ and $b < c$, then $a < c$.

Since ϕ_4 is not reflexive, it is neither a partial ordering nor an equivalence relation.

(5) ϕ_5 is reflexive: For any integer a, $a - a = 0$; since 7 divides 0, $a - a$ is a multiple of 7, so $(a, a) \in \phi_5$.

ϕ_5 is symmetric: For integers a and b, if $a - b = 7q$ for some integer q, then $b - a = 7(-q)$. Thus, if $(a, b) \in \phi_5$, then $(b, a) \in \phi_5$ also.

ϕ_5 is transitive: Suppose that a, b, and c are integers. If $(a, b) \in \phi_5$, then $a - b = 7q$ for some integer q; likewise, if $(b, c) \in \phi_5$, then $b - c = 7p$ for some integer p. Therefore,

$$a - c = (a - b) + (b - c) = 7q + 7p = 7(q + p).$$

Since $q + p$ is an integer, $(a, c) \in \phi_1$.

Thus, ϕ_5 is an equivalence relation. In fact, ϕ_5 is an example of a special type of equivalence relation, called a **congruence modulo** n (where, in this case, $n = 7$). Congruence relations will be explored much more fully in Chapter 7. □

Equivalence relations are useful in mathematics because they provide a convenient way to group elements that are "similar". Given an equivalence relation R on a set A and some $a \in A$, the set of all elements related to a under R is called the **equivalence class containing** a, denoted by $[a]$.[2] That is,

$$[a] = \{x \in A : aRx\} \quad (= \{x \in A : xRa\}).$$

Example 54. In the opening hook problem for this chapter, two open necklaces, each with four red beads and four blue beads, were considered to be identical (i.e., equivalent) if one could be obtained from the other by a combination of rotating the beads and reversing their order. The necklace patterns, described as a sequence of eight red or blue beads, form an equivalence relation, because the relationship of being identical is reflexive, symmetric, and transitive. List all patterns in the same equivalence class as the pattern BBBRRBRR.

Solution. By repeatedly rotating the first bead to the end of the necklace, we obtain the additional patterns BBRRBRRB, BRRBRRBB, RRBRRBBB, RBRRBBBR, BRRBBBRR, RRBBBRRB, and RBBBRRBR. By inspection, reversing the order of each of these eight patterns does not yield any new ones, so these eight patterns are the only elements of the equivalence class. □

In the equivalence relation ϕ_5 in Example 53, the equivalence class containing 0 consists of all multiples of 7. The equivalence class containing a consists of all integers

---

[2]Here we have yet another unfortunate instance where the same notation is used to describe two different mathematical entities. One must use context to determine whether $[n]$ denotes an equivalence class or the set of integers from 1 to n.

m for which m mod $7 = a$ mod 7. The seven equivalence classes, more specifically known as **congruence classes** in this context, are shown below:

$$[0] = \{\dots, -7, 0, 7, 14, \dots\},$$
$$[1] = \{\dots, -6, 1, 8, 15, \dots\},$$
$$[2] = \{\dots, -5, 2, 9, 16, \dots\},$$
$$[3] = \{\dots, -4, 3, 10, 17, \dots\},$$
$$[4] = \{\dots, -3, 4, 11, 18, \dots\},$$
$$[5] = \{\dots, -2, 5, 12, 19, \dots\},$$
$$[6] = \{\dots, -1, 6, 13, 20, \dots\}.$$

Notice, for example, that $[-5] = [2] = [9] = \dots = [100] = \dots$. In general, related elements generate identical equivalence classes, as stated in the following theorem.

Theorem 5.1. *Let R be an equivalence relation on a set S, with $a \in S$ and $b \in S$. The following statements are equivalent.*

(1) $[a] = [b]$.

(2) $[a] \cap [b] \neq \varnothing$.

(3) aRb.

Proof. You are asked to show that $(2) \Rightarrow (1)$ in Problem 23. Clearly, $(1) \Rightarrow (2)$, and this proves that (1) and (2) are equivalent statements. Now assume that $[a] = [b]$. Then, since $b \in [b]$, $b \in [a]$. By definition $[a]$ contains all elements x for which aRx, so aRb, which proves that $(1) \Rightarrow (3)$. On the other hand, if aRb, then since R is an equivalence relation, bRa. Let x be an arbitrary element in $[a]$. Then aRx and, by transitivity, bRx, so $x \in [b]$. This proves that $[a] \subseteq [b]$. Similarly, $[b] \subseteq [a]$, so $[a] = [b]$. Thus $(3) \Rightarrow$ (1). Together these implications prove that (1), (2), and (3) are equivalent. $\qquad\square$

Theorem 5.1 asserts that the equivalence classes are either identical or disjoint. Furthermore, since every element of S belongs to the equivalence class it generates, the union of all equivalence classes will be S itself. These are the attributes of a partition: any collection of nonempty, pairwise disjoint subsets of a set S whose union is all of S is called a **partition** of S. That is, a partition of a set S is a collection of subsets A_i, $i \in I$ for some indexing set I, for which

$$A_i \neq \varnothing \text{ for all } i \in I, \quad A_i \cap A_j = \varnothing \text{ for } i \neq j, \quad \text{and} \quad \bigcup_{i \in I} A_i = S.$$

The subsets A_i are called the **cells** of the partition. For example, if $S = \{1, 2, \dots, 10\}$, one partition of S is

$$\{\{1, 2, 3\}, \{4, 5, 6, 7\}, \{8, 9, 10\}\}.$$

A partition of S with different cells is $\{A_1, A_2, A_3, A_4\}$, where $A_1 = \{1, 4, 5, 7, 9\}$, $A_2 = \{6\}$, $A_3 = \{2, 10\}$, and $A_4 = \{3, 8\}$.

While the equivalence classes of an equivalence relation R on a set S form the cells of a partition of S, the converse is also true. That is, any partition of S defines an equivalence relation on S (two elements are defined to be related if and only if they lie in the same cell of the partition).

Exercises and Problems.

(1) Graph the points of the relation $\phi \subseteq \mathbb{R} \times \mathbb{R}$ in each case.

    (a) $\phi = \{(x, y) : |x| = |y|\}$.

    (b) $\phi = \{(x, y) : -2 \leq x < 2,\ x + y = 1\}$.

    (c) $\phi = \{(x, y) : 2 \leq x \leq 4,\ 2x + 1 \leq y \leq x^2 + 12\}$.

(2) Define a relation R on the set $A = \{1, 2, \dots, 6\}$ whereby aRb if and only if $a|b$ (a is a divisor of b).

    (a) List all ordered pairs in R.

    (b) Sketch the digraph of R.

    (c) Determine the matrix of R.

(3) Define the relation $\succ$ on the set consisting of the elements Rock, Paper, and Scissors as follows: $x \succ y$ if x beats y in the game Rock-Paper-Scissors.

    (a) List all ordered pairs in the relation.

    (b) Sketch the digraph of the relation.

    (c) Does the digraph have any loops? Describe how the rules of the game dictate your answer.

    (d) Is the relation transitive?

(4) Let $A = \{1, 2, \dots, 9\}$. Define the relation R on A by xRy if x and y have the same number of letters when written as English words. For example, $(9, 4) \in R$ because "nine" and "four" are both four-letter words. It may be helpful to know that $|R| = 18$.

    (a) Determine the matrix of R.

    (b) Sketch the digraph of R.

    (c) Is R an equivalence relation?

(5) Let H be the set of all people in the world. Define a "cousin" relation κ on H via $a\kappa b$ if and only if Person a and Person b are descended from a same grandmother.

    (a) Explain why it is reasonable to conclude κ is reflexive.

    (b) Is κ symmetric, asymmetric, or antisymmetric?

    (c) Is κ transitive?

    (d) Is κ an equivalence relation?

    (e) Explain why κ is not truly a cousin relationship in the usual understanding of the word.

(6) Consider the equivalence relation β defined on the set H of all people in the world, whereby $a\beta b$ if and only if Person a and Person b share the same blood type. Give a detailed description of the equivalence classes of β.

(7) Let ϕ be the relation on the set of integers, where $(a, b) \in \phi$ if $a + b$ is even.

    (a) List five elements of ϕ.

    (b) Is ϕ reflexive?

    (c) Is ϕ symmetric?

    (d) Is ϕ transitive?

    (e) Is ϕ an equivalence relation? If so, describe the equivalence classes.

(8) Do the four quadrants of the Cartesian plane partition its points? Why or why not?

(9) The prime numbers, P, consist of all positive integers that have exactly two positive divisors, and the composite numbers, C, are those with more than two positive divisors. Does $\{P, C\}$ form a partition of $\mathbb{N}$? Briefly explain.

(10) Let $S = \{1, 2, 3, 4, 5\}$. Determine which of the following are partitions of S.

    (a) $\{\{1, 3, 5\}, \{2\}\}$

    (b) $\{\{1\}, \{2\}, \{3, 4\}, \{5\}\}$

    (c) $\{\{1, 2, 3\}, \{3, 4, 5\}\}$

    (d) $\{\{1, 2, 3, 4, 5\}\}$

(11) Create partitions of the set $\mathbb{Z}$ that meet the following criteria.

    (a) The partition consists of two cells.

    (b) The partition consists of three cells, one with finite cardinality and two with infinite cardinality.

    (c) The partition consists of three cells, each with infinite cardinality.

(12) Consider the relation $R = \{(a, a), (a, b), (a, c), (b, a), (b, b), (b, c), (c, a), (c, b), (c, c), (d, d), (d, e), (e, d), (e, e), (f, f)\}$ on $A = \{a, b, c, d, e, f\}$.

    (a) Verify that R is an equivalence relation.

    (b) Determine the equivalence classes for R and verify that the equivalence classes form a partition of A.

    (c) Sketch the digraph of R in such a way that the partition of equivalence classes is clearly shown.

(13) Consider the partition $\mathcal{P} = \{\{1, 3\}, \{2\}, \{4, 5\}\}$ of $A = \{1, 2, 3, 4, 5\}$. Let R be the relation on A defined by aRb if and only if a and b belong to the same cell of $\mathcal{P}$. List all ordered pairs in R.

(14) Let R be a relation on the prime numbers, P, where $(a, b) \in R$ if $a + b$ is odd.

    (a) List five elements of R. Do you think R is an infinite set?

    (b) Determine which of these properties R has: reflexive, irreflexive, symmetric, antisymmetric, transitive.

    (c) True or false: For all a, b, c in P, if $(a, b) \in R$ and $(b, c) \in R$, then $(a, c) \notin R$.

(15) Let R be a relation on the prime numbers, P, where $(a, b) \in R$ if $a - b$ is prime. List five elements of R. Do you think R is an infinite set?

---

(16) Let R be the relation on $\mathbb{N} \times \mathbb{N}$ (where $\mathbb{N}$ is the set of natural numbers) given by $((a, b), (c, d)) \in R$ if $a \leq c$ and $b \leq d$. Determine which of these properties R has: reflexive, irreflexive, symmetric, antisymmetric, transitive.

(17) Let α be the relation on the set of positive integers where $(a, b) \in \alpha$ if a and b share a common divisor greater than 1.

   (a) List five elements of α.

   (b) Is α reflexive?

   (c) Is α symmetric?

   (d) Is α transitive?

   (e) Is α an equivalence relation?

   (f) Is α a partial ordering on $\mathbb{N}$?

(18) Explain why the "divides" relation ($a|b$ if a is a divisor of b) is neither a partial ordering nor an equivalence relation on the set of nonzero integers.

(19) Let R be a binary relation on A. Each of the following sentences would be an appropriate first line of a proof for proving one or more of the following properties about R: symmetric, asymmetric, antisymmetric, transitive, or reflexive. In each case, list the properties.

   (a) Let (a, b) be an arbitrary element of R.

   (b) Suppose that (a, b) and (b, a) are both elements of R.

   (c) Suppose that (a, b) and (b, c) are both elements of R.

(20) Let R be a binary relation A. Each of the following sentences would be an appropriate penultimate line of a proof for proving one or more properties about R: symmetric, asymmetric, antisymmetric, transitive, reflexive, not reflexive, irreflexive, or a partial ordering. In each case, write a final sentence for the proof.

   (a) We have found an element $a \in A$ such that $(a, a) \notin R$.

   (b) Therefore, the only way that two elements of A can be related to each other is if the two elements are actually equal.

   (c) We have shown that each element of A is related to itself, that two distinct elements cannot be related to each other, and that the relation R is transitive.

   (d) This proves that no element of A is related to itself.

   (e) Therefore, for arbitrary elements a and b in A, if aRb, then bRa.

   (f) Thus each element of A is related to itself.

   (g) This proves that $(b, a) \notin R$.

(21) Define the "approximate" relation on the real numbers via $x \approx y$ if $|x - y| \leq 0.1$. Is $\approx$ reflexive? Symmetric? Transitive? Prove your answer in each case.

(22) Suppose R is a relation on a set A with $|A| = n$. Describe how each of the following facets of R and its digraph are detected or computed from the matrix M_R.

   (a) The in-degree of any vertex.

   (b) The out-degree of any vertex.

   (c) The total number of pairs in R or edges in its digraph.

   (d) Loops in the digraph.

   (e) Whether R is reflexive, irreflexive, or neither.

   (f) Whether R is symmetric.

   (g) Whether R is asymmetric.

   (h) Whether R is antisymmetric.

(23) Let R be an equivalence relation on a set A, and let $a \in A$. Prove that if $b \in [a]$, then $[a] = [b]$.

(24) In a popular word game, one is given a starting word and then must change one letter at a time, forming a valid word at each step, and end with another given word. For example, given the words DOG and CAT, a valid sequence of words is DOG-COG-COT-CAT. Define a relation on the set of all words in the English language as follows: (a, b) is in the relation if there is a valid sequence of moves that takes word a to word b. Is this an equivalence relation? Why or why not?

(25) Suppose R is a relation on a finite set A, and let D denote the digraph of R. Prove that the sum of the in-degrees plus the sum of the out-degrees for the vertices in D equals twice the number of edges in D.

---

(26) Is it possible for a relation on a set A to be both symmetric and antisymmetric? If so, give an example.

(27) Consider the relation $\sim$ on $\mathbb{N} \times \mathbb{N}$ where $(a, b) \sim (c, d)$ if $ad = bc$.

    (a) List five elements of $\sim$.

    (b) Prove that $\sim$ is an equivalence relation.

    (c) List five members of the equivalence class $[(3, 4)]$.

    (d) List five members of the equivalence class $[(2, 5)]$.

    (e) Explain the connection between the equivalence classes of $\sim$ and the positive rational numbers.

    (f) Explain why the same relation on $\mathbb{Z} \times \mathbb{Z}$ is not an equivalence relation.

(28) Let L be the set of all lines in a Euclidean plane. Define a relation $\parallel$ on L such that $m \parallel n$ if line m is parallel to line n. Remember that a line is a set of points of the plane. We assume it is known that $\parallel$ is an equivalence relation.

    (a) Given $m \in L$, what is $[m]$? What is $\bigcup_{l \in [m]} l$?

    (b) For fixed lines m and n, $m \nparallel n$, what is $\bigcup_{k \in [m]} (n \cap k)$?

5.3 Functions

Let A and B be arbitrary sets. A relation f from A to B is called a **function** from A to B if $\mathrm{Dom}(f) = A$ and f contains no two pairs with the same first coordinate but different second coordinates. If $(a, b) \in f$, we will also say that f **maps** a to b and that b is the **image** of a under f.

The fact that f is a function from A to B is often denoted by

$$f : A \to B,$$

and instead of using the form afb or $(a, b) \in f$, we usually write

$$f(a) = b \qquad \text{or} \qquad f : a \mapsto b.$$

If f is a function, then the graph of f contains exactly one point on each vertical line through a point of A, and in the digraph of f, a unique arrow initiates from every point of A.

Of the four relations in Example 48, only $f = \{(1, x), (5, z)\}$ is a function; its domain is $\{1, 5\}$ and its range is $\{x, z\}$. Relations α and β have 1 related to both x and y, while θ has 5 related to x, y, and z.

Some functions are described more succinctly in terms of a formula, such as $f(a) = a^2$ for $a \in \mathbb{R}$. This function is the set of ordered pairs $f = \{(a, a^2) : a \in \mathbb{R}\}$, which when graphed in the Cartesian plane $\mathbb{R} \times \mathbb{R}$ is a familiar parabola.

Although it is rarely done, the definition of a function f from A to B can also be written as

$$[\text{Dom}(f) = A] \wedge [\forall a \in A ((a \, f \, b_1) \wedge (a \, f \, b_2)) \Rightarrow (b_1 = b_2)].$$

We call two functions f and g **equal** if they have the same domain, A, and for all $x \in A$, $f(x) = g(x)$.

Very little can be said about functions in general; usually special classes (sets) of functions are considered in which all functions in a class satisfy a certain property. The reader has probably already studied linear, quadratic, polynomial, exponential, and trigonometric functions. Other classes include continuous, differentiable, or monotone functions. The following properties give yet another way to classify functions and are commonly used in higher level mathematics.

5.3.1 One-to-one and onto functions. A function from A to B is called an **injective** function, an **injection**, or a **one-to-one** function if distinct elements of A are mapped to distinct elements of B. Symbolically, this is expressed as

$$\forall a_1 \in A, \forall a_2 \in A, (a_1 \neq a_2) \Rightarrow (f(a_1) \neq f(a_2)),$$

or, by using the contrapositive,

$$\forall a_1 \in A, \forall a_2 \in A, (f(a_1) = f(a_2)) \Rightarrow (a_1 = a_2).$$

For an injective function $f : A \to B$, each $b \in B$ is paired with at most one $a \in A$. Hence, the graph of f contains at most one point on each horizontal line through a point of B, and in the digraph of f, no two arrows terminate at the same point of B.

Example 55. Suppose $f : D \to C$ is a function mapping a date in the 20th century to the closing price of the Dow Jones Industrial Average (DJIA) on that date. What are the domain and range? Is the function an injection?

Solution. The domain of this function is a set whose elements are ordered triples of the form (*month, day, year*), where "year" is an integer between 1900 and 1999. The range is the nonnegative real numbers to two decimals. For example, $f(7, 8, 1932) = 41.22$ and $f(10, 19, 1987) = 1{,}738.74$ indicates that the closing price of the DJIA was 41.22 on July 8, 1932 and 1738.74 on October 19, 1987.[3]

Since the DJIA has exactly one closing price each day, f is a function. However, since there happen to be two different domain values that get mapped to the same range value (the DJIA closing price was 2,306.25 on both March 13, 1989, and March 14, 1989), the function is not an injection.   □

---

[3]The observant reader may notice that since elements of the domain are ordered triples, the correct way to denote the fact that the closing price of the DJIA was 41.22 on July 9, 1932, should be $f((7, 9, 1932)) = 41.22$. Although this is technically correct, the convention is to use only a single set of parentheses, and we will follow the convention.

A function f from A to B is called a **surjective** function, a **surjection**, or an **onto** function if Range$(f) = B$. In other words, every element of B is the image of an element of A. Symbolically, this is expressed as

$$\forall b \in B,\ \exists a \in A,\ f(a) = b.$$

For a surjection $f : A \to B$, each $b \in B$ is paired with at least one $a \in A$. Hence, the graph of f contains at least one point on each horizontal line through a point of B; in the digraph of f, every element of B is an endpoint of at least one arrow.

A function that is both injective and surjective is called a **bijection** or a **one-to-one correspondence**.

Often we will want to show that a function is not an injection or is not a surjection. Using the rules of logic to form negations of the definitions we just introduced, we obtain the following: $f : A \to B$ is

- not injective if there exist distinct a_1 and a_2 in A such that $f(a_1) = f(a_2)$;
- not surjective if Range$(f) \neq B$, or, equivalently, there exists $b \in B$ such that for each $a \in A$, $f(a) \neq b$;
- not bijective if f is not injective or f is not surjective.

Example 56. Which of the following functions are injective? Which are surjective? Note any that are bijections.

(1) $\phi : B \to C$, where $B = \{x, y, z\}$, $C = \{5, 7, 14\}$, and $\phi = \{(x, 7), (y, 5), (z, 14)\}$.

(2) $f : \mathbb{R} \to [-1, 1]$, where $f(x) = \sin x$.

(3) $g : \mathbb{R} \to \mathbb{R}$, where $g(x) = 2x - 3$.

(4) $h : \mathbb{R} \to \mathbb{R}$, where $h(x) = x$ for $x < 0$ and $h(x) = x + 1$ for $x \geq 0$.

(5) $p : S \to G$, where S is the set of Discrete Math students in **your** class, G is the set of letter grades $\{A, B, C, D, F\}$, and p assigns a course grade to each student.

Solution. (1) In a finite example like this one, it is easy to determine by inspection whether the given function is injective, surjective, both, or neither. Here, for any distinct b_1 and $b_2 \in B$, we see that b_1 and b_2 are also distinct; hence, by definition, ϕ is injective.

Additionally, since each element in C is the image of some element of B,

$$\text{that is },\ \forall c \in C,\ \exists b \in B,\ \phi(b) = c,$$

ϕ is surjective. Since ϕ is both injective and surjective, ϕ is a bijection.

(2) In order to see whether f is injective, we check the definition. Does $a_1 \neq a_2$ imply $\sin(a_1) \neq \sin(a_2)$ for all $a_1, a_2 \in \mathbb{R}$? Of course not: for example, $0 \neq \pi$, but $\sin 0 = \sin \pi$. Hence, f is not injective.

On the other hand, for any $y \in [-1, 1]$, there exists an $x \in \mathbb{R}$ such that $\sin x = y$. (We assume this trigonometric fact is well known.) Therefore, by definition, f is surjective.

(3) Suppose $x_1, x_2 \in \mathbb{R}$ and $g(x_1) = g(x_2)$. Then $2x_1 - 3 = 2x_2 - 3$, from which it follows that $x_1 = x_2$. Thus, g is injective.

Given any $y \in \mathbb{R}$, let $x = (y + 3)/2 \in \mathbb{R}$. Then $g(x) = 2((y + 3)/2)) - 3 = y$. Thus, g is surjective.

Since g is both injective and surjective, g is a bijection. This can be seen in Figure 5.6(a): since the graph passes a "horizontal line test", g is injective, and since the projection of the graph onto the y-axis is all of $\mathbb{R}$, g is surjective.

(4) Suppose $h(x_1) = h(x_2)$ for some $x_1, x_2 \in \mathbb{R}$. As before, we wish to show that this condition forces $x_1 = x_2$. Because h is piecewise defined, we need to consider several cases. Suppose that one of the values (say, x_1) is negative while the other is nonnegative; then $h(x_1) = h(x_2)$ means that $x_1 = x_2 + 1$, which is an impossibility. If both x_1 and x_2 are negative, then $h(x_1) = h(x_2)$ means that $x_1 = x_2$. If both x_1 and x_2 are nonnegative, then $h(x_1) = h(x_2)$ means that $x_1 + 1 = x_2 + 1$, which again implies that $x_1 = x_2$. Thus, h is injective.

However, note that if $y \in (0, 1)$, there is no $x \in \mathbb{R}$ such that $h(x) = y$. Therefore, h is not surjective, since $\text{Range}(h)$ contains no elements from the interval $(0, 1)$.

The fact that h is injective but not surjective is evident from the graph on the right-hand side of Figure 5.6; in particular, the gap from $y = 0$ to $y = 1$ in the projection of h onto the y-axis indicates that the function fails to be surjective.

(5) If each student in S receives a different grade, then p is injective. Note that if $|S| > 5$, then p cannot be injective. The function will be surjective only if all five grades are collectively achieved by the students in the class. This could fail to happen in a number of ways: for example, there might be fewer than five students in the class, and even if $|S| > 5$, all students could receive the same grade. The only scenario in which p is a bijection is when $|S| = 5$ and each grade in G is assigned to precisely one student.   □

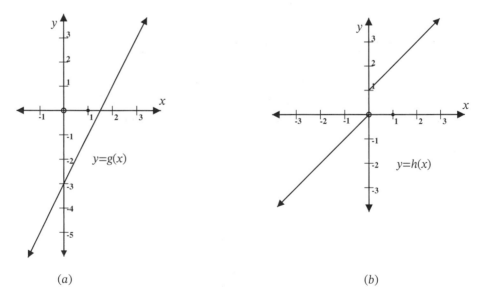

(a)   (b)

Figure 5.6.   Graphs of $y = g(x)$ and $y = h(x)$, Example 56

As illustrated in the previous example, injective or surjective functions with finite domains and ranges have limitations on the cardinalities of those sets.

Theorem 5.2. *Suppose $f : A \to B$ is a function and A and B are finite sets with $|A| = n$ and $|B| = m$.*

(1) *If f is injective, then $n \leq m$.*
(2) *If f is surjective, then $n \geq m$.*
(3) *If f is a bijection, then $n = m$.*

Proof. (1) Using the definition of injective, for each element $b \in B$ there is at most one $a \in A$ for which b is the image of a. Since every element of A is paired with exactly one element of B, we see $|A| \leq |B|$.

(2) This proof is similar to the one for part (1) and is left as an exercise.

(3) If f is both injective and surjective, then $n \leq m$ and $n \geq m$, respectively. Therefore, $n = m$.   □

An immediate consequence of Theorem 5.2 is that an injection (or surjection) from A to B is necessarily a bijection when A and B are equicardinal, finite sets.

Corollary 5.3. *Suppose $f : A \to B$ is a function and A and B are finite sets with $|A| = n = |B|$. Then, f is injective if and only if f is surjective if and only if f is a bijection.*

5.3.2 Inverses. For any relation R from A to B, the **inverse** of R, denoted R^{-1}, is the relation from B to A given by $R^{-1} = \{(b, a) : (a, b) \in R\}$. For example, the inverse of $\beta = \{(1, x), (1, y)\}$ in Example 48 is $\beta^{-1} = \{(x, 1), (y, 1)\}$. Notice that $\text{Dom}(R^{-1}) = \{b \in B : (b, a) \in R^{-1}\} = \{b \in B : (a, b) \in R\} = \text{Range}(R)$, and similarly $\text{Range}(R^{-1}) = \text{Dom}(R)$. As R^{-1} simply contains the transposed ordered pairs from R, the digraph and matrix of R^{-1} are easily obtained from the digraph and matrix of R; see Exercise 18.

Inverse functions are a familiar topic to algebra and calculus students, and any understanding gained in those courses on this topic is applicable here. A function $f : A \to B$ is said to be **invertible** if the relation f^{-1} from $\text{Range}(f)$ to A is a function. If f is invertible, then f^{-1} is called the **inverse** function of f and

$$f(a) = b \quad \text{if and only if} \quad f^{-1}(b) = a. \tag{5.1}$$

Example 57. Which of the following functions are invertible?

(1) $\phi = \{(x, 7), (y, 5), (z, 14)\}$ from Example 56.
(2) $f = \{(1, x), (5, z)\}$ from Example 48.
(3) $g = \{(1, a), (2, a), (3, b)\}$.

Solution. (1) $\phi^{-1} = \{(7, x), (5, y), (14, z)\}$. As this relation is a function, ϕ is invertible.
(2) $f^{-1} = \{(x, 1), (z, 5)\}$. Hence, f is invertible.
(3) $g^{-1} = \{(a, 1), (a, 2), (b, 3)\}$. Since g^{-1} is not a function, g is not invertible.   □

In order for f^{-1} to be a function, by definition, for each $b \in \text{Dom}(f^{-1})$, there is a unique $a \in \text{Range}(f^{-1})$ such that $(b, a) \in f^{-1}$. Equivalently, for each $b \in \text{Range}(f)$, there is a unique $a \in \text{Dom}(f)$ such that $(a, b) \in f$. This is precisely the definition for f to be injective; this proves the following theorem.

Theorem 5.4. *Let $f : A \to B$. Then f is invertible if and only if f is one-to-one.*

This matches the pre-calulus criterion for f^{-1} to exist: a function $f(x)$ is invertible if and only if the graph of $y = f(x)$ passes the Horizontal Line Test.

In Example 57, notice that ϕ and f are one-to-one functions, while g is not one-to-one. For the functions from $\mathbb{R}$ to $\mathbb{R}$ in Example 56, $f(x) = \sin x$ is not injective and thus is not invertible, while $g(x) = 2x - 3$ and the piecewise function $h(x)$ are injective and thus are invertible. The inverse of g is given by the formula $g^{-1}(x) = (x + 3)/2$; the inverse of h is left as an exercise.

The familiar function $f(x) = x^2$ from $\mathbb{R}$ to $\mathbb{R}$ is not one-to-one, as $f(-2) = f(2)$; hence it is not invertible. However, the function $f(x) = x^2$ with domain $\{x \in \mathbb{R} : x \geq 0\}$ is one-to-one, and its inverse is $f^{-1}(x) = \sqrt{x}$; see Exercise 19.

In the event that $f : A \to B$ and $B = \text{Range}(f)$, then, by Theorem 5.4, f is invertible if and only if f is a bijection. Therefore, a function f is a one-to-one correspondence from $\text{Dom}(f)$ to $\text{Range}(f)$ if and only if f is invertible. In Example 57, $f = \{(1, x), (5, z)\}$ is an invertible function from $\{1, 5\}$ to $\{x, y, z\}$, but it is not a bijection. However, f is a bijection from $\{1, 5\}$ to $\text{Range}(f) = \{x, z\}$. We conclude that if $f : A \to B$ and f^{-1} is a function from B to A, then f (as well as f^{-1}) is a one-to-one correspondence between A and B.

In the language of Section 3.7, sets A and B are equivalent (have the same cardinality) if and only if there is a function from A to B with an inverse function defined from B to A. For instance, the function $f(n) = 2n$ maps $\mathbb{N}$ to $\mathbb{E}$ and its inverse $f^{-1}(e) = e/2$ is a function from $\mathbb{E}$ to $\mathbb{N}$. Therefore, $\mathbb{N}$ and $\mathbb{E}$ are equivalent.

5.3.3 Function Compositions.
Under certain circumstances, functions can be combined by applying one function to the range of another. This can happen only if an element of the range of one function is also an element of the domain of the other. Formally, if $f : B \to C$ and $g : A \to B$ are functions, then the **composition** of f and g, denoted $f \circ g$, is a function from A to C such that $(f \circ g)(a) = f(g(a))$ for each $a \in A$. (See Figure 5.7.)

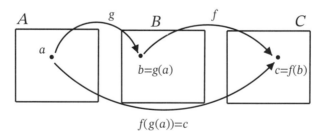

$$f(g(a)) = c$$

Figure 5.7. $f \circ g$ is a function from A to C

Example 58. For each pair of functions given below, determine whether $g \circ f$ or $f \circ g$ is defined. For those compositions which are defined, determine their ranges. If both $g \circ f$ and $f \circ g$ are defined, determine whether they are equal functions.

(1) Let $f : \mathbb{R} \to \mathbb{R}$ be defined by $f(x) = 2x + 3$ and let $g : \mathbb{R} \to \mathbb{R}$ be defined by $g(x) = -x^2 + 5$.

(2) Let $f : A \to B$, where $A = \{a, b, c, d\}$ and $B = \{q, r, s, t, u, v\}$, be defined by $f = \{(a, s), (b, q), (c, v), (d, s)\}$. Let $g : C \to D$, where $C = \{q, r, s, v\}$ and $D = \{x, y, z\}$, be defined by $g = \{(q, z), (r, y), (s, z), (v, x)\}$.

(3) Let $f : \mathbb{R} \to \mathbb{R}$ be defined by $f(x) = x^2$ and let $g : \mathbb{R}^+ \to \mathbb{R}^+$ be defined by $g(x) = \sqrt{x}$.

(4) Let $f : \mathbb{R} \to \mathbb{R}$ be defined by $f(x) = x^2$ and let $g : \mathbb{R} \to \mathbb{R}$ be defined by $g(x) = x^3$.

Solution. (1) $f(g(x)) = 2(-x^2 + 5) + 3 = -2x^2 + 13$; the range is $(-\infty, 13]$.

$$g(f(x)) = -(2x + 3)^2 + 5 = -(4x^2 + 12x + 9) + 5$$
$$= -4x^2 - 12x - 4 = -4(x + 3/2)^2 + 5.$$

The range of $g \circ f$ is $(-\infty, 5]$.

Clearly, $f \circ g$ and $g \circ f$ are not the same function.

(2) $g \circ f = \{(a, z), (b, z), (c, x), (d, z)\}$.

On the other hand, $f \circ g$ is undefined, since the range of g is not a subset of the domain of f; that is, D is not a subset of A.

(3) $f(g(x)) = (\sqrt{x})^2 = x$; the range is $\mathbb{R}^+$. (Note that 0 is not in the range due to the given domain of g.)

$g(f(x)) = \sqrt{x^2} = |x|$; the range is $\mathbb{R}^+$. The two compositions are not equal since, for example, they have different images at $x = -1$.

(4) Since $(x^3)^2 = (x^2)^3 = x^6$ for all real numbers x, the compositions $f \circ g$ and $g \circ f$ are equal functions. $\qquad\square$

Suppose $f : A \to B$ is invertible. Then, by (5.1), $f^{-1}(f(a)) = f^{-1}(b) = a$ for all $a \in \text{Dom}(f)$ and $f(f^{-1}(b)) = f(a) = b$ for all $b \in \text{Range}(f)$. For example, with $g(x) = 2x - 3$ and $g^{-1}(x) = (x + 3)/2$, $g^{-1}(g(x)) = g^{-1}(2x - 3) = ((2x - 3) + 3)/2 = x$ and $g(g^{-1}(x)) = g((x + 3)/2) = 2((x + 3)/2) - 3 = x$ for all $x \in \mathbb{R}$. The function $I_A(a) = a$ for all $a \in A$ is called the **identity function** on A. Thus, we have shown $f^{-1} \circ f = I_A$ and $f \circ f^{-1} = I_B$.

By taking the collection of all bijections from set A to A and applying the operation of composition to functions in this collection, we see the role of identities and inverses appearing in precisely the same manner as in Property 2.1 (for numbers) and Property 2.7 (for matrices). The revelation of such similarities is the primary step in the most pleasant mathematical task known as *abstraction*.

Exercises and Problems.

(1) For each of the following relations, write down the domain and range of the relation and determine whether the relation is a function.

(a) $f_1 = \{(1, 1), (1, 2), (2, 1), (2, 2)\}$.

(b) $f_2 = \{(1, 1), (2, 1), (3, 2)\}$.

(c) $f_3 = \{(1, 3), (2, 2), (3, 1)\}$.

(d) $f_4 = \{(x, y) : x = y^2 \text{ for } y \in \mathbb{R} \}$.

(2) For each of the following functions, determine if the function is injective or surjective. If the function is both, label it as a bijection.

(a) $f_1 : \mathbb{R} \to \mathbb{R}, f_1(x) = x^3$.

(b) $f_2 : [3] \to [3], f_2 = \{(1, 1), (2, 1), (3, 2)\}$.

(c) $f_3 : [3] \to [3], f_3 = \{(1, 3), (2, 2), (3, 1)\}$.

(d) $f_4 : \mathbb{R} \to \mathbb{R}$, $f_4(x) = |x|$.

(e) $f_5 : \mathbb{R} \to \mathbb{Z}$, $f_5(x) = \lfloor x \rfloor$ (floor function).

(3) For each function in Exercise 2, determine whether the function is invertible. If the function is invertible, find its inverse; write the answer as a set of ordered pairs or with a formula.

(4) Using the functions f_2 and f_3 in Exercise 2, write the function $f_2 \circ f_3$ as a set of ordered pairs. Do the same for $f_3 \circ f_2$ and compare the results.

(5) Explain why $f(x) = 1/(x + 1)$ and $g(x) = (x - 1)/(x^2 - 1)$ are different functions.

(6) Consider functions $f, g : \mathbb{R} \to \mathbb{R}$ given by $f(x) = \sqrt{x^2 + 1} - x$ and $g(x) = 1/(\sqrt{x^2 + 1} + x)$. Show that f and g are equal functions.

(7) Let $A = \{X, Y\}$ be a set consisting of two people, and let $B = \mathbb{R}$ be the set of real numbers. Let $f, g : A \to B$ be two functions, where f is the number of cents representing the amount of money in the person's wallet and g is the number of inches representing the person's height. Can f and g be equal?

(8) Although a domain is necessary in order for a function to be defined, in practice the domain is not always explicitly stated. In such cases, the reader must determine the **natural domain** of the function. Typically, the natural domain is the largest subset of real numbers for which the function is defined. Give the natural domain and range of each of the following functions. Also determine whether the function is injective and whether it is surjective on the set of real numbers.

(a) $f(x) = x^2$.

(b) $g(x) = 1/\sqrt{x + 1}$.

(c) $h(x) = \dfrac{x}{x^2 - 1}$.

(d) $k(x) = \dfrac{1}{1 - 1/\sqrt{x + 1}}$.

(9) For each of the following, suppose $A = \{a, b, c\}$ and f is a relation from A to B. Provide an example of a function f and set B for which f has the desired property. (Your examples should be different from the ones found in this chapter.)

(a) f is an onto function, but not one-to-one.

(b) f is an injective function, but not surjective.

(c) The domain of f is A, but f is not a function.

(d) The range of f is B, but f is not a function.

(10) Consider a function $f : A \to B$. Each of the following lines would be an appropriate first line of a proof of a property of f discussed in this section. In each case, name that property.

(a) Suppose $a_1 \in A$ and $a_2 \in A$ such that $f(a_1) = f(a_2)$.

(b) Let b be an arbitrary element of B.

(c) Let $a_1 \in A$ and $a_2 \in A$ such that $a_1 \neq a_2$.

(11) Consider a function $f : A \to B$. Each of the following lines would be an appropriate penultimate line of a proof of a property of f discussed in this section. In each case, name that property.

(a) We have found an element $a \in A$ such that $f(a) = b$.

(b) We have shown that f is both injective and surjective.

(c) Therefore, $a_1 = a_2$.

(d) We have shown that f is both one-to-one and onto.

(e) This proves that $f(a_1) \neq f(a_2)$.

(12) Prove that the function $f(x) = 3x + 5$ from $\mathbb{Z}$ to $\mathbb{Z}$ is not a bijection but that a function given by the same formula from $\mathbb{R}$ to $\mathbb{R}$ is a bijection.

(13) Suppose $f : A \to B$ is a function and A and B are finite sets, say $|A| = n$ and $|B| = m$. Determine whether the following statements are true or false. If the statement is true, provide a brief explanation of why it is true. If the statement is false, provide an example of a function that does not satisfy the statement; either list the ordered pairs or draw the digraph.

(a) If $n > m$, then f is not one-to-one.

(b) If f is one-to-one, then $n < m$.

(c) If $n \leq m$, then f is one-to-one.

(d) If $n > m$, then f is onto.

(e) If $n < m$, then f is not onto.

(f) If f is onto, then $n \geq m$.

(g) If $n = m$, then f is a bijection.

(h) If f is one-to-one and onto, then $n = m$.

(i) If f is a bijection, then $A = B$.

---

(14) Let $f(y)$ be the cost (in dollars and cents) of y gallons of gasoline. Let $g(x)$ be the number of gallons of gasoline needed to drive x miles. Which composition makes sense, $f \circ g$ or $g \circ f$? Interpret the composition.

(15) Find the range for the following functions defined over $\mathbb{N}$. Justify your answers.

(a) The function f that assigns to each natural number its last digit (base-10 representation).

(b) The function g that assigns to each natural number its number of digits (base-10 representation).

(c) $f \circ g$.

(d) $g \circ f$.

(16) (a) Let $f, g : \mathbb{R} \to \mathbb{R}$ be two functions such that $f(x) = 5x - 2$ and $g(x) = (x + 2)/5$. Prove that $g \circ f = f \circ g$.

(b) Let $f : \mathbb{R} \to \mathbb{R}$ via $f(x) = 2x + 1$. Find all functions $g : \mathbb{R} \to \mathbb{R}$ of the form $g(x) = ax + b$, $a, b \in \mathbb{R}$, such that $g \circ f = f \circ g$.

(c) Let $f : \mathbb{R} \to \mathbb{R}$ via $f(x) = x^2$. Find all functions $g : \mathbb{R} \to \mathbb{R}$ of the form $g(x) = ax + b$, $a, b \in \mathbb{R}$, such that $g \circ f = f \circ g$.

(17) Suppose R is a relation from A to B. Explain why $\text{Range}(R^{-1}) = \text{Dom}(R)$ using the definition of each.

(18) Let R be a relation on a set A with inverse relation R^{-1}.

   (a) How are the digraphs of R and R^{-1} related? How is one easily obtained from the other?

   (b) How are the matrices of R and R^{-1} related? How is one easily obtained from the other?

(19) Consider the function $q(x) = x^2$ with domain $A = \{x \in \mathbb{R} : x \geq 0\}$.

   (a) Show that q is a one-to-one function on A and determine the range of q.

   (b) Show that $q^{-1}(x) = \sqrt{x}$ using the definition of an inverse relation. Determine the domain and range of q^{-1}.

   (c) Show that $q^{-1}(q(x)) = x$ and $q(q^{-1}(x)) = x$ for all $x \geq 0$. Identify the step that requires x to be nonnegative.

   (d) Sketch the graphs of $y = q(x)$ and $y = q^{-1}(x)$ in the xy-plane and describe the relationship between the two graphs.

(20) Describe the inverse function of the piecewise function $h(x)$ found in Example 56. The answer could be a formula, a graph, or a set of ordered pairs.

(21) Show that there is a bijection between the set of all positive integers, $\mathbb{N}$, and the set of all squares of positive integers, $S = \{1, 4, 9, 16, ...\}$, by finding an appropriate invertible function.

(22) Let S be a set, with power set $\mathcal{P}(S)$, and suppose that $x \notin S$. Define the function

$$f : \mathcal{P}(S) \to \mathcal{P}(S \cup \{x\}) \quad \text{by} \quad f(A) = A \cup \{x\}$$

for all $A \in \mathcal{P}(S)$. Is this function f injective? Is f surjective? Explain your answers.

(23) Find a bijection between each pair of sets.

   (a) The two closed intervals of real numbers $[1, 2]$ and $[5, 20]$.

   (b) The two open intervals of real numbers $(0, 1)$ and $(1, \infty)$.

   (c) The open interval $(0, 1)$ of real numbers and the set of real numbers, $\mathbb{R}$.

   (d) $\mathbb{N}$ and its subset $A = \{x : x \in \mathbb{N} \text{ and } x \geq 5\}$.

(24) Show that there exists a bijection between $\mathbb{N}$ and $\mathbb{Z}$.

(25) Provide a proof of Theorem 5.2(2).

---

(26) Let $f : A \to B$ and $g : B \to C$ be one-to-one functions. Prove that their composition, $g \circ f$, is a one-to-one function from A to C.

(27) Let $f : A \to B$ and $g : B \to C$ be functions, and assume that their composition $g \circ f : A \to C$ is one-to-one. Does this imply that both f and g are one-to-one? Prove your answer.

(28) Let $f : A \to B$ and $g : B \to C$ be one-to-one functions, and assume that their composition $g \circ f : A \to C$ is onto. Does this imply that both f and g are onto? Prove your answer.

(29) Show that there exists a bijection between $\mathbb{N}$ and $\mathbb{N} \times \mathbb{N}$.

(30) In Section 3.7, sets A and B are called *equivalent* if there exists a bijection from A to B. Prove that equivalence of sets is an equivalence relation on the collection of all possible sets. Also, describe the equivalence classes of this equivalence relation.

(31) Show that there exists a bijection between the set of points of a circle with one point removed and the set of all real numbers.

(32) Let A be a nonempty set, and let $\mathcal{F}$ denote the set of all bijections from A to A. Which of the following properties hold for the operation of function composition on elements of $\mathcal{F}$?

    (a) closure of $\circ$

    (b) commutativity of $\circ$

    (c) associativity of $\circ$

    (d) identity for $\circ$

    (e) inverses for $\circ$

5.4 Sequences

Sequences form a special category of functions, worthy of their own section. Anyone who has created a "prioritized" to-do list or has stood in line at a grocery store checkout counter is familiar with everyday occurrences of sequences. Simply put, any arrangement of finitely or infinitely many items in a particular order is a sequence. This is in contrast to sets, where the order in which elements are listed is of no consequence. Moreover, elements are never repeated in a set, while items may occur many times in a sequence. In mathematics, to define a sequence precisely, we will utilize the concept of function.

Let A be a set and m a positive integer. A **finite sequence f of elements of A** is a function $f : [m] \to A$. An **infinite sequence f of elements of A** is a function $f : \mathbb{N} \to A$. Sometimes the domain of a sequence is taken to be the set of consecutive integers $\{n, n+1, \ldots, m\}$ or $\{n, n+1, n+2, \ldots\}$ for some integer $n \neq 1$, with $n = 0$ being a frequent alternative.

For n in the domain of f, we call $f(n)$ the **nth term** of the sequence f, and we denote it by a_n. Using this notation, a finite sequence is often denoted by $(a_n)_{n=1}^{m}$, and an infinite sequence by $(a_n)_{n \geq 1}$. Though f is not present in these notations, we always remember that $a_n = f(n)$. When listing the specific terms of a sequence, we will typically enclose them in parentheses to distinguish the sequence from the set with the same elements. For example, the sequence $(2, 6, 4, 9)$ is a 4-tuple, which is a different mathematical entity than the set $\{2, 6, 4, 9\}$. For convenience, the sequence is sometimes displayed without parentheses: $2, 6, 4, 9$.

Example 59. The following examples illustrate a variety of ways (lists, formulas, descriptions) that sequences can be described.

(1) For $n \geq 1$, let a_n be the nth perfect square. That is, $a_1 = 1^2 = 1, a_2 = 2^2 = 4, a_3 = 3^2 = 9, a_4 = 4^2 = 16, \ldots$. This is an infinite sequence, with $a_n = f(n) = n^2$.

(2) The sequence $(5, -1, 3, 3, 4)$ is a finite sequence with just five terms: $a_1 = 5, a_2 = -1, a_3 = 3, a_4 = 3$, and $a_5 = 4$. The only obvious function f here is $f : \{1, 2, 3, 4, 5\} \to \mathbb{Z}$ such that

$$f(1) = 5, \quad f(2) = -1, \quad f(3) = 3, \quad f(4) = 3, \quad f(5) = 4.$$

(3) Consider the sequence in which a_n is the sum of the measures (in degrees) of the interior angles of a (non-self-intersecting) n-sided polygon. In this case, a function

is easily defined, although the usual definition generates a sequence that begins with $n = 3$: $a_n = (n - 2)180°$ for $n \geq 3$, and the sequence is $(180°, 360°, 540°, ...)$.

(4) Let a_n denote the nth digit after the decimal point in the decimal expansion of π. This infinite sequence is easily defined, although no one knows an explicit formula for its nth term. The sequence is $(1, 4, 1, 5, 9, ...)$.

Recognition of patterns is a key mathematical skill; in fact, often mathematics is described as the science of patterns. Pattern recognition can lead to important discoveries.

Suppose we have a sequence whose first three terms are $1, 2, 3$. What is the next term? Any child will say "4". However, we could certainly find a function f for which $f(1) = 1, f(2) = 2$, and $f(3) = 3$, but $f(4) \neq 4$. For example, take

$$f(x) = x + (x - 1)(x - 2)(x - 3).$$

Then $f(n) = n$ for $n = 1, 2, 3$, but $f(4) = 10 \neq 4$. Another possibility is $f(x) = x + x^{2,018}(x - 1)(x - 2)(x - 3)$. In general, a sequence cannot be determined based on its first few terms, a point that is illustrated again in the following example.

Example 60. Take $n \geq 1$ points on a circle and connect every two of them by a chord. Let a_n denote the greatest number of regions into which these chords can divide the disc. Assuming $a_1 = 1$ (no chords "divide" the disc into one region), find $(a_n)_{n \geq 1}$.

Solution. We are given that $a_1 = 1$. Experimenting with pictures, one immediately realizes that in order to maximize the number of regions, no three of these chords should meet at a point inside the circle. The count gives $a_2 = 2$, $a_3 = 4$, $a_4 = 8$. (See Figure 5.8.) Continuing this way, one gets $a_5 = 16$. Hence, the sequence begins like this:

$$1, 2, 4, 8, 16, ...$$

The pattern seems to be clear: $a_n = 2^{n-1}$. However, counting (and recounting) gives $a_6 = 31$, rather than the anticipated $2^5 = 32$. In Problem 25 of Section 13.2, we will determine a not-too-complicated formula for a_n. See if you can find it for yourself. □

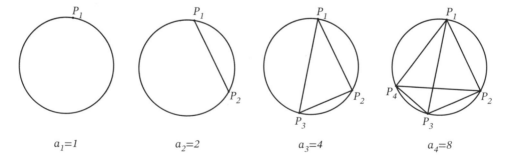

Figure 5.8. Example 60

Example 61. Find a pattern for each of the finite sequences below and use your pattern to extend the sequence for two more terms.

(1) $(2, 5, 8, 11, 14)$

(2) $(1, -1, 1, -1, 1)$

(3) $(3, 5, 7)$

(4) $(1, 2, 6, 24, 120)$

(5) $(10, 5, 5/2, 5/4, 5/8)$

(6) $(22, 16, 10, 4, -2)$

(7) $(1, 1, 2, 3, 5, 8, 13)$

Solution. (1) One pattern that fits the sequence is that each term of the sequence is obtained by adding 3 to the immediately preceding term. Thus, the next two terms would be 17 and 20.

(2) The pattern of this sequence appears to be simply alternating between 1 and -1, so the next two terms would be 1 and -1.

(3) There are at least two obvious possibilities for this sequence: it could be a list of the positive odd integers, starting with 3; alternatively, a_n could be the nth odd prime. The sequence continues with $a_4 = 9$ and $a_5 = 11$ or with $a_4 = 11$ and $a_5 = 13$, respectively.

(4) This appears to be the sequence of **factorials**: $a_1 = 1! = 1$, $a_2 = 2! = 2$, $a_3 = 3! = 6$, and so on, where $n!$ is defined as the product of integers from 1 to n for $n \geq 1$. The next two terms of the sequence would be $a_6 = 6! = 720$ and $a_7 = 7! = 5,040$.

(5) Notice that each term is half of the preceding term. Continuing this pattern, the next two terms of the sequence are $5/16$ and $5/32$.

(6) Note that for the terms shown, a_n is obtained by adding -6 to a_{n-1}. This gives $a_6 = -8$ and $a_7 = -14$.

(7) The pattern for this sequence may be more difficult to determine than in the previous examples. After some experimentation, you might see that after the first two terms, each term can be obtained by adding the two terms that immediately precede it. The next two terms would be 21 and 34. This is an example of a *recursive sequence*, to be defined below.   □

A **recursively-defined sequence** is one in which the first several terms (called the **initial terms**) are given, and each subsequent term is determined from some previous ones.

Example 62. Define $a_1 = 1$ and $a_{n+1} = (n+1) \cdot a_n$ for $n \geq 1$. This recursive definition yields the following terms for $n = 1, 2, \dots, 6$.

$$a_1 = 1, \quad a_2 = 2 \cdot 1 = 2, \quad a_3 = 3 \cdot 2 = 6, \quad a_4 = 4 \cdot 6 = 24,$$

$$a_5 = 5 \cdot 24 = 120, \quad a_6 = 6 \cdot 120 = 720.$$

Thus, a_n is simply $n!$, as described in Example 61(4).

Example 63. (1) Define $a_1 = 1$, $a_{n+1} = 2a_n + 1$ for $n \geq 1$.

The sequence is $(1, 3, 7, 15, 31, \dots)$.

(2) Define $a_0 = 0$, $a_1 = 1$, and $a_{n+1} = a_n - a_{n-1}^2$ for $n \geq 1$.

The sequence is $(0, 1, 1, 0, -1, -1, -2, -3, -7, -16, \dots)$.

(3) Define $a_1 = 4$, $a_2 = -3$, $a_3 = 1$, $a_4 = 5$, and $a_{n+1} = a_{n-3} + 2a_{n-2}$ for $n \geq 4$.
The sequence is $(4, -3, 1, 5, -2, -1, 11, 1, -4, 21, 13, \dots)$.

Example 64. Define $f_1 = 1$, $f_2 = 1$, and $f_{n+1} = f_n + f_{n-1}$ for $n \geq 2$. This is the recursive sequence from Example 61(7), called the *Fibonacci sequence*. It was named after Leonardo de Pisa, also known as Fibonacci, and the numbers in the sequence are called Fibonacci numbers. These numbers arise surprisingly often as answers in a variety of counting problems, but the sequence originated as the solution to a problem posed by him in his 1202 book, *Liber Abaci*. The statement of Fibonacci's problem can be paraphrased as follows:

> *How many pairs of rabbits can be produced in a year from a single pair if in every month each pair bears a new pair of young rabbits, each new pair reproduces starting at the age of one month, the gestation period is one month, and the rabbits never die?*

Let f_n denote the number of pairs of rabbits at the beginning of Month n. At the beginning of this scenario, there is just one pair of rabbits, so $f_1 = 1$.

At the beginning of Month 2, the first pair of rabbits has matured but not yet given birth to baby rabbits. Thus, $f_2 = 1$.

At the beginning of Month 3, the first pair of rabbits has given birth to a new pair, so $f_3 = 2$.

At the beginning of Month 4, the first pair of rabbits gives birth to a second pair of babies; their first pair of offspring are not yet old enough to bear young. Thus, $f_4 = 3$.

At the beginning of Month 5, both the original pair of rabbits and their first pair of offspring give birth to babies, so $f_5 = 5$. And, so on. □

Two straightforward and widely applicable types of sequences are arithmetic sequences and geometric sequences. Each of these types of sequences are defined recursively below and explored more fully in the problem set. Often an arithmetic (or a geometric) sequence is called an arithmetic (a geometric) **progression**. The term progression comes from the Latin words *progredior* (to walk forward) and *progressio* (movement forward, success). Problems about progressions go back to the *Rhind Papyrus*, c. 1550 BC, and Babylonian astronomical tables, c. 2500–2000 BC.

A sequence is called **arithmetic** if each term after the first one can be obtained by adding a fixed number to the preceding term. In this case, the terms can be expressed as $a_1 = a$, $a_2 = a + d$, $a_3 = a + 2d$, ..., $a_n = a + (n-1)d$, where a and d are fixed real numbers. That is, a sequence with first term a_1 is arithmetic if

$$\exists d \, \forall n \in \mathbb{N}, \; a_{n+1} = a_n + d$$

In an arithmetic sequence, $a_1 = a$ is called the **initial term** and d is called the **common difference**.

Example 65. Define $a_1 = 2$ and $a_{n+1} = a_n + 3$ for $n \geq 1$. This recursive definition yields the following terms for $n = 1, 2, \dots, 5$.

$$a_1 = 2, \quad a_2 = 2 + 3 = 5, \quad a_3 = 5 + 3 = 8, \quad a_4 = 8 + 3 = 11,$$

$$a_5 = 11 + 3 = 14.$$

Since the nth term of the sequence is $a_n = 2 + 3n$ for $n \geq 1$, this is an arithmetic sequence with initial term $a = 2$ and common difference $d = 3$. (This is the same sequence given in Example 61(1).) $\qquad\square$

Below are additional examples of arithmetic sequences:

$$
\begin{aligned}
a_1 &= 1, d = 1: & (1, 2, 3, 4, 5, \ldots), \\
a_1 &= 12, d = 8.5: & (12, 20.5, 29, 37.5, \ldots), \\
a_1 &= 5, d = -4: & (5, 1, -3, -7, \ldots), \\
a_1 &= 5, d = 0: & (5, 5, 5, 5, \ldots).
\end{aligned}
$$

A sequence is called **geometric** if each term after the first one can be obtained by multiplying the preceding term by a fixed number. The terms of a geometric sequence have the form

$$b_1 = b, b_2 = br, b_3 = br^2, \ldots, b_n = b*r^{n-1},$$

where b and r are fixed real numbers. Thus, a sequence with first term b_1 is geometric if

$$\exists r \, \forall n \in \mathbb{N}, \, b_{n+1} = b_n r.$$

In a geometric sequence, $b_1 = b$ is called the **initial term** and r is called the **common ratio**.

The following are all examples of geometric sequences:

$$
\begin{aligned}
b_1 &= 2, r = 1: & (2, 2, 2, 2, \ldots), \\
b_1 &= 1, r = 2: & (1, 2, 4, 8, \ldots), \\
b_1 &= 3, r = 5: & (3, 15, 75, 375, \ldots), \\
b_1 &= -3, r = -1/2: & (-3, 3/2, -3/4, 3/8, \ldots).
\end{aligned}
$$

Exercises and Problems.

(1) For each sequence in Example 61, determine whether it is arithmetic, geometric, or neither.

(2) Find a pattern for each of the sequences below and use the pattern to extend the sequence two more terms. Classify each as arithmetic (find the common difference), geometric (find the common ratio), or neither.

(a) $2, 2, 2, 2, 2, \ldots$

(b) $1, \dfrac{1}{2}, \dfrac{1}{3}, \dfrac{1}{4}, \dfrac{1}{5}, \ldots$

(c) $1, \dfrac{-1}{3}, \dfrac{1}{9}, \dfrac{-1}{27}, \ldots$

(d) $\dfrac{1}{2}, \dfrac{2}{3}, \dfrac{3}{4}, \dfrac{4}{5}, \ldots$

(e) $12, 9, 6, 3, \ldots$

(f) $1, 10, 100, 1,000, \ldots$

(3) Given a positive integer N, let (r_n) be the (finite) sequence of remainders when N and its subsequent quotients are divided by 2. The sequence ends when a quotient of 0 is obtained. For example, when $N = 12$, the quotients and remainders are 6 remainder 0, 3 remainder 0, 1 remainder 1, and 0 remainder 1. The sequence of

remainders is thus 0, 0, 1, 1. Find the corresponding sequences for $N = 75$ and $N = 128$.

Note: These sequences are utilized when converting integers to their binary form, which we discuss in Chapter 12.

(4) Let $(a_n)_{n \geq 1}$ be an arithmetic sequence with terms $3, 8, 13, \ldots$.

    (a) Find a formula for a_n.

    (b) Compare the three sums $a_1 + a_6$, $a_2 + a_5$, and $a_3 + a_4$.

    (c) What is the sum of the first six terms of the sequence? Ten terms? Twenty terms?

---

(5) Given a sequence $(a_n)_{n \geq 0}$, define a new sequence $(b_n)_{n \geq 0}$ via

$$b_n = \sum_{k=0}^{n} a_k a_{n-k}.$$

    (a) Find the first nine terms of (b_n) if $(a_n) = (1, 3, -2, 5, -1, 0, 0, 0, \ldots)$.

    (b) Explain the relationship between part (a) and the expansion of the polynomial $(1 + 3x - 2x^2 + 5x^3 - x^4)^2$.

(6) Given two numbers a, b, the number $(a + b)/2$ is called their **arithmetic mean**.

    (a) Show that every term of an arithmetic sequence, except the first term and the last term (in case of a finite sequence), is the arithmetic mean of the preceding term and the following term. That is, show that for $n \geq 2$, $a_n = \dfrac{a_{n-1} + a_{n+1}}{2}$.

    (b) Generalize the statement in (a) by proving that the kth term of an arithmetic sequence is the arithmetic mean of the $(k - i)$th term and the $(k + i)$th term for all i such that these terms exist.

(7) Consider the sequence $(a_n)_{n \geq 1}$, where for $n \geq 2$,

$$a_n = \begin{cases} a_{n-1}/2 & \text{if } a_{n-1} \text{ is even,} \\ 3a_{n-1} + 1 & \text{if } a_{n-1} \text{ is odd.} \end{cases}$$

Clearly, the sequence depends on the initial term a_1. For example, when $a_1 = 17$, the sequence is $(17, 52, 26, 13, 40, 20, \ldots)$.

    (a) Find the first 15 terms of the sequence when $a_1 = 17$.

    (b) If $a_1 = 2^k$, what is the value of a_{k+1}?

    (c) Choose five different values of a_1 and compute enough terms of the resulting sequence to notice a pattern. What is it?

    Note: The pattern that you hopefully found is not one that is easy to prove. The problem was proposed by Lothar Collatz in 1937, and mathematicians have been working to solve it since that time, without success.

(8) Let $(a_n)_{n \geq 1}$ be the sequence of **triangular numbers**, where a_n is defined as the number of dots used to create a triangle built upon a base of n dots (see Figure 5.9), where each row has one less dot than the one above it. Find the first ten terms of the sequence and the general form for a_n.

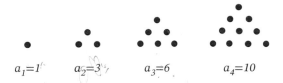

$a_1=1$ $a_2=3$ $a_3=6$ $a_4=10$

Figure 5.9. Triangular numbers

(9) Let $(a_n)_{n\geq1}$ be an arithmetic sequence with first term a and common difference d.

    (a) Give the value of a_n in terms of a, d, and n.

    (b) For integers k and M, with $1 \leq k < M$, prove that $a_k + a_{M+1-k} = a_1 + a_M$.

    (c) Find the sum of the series $a_1 + a_2 + \cdots + a_m$. Prove your answer.

(10) Suppose that the sum of the first n terms of an arithmetic sequence is given by the formula $S_n = 4n^2 - 3n$ for every $n \geq 1$. Find the first three terms of the arithmetic sequence and its common difference.

(11) Let $\{b_n\}_{n\geq1}$ be a geometric sequence with ratio r. Prove the following statements.

    (a) $b_n = b_1 r^{n-1}$, for all $n \geq 2$.

    (b) $b_1 b_n = b_2 b_{n-1} = b_3 b_{n-2} = \dots$.

    (c) Let $n \geq 2$, and let $S_n = b_1 + b_2 + \cdots + b_n$. Then

$$S_n = \sum_{i=1}^{n} b_i = \begin{cases} \frac{b_1 - b_n r}{1-r} = \frac{b_1(1-r^n)}{1-r} & \text{if } r \neq 1, \\ n b_1 & \text{if } r = 1. \end{cases}$$

    (d) If $|r| < 1$, then $\lim_{n\to\infty} r^n = 0$ and

$$\sum_{i=1}^{\infty} b_i = \frac{b_1}{1-r}.$$

(12) Given two numbers a, b, $ab \geq 0$, the number $\sqrt{ab}$ is called their **geometric mean**.

    (a) Show that every term of a geometric sequence with nonnegative terms, except the first term and the last term (in case of a finite sequence), is the geometric mean of the preceding term and the following term.

    (b) Show that if a sequence has the property above, it must be a geometric sequence.

    (c) Generalize the statement in (a) by proving that the kth term is the geometric mean of the $(k-i)$th term and the $(k+i)$th term for all i such that these terms exist.

(13) Consider the sequence

$$f_n = \left(\frac{1}{2^n \cdot \sqrt{5}} \right) \left((1 + \sqrt{5})^n - (1 - \sqrt{5})^n \right), \qquad n \geq 1.$$

    (a) Evaluate f_1, f_2, and f_3 by hand.

    (b) Determine f_{10} using a calculator.

    (c) Repeat the previous calculation, but only use the first term (leave out the term with $(1 - \sqrt{5})$. How close is the value to f_{10}?

(d) Show that f_n satisfies the recursive relation $f_n = f_{n-1} + f_{n-2}$ for $n \geq 3$, and explain why this implies that f_n is the nth Fibonacci number.

(14) Given that each sum below is the sum of a part of an arithmetic or geometric progression, find (or simplify) each sum.

  (a) $75 + 71 + 67 + \cdots + (-61)$.

  (b) $75 + 15 + 3 + \cdots + \frac{3}{5^7}$.

  (c) $1 + 2 + 3 + \cdots + (n - 1) + n$.

  (d) $i + (i + 1) + (i + 2) + \cdots + j$, where $i, j \in \mathbb{Z}, i < j$.

  (e) $1 + 3 + 5 + \cdots + (2n - 1)$, where $n \in \mathbb{N}$.

  (f) $x^i + x^{i+1} + x^{i+2} + \cdots + x^j$, where $i, j \in \mathbb{Z}, i < j, x \neq 1$.

---

(15) In how many ways could one climb a 15-step staircase if only one or two steps may be taken at a time? For example, one could climb the fifteen steps one step at a time. Alternatively, the numbers of steps taken could be 2, 1, 2, 1, 2, 1, 2, 1, 2, and 1, in that order.

(16) In how many 0-1 sequences of length 15 are no two 1's next to each other?

(17) How many subsets of letters of the alphabet do not contain consecutive letters? Hint: Find a connection to Problem 16.

(18) The following construction leads to an object called the **Koch snowflake**. Begin with an equilateral triangle K_1 having unit side lengths. Then divide each side into three congruent segments and on each middle segment build an equilateral triangle (exterior to the original triangle) with the middle segment as the base. Now delete each of the bases to obtain a 12-gon, K_2, with every side having length 1/3. We continue by dividing each side of K_2 into three congruent segments, building an equilateral triangle on each middle segment. Then delete the base of each of these triangles to obtain a polygon K_3. Applying a similar procedure to K_3 leads to the polygon K_4, and so on. Polygons K_1, K_2, and K_3 are shown in Figure 5.10.

Figure 5.10. K_1, K_2, and K_3

Compute the following:

  (a) the number of sides of K_n,

  (b) the perimeter p_n of K_n,

  (c) the area a_n of K_n,

  (d) $\displaystyle\lim_{n\to\infty} p_n$ and $\displaystyle\lim_{n\to\infty} a_n$.

One can show that the sequence of figures K_n "approaches" a distinguished set of points of the plane, called the Koch snowflake, in honor of N. F. H. von Koch (1870–1924).

(19) The construction of the **Sierpiński triangle** begins with an equilateral triangle of unit side length, which we denote by T_1. The triangle is cut into four congruent equilateral triangles, and the central subtriangle is removed. We denote the obtained figure by T_2. The same procedure is then applied to the remaining three subtriangles, leading to a figure T_3, and so on. See Figure 5.11.

Figure 5.11. $T_1, T_2, T_3, T_4,$ and T_5

One can show that the sequence T_n "approaches" a specific set of points in the plane, called the Sierpiński triangle, in honor of W. Sierpiński (1882–1969).

Compute the area of T_n, and show that it tends to zero as n increases. (This shows that the Sierpiński triangle has no area.)

Remark. Sets of points like the Koch snowflake and the Sierpiński triangle are examples of objects called **fractals**. Informally, a fractal is a figure that can be split into parts, each of which is a reduced-size copy of the whole, a property called **self-similarity**. Fractal-like objects are found throughout nature. For more on fractals, including beautiful pictures, see http://en.wikipedia.org/wiki/Fractal. ◇

Each of two bright math students, Terence and Srinivasa, has a number on his forehead. Each can see the number on the other person's forehead, but not the number on his own. They are told that the two numbers are consecutive positive integers. They start the conversation with Terence saying, "I don't know my number." Srinivasa thinks for a moment and replies, "I don't know my number either." They then repeat this exact same conversation, for a total of 53 times, at which point Terence exclaims, "Hey, I just figured out my number!" Srinivasa immediately follows, "Yeah? Me too!" What are the possible numbers on Terence's forehead?

6

Induction

Induction makes you feel guilty for getting something out of nothing,
and it is artificial, but it is one of the greatest ideas of civilization.

—Herbert Wilf (1931–2012)

Recall that in Example 60 of Chapter 4, we encountered a sequence (namely, 1, 2, 4, 8, 16, ...) that could be obtained by letting a_n denote the number of regions created when n points are placed on a circle and connected, with no three segments meeting at a point. Although the first five terms can be expressed as powers of 2 ($a_1 = 1 = 2^0$, $a_2 = 2 = 2^1$, $a_3 = 4 = 2^2$, $a_4 = 8 = 2^3$, and $a_5 = 16 = 2^4$), the pattern does not hold beyond a_5, since $a_6 = 31$. On the other hand, we saw in Chapter 3 that a sequence with the same first five terms can be created by letting b_n represent the number of subsets of a set with n elements; in that case, however, $b_n = 2^n$ for *all* positive integer values of n.

This raises the question of how one verifies that a statement about integers holds for infinitely many consecutive positive integers. Proof by mathematical induction is a technique that is frequently used for establishing that a conjecture is true for all positive integers beyond a particular starting value. We will see that this proof technique involves two steps: first, showing that the statement is true for a smallest value; second, demonstrating that if the statement holds for a given positive integer, it must also hold for the next integer. Before looking at the Principle of Mathematical Induction in detail, though, we will examine the role of induction more generally in science and mathematics.

6.1 Inductive and Deductive Thinking

Within the physical sciences (biology, chemistry, physics, astronomy), inductive reasoning is used on an everyday basis to derive a general conclusion after examining many specific cases. If after many observations, it becomes clear that objects that go up in the air (a ball, a skateboarder, a frisbee) return to the earth, one might conclude that "what goes up must come down"—i.e., the law of gravity holds. Similarly, noting that plant growth rates are directly related to the amount of water the plant absorbs leads (inductively) to the conclusion that water is critical to plant growth. Inductive reasoning is used outside of science also. After listening many times to older speakers of English, a young child concludes that the way to make a verb past tense is to add "d" or "ed": played, walked, pushed, and so on. Whenever a general principle is reached after observing multiple specific instances, induction has been used.

Deductive reasoning, on the other hand, is the application of general theory to a specific situation. That is, the use of deduction involves determining logical consequences of an established fact. The Pythagorean Theorem can be proven in complete generality: *A triangle is a right triangle if and only if the square of the longest side is equal to the sum of the squares of the two shorter sides.* Because this general statement can be proven, any specific example of a triangle in the plane must follow this rule; if the shorter sides of a right triangle have lengths 3 and 4, the longest side must have length 5. Many other examples of deductive reasoning were presented in Chapter 4.

In deductive reasoning, if all of the premises are true and valid rules of inference are applied, then the conclusion must also be true. This is not the case with inductive reasoning, where regardless of how many previous observations have been made for which the pattern holds, an exception may still exist. For example, the preschooler applying inductive reasoning to verb conjugation may announce, "I runned fast." Although gravity does cause most things that leave Earth's surface to return (so that what goes up does come down), scientists have succeeded in building a space craft that is capable of escaping the pull of gravity.

In the mid-1700s, German mathematician Christian Goldbach conjectured that any even integer n greater than 2 can be expressed as the sum of two primes. Although the use of computers has shown this to be true up through $n = 4 \times 10^{18}$, because it has not been proven in generality, it remains the "Goldbach conjecture"[1] rather than the "Goldbach theorem"—and rightly so, because within mathematics, there are many classic examples of long patterns that break eventually.

Example 66. (1) The polynomial n^2+n+41 found by Swiss mathematician Leonhard Euler (1707–1783) in the 18th century produces a prime number for $n = 0, 1, \ldots, 39$, but fails to do so when $n = 40$ and $n = 41$.

(2) The French mathematician Pierre de Fermat (1601–1665) conjectured that any integer of the form $2^{2^n} + 1$ would be prime, but Euler proved this to be false for $n = 5$. In fact no primes of this form have been found for $n > 4$.

(3) For every n let $\pi_1(n)$ be the number of primes less than or equal to n that are congruent to 1 mod 4. Similarly, define $\pi_3(n)$. For $n < 26{,}861$ and $26{,}861 < n < 616{,}841$, $\pi_1(n) \leq \pi_3(n)$. However, $\pi_1(26{,}861) > \pi_3(26{,}861)$ [10]. While data supports that $\pi_1(n)$ is "usually" less than $\pi_3(n)$, English mathematician John Littlewood (1885–1977) showed that there are arbitrarily large values of n for which $\pi_1(n) > \pi_3(n)$ [16]. □

In short, induction goes from the specific to the general, whereas deduction goes from the general to the specific. While mathematical truths are often discovered via inductive reasoning, their proofs are established deductively; that is, the statement is proven in general, even if it was derived through observation of many specific cases.

---

[1]See http://mathworld.wolfram.com/GoldbachConjecture.html.

The method of mathematical induction, however, is unusual in that it combines the examination of a special case with a deductive argument. In order to proceed, we need an important property of integers that has not been mentioned yet.

6.2 Well-Ordering Principle

The so-called Well-Ordering Principle is simple to state and understand.

Principle 6.1. *Any nonempty subset of positive integers has a (unique) smallest element.*

As obvious as this statement may seem, it cannot be proven using the other properties that were provided in Chapter 2. Therefore, the Well-Ordering Principle must also be accepted as an axiom. Notice that the Well-Ordering Principle is a property of the positive integers; it may not apply to a set that includes negative integers, nor does it apply to all subsets of the positive rational numbers.

Example 67. For each set given below, determine if the Well-Ordering Principle can be applied:
(1) the set of integers greater than 7,
(2) the set of real numbers greater than 7,
(3) the set of all multiples of 5,
(4) $\{1 - 2^{-n} : n \in \mathbb{Z}^+\}$.

Solution. (1) The set of integers greater than 7 is a nonempty set of positive integers: $\{8, 9, 10, 11, ...\}$. Therefore, the conclusion of the Well-Ordering Principle applies. In this case, the unique smallest element is 8.

(2) The set of real numbers greater than 7 is nonempty and positive, but because the set does not consist of integers only, the conclusion of the Well-Ordering Principle may not hold. In particular, it is impossible to find a smallest real number greater than 7.

(3) The set of all multiples of 5 is a nonempty set of integers, but they are not all positive. Once again, the conclusion of the Well-Ordering Principle does not hold; a smaller (negative) multiple of 5 can always be found.

(4) Notice that $\{1 - 2^{-n} : n \in \mathbb{Z}^+\} = \{1 - \frac{1}{2}, 1 - \frac{1}{4}, 1 - \frac{1}{8}, ...\} = \{\frac{1}{2}, \frac{3}{4}, \frac{7}{8}, ...\}$. The Well-Ordering Principle cannot be applied since this is not a set of integers—not every set with a smallest element meets the hypotheses of the principle. $\square$

In the next example we will utilize the so-called "method of smallest counterexample", which relies on the Well-Ordering Principle. In fact, the argument is logically equivalent to a proof by induction.

Example 68. Prove that for all $n \in \mathbb{N}$, $27|(10^n + 18n - 1)$.

First Thoughts. When $n = 1$ and $n = 2$, the statement says $27|27$ and $27|135$, which are both true. We can continue verifying that the statement holds for integers $n = 3, 4, 5, ...$, but even if we never find a counterexample before exhausting all the computer power on earth, we still will not have *proven* the statement true for all positive integers! On the other hand, if the statement is not true in general, we will eventually reach a specific value of n for which the statement fails.

Suppose this happens; that is, after showing the statement is true for $n = 1$, let us assume that there is some integer $k \geq 1$ for which the given statement is true for $n = 1, 2, 3, \ldots, k$ but fails when $n = k + 1$. In other words, let us assume that $n = k + 1$ is the first positive integer for which the statement $27|(10^n + 18n - 1)$ is false, and see if any contradiction arises. The legitimacy of making such an assumption is grounded in the Well-Ordering Principle; in fact, this is why we need to show that the statement is true for $n = 1$ (or for some other value), because the Well-Ordering Principle only applies to nonempty sets.   ◇

Proof. The statement is true when $n = 1$. Under the assumption that $n = k + 1$ is the first positive integer for which the statement fails, the integer $A = 10^k + 18k - 1$ is divisible by 27 but the integer $B = 10^{k+1} + 18(k + 1) - 1 = 10^{k+1} + 18k + 17$ is *not* divisible by 27. Consider then

$$C = B - 10A$$
$$= (10^{k+1} + 18k + 17) - (10^{k+1} + 180k - 10)$$
$$= -162k + 27$$
$$= 27(-6k + 1).$$

It follows that $27|C$. Since $B = 10A + C$, by Theorem 2.5(3), $27|B$. This contradicts our assumption that B is *not* divisible by 27, so our proof is complete.   □

Accepting the Well-Ordering Principle as an axiom allows us to prove many other useful mathematical statements; we give an example below. The Well-Ordering Principle will also be foundational in establishing the Principle of Mathematical Induction, to be stated in the next section.

Theorem 6.2. *Every integer greater than 1 is either prime or the product of finitely many primes.*

Proof. Suppose the statement is false; that is, suppose the set S of integers greater than 1 that are neither prime nor the product of primes is not empty. By the Well-Ordering Principle, S has a unique smallest element, n. If n is not a prime, it is composite. Therefore, n can be expressed as the product of smaller integers, x and y, where $1 < x, y < n$. Since n is the smallest element of S, both x and y are either prime or the product of a finite number of primes. In either case, this means that $n = xy$ is also the product of a finite number of primes, contradicting our assumption. Hence, S is empty.   □

Exercises and Problems.

(1) Use the method of smallest counterexample to prove

$$P(n) : 1 + 2 + 4 + \cdots + 2^n = 2^{n+1} - 1 \quad \text{for } n \geq 1.$$

(2) Use the method of smallest counterexample to prove that $n^3 + 2n$ is divisible by 3 for all positive integers n.

(3) Consider a sequence of 0's and 1's of length $n \geq 2$ in which the first element in the string is 0 and the last element is 1. Prove by the method of smallest counterexample that there must be some point in the sequence where a 0 immediately precedes a 1.

---

(4) A Fermat number is an integer of the form $F_n = 2^{(2^n)} + 1$. (Only five prime Fermat numbers are known, with the first two being $F_0 = 3$ and $F_1 = 5$.) Use the method of smallest counterexample to prove that $F_{n+1} = F_0 \cdot F_1 \cdot F_2 \cdots \cdot F_n + 2$.

6.3 Method of Mathematical Induction

As we noted, the proof by method of smallest counterexample used in Example 68 is equivalent to the proof method that utilizes mathematical induction. However, the method of mathematical induction is more commonly used, perhaps because some mathematicians try to avoid proof by contradiction when possible. We will see that mathematical induction provides a two-step method for proving statements of the type $\forall n \in \mathbb{N}, P(n)$, where $P(n)$ is a predicate, often conjectured from experimentation.

The earliest evidence of the use of reasoning akin to mathematical induction is found in the writings of Sicilian priest and mathematician Francesco Maurolico (1494–1575). Maurolico's *Arithmeticorum Libri Duo*, published in 1575, appears to have influenced the work of French mathematician Blaise Pascal, who utilized a more polished version of mathematical induction throughout his *Triangle Arithmétique*. Published in 1665, the *Triangle Arithmétique* played an important role in disseminating the technique to other mathematicians. At approximately the same time, Pierre de Fermat used a technique that he called the "method of infinite descent", which was roughly equivalent to mathematical induction (though in reverse) [6]. In an article published in 1838, Augustus De Morgan carefully described mathematical induction in the following form, which is the form now commonly used.[2]

Principle 6.3 (Principle of Mathematical Induction). *Let $P(n)$ be a predicate that is defined for all $n \in \mathbb{N}$. Suppose the following two statements are true.*

(1) *$P(1)$ is true.*

(2) *For all $k \geq 1$, if $P(k)$ is true, then $P(k + 1)$ is true.*

Then $P(n)$ is true for all $n \in \mathbb{N}$.

Proving part (1) of the Principle of Mathematical Induction is called the **basis step**, while proving part (2) is called the **inductive step**. Often the process of using the Principle of Mathematical Induction is likened to the process of establishing that a ladder containing any finite number of steps can be climbed: proving $P(1)$ establishes that one can step onto the first rung of the ladder, while proving the inductive step verifies that if one has climbed to the kth rung of the ladder, it is possible to step onto the $(k + 1)$st rung. If both of these are true, then one can climb to the top of a ladder of height n for any $n \in \mathbb{N}$.

In some instances, $P(1)$ may not be true or might not even make sense. Therefore, depending on the statement to be proven, the basis statement might require proving

---

[2]http://education.lms.ac.uk/2011/10/augustus-de-morgan-mathematical-induction/.

$P(a)$ for some integer a other than 1. In such cases, the conclusion will be that $P(n)$ is true for all integers $n \geq a$.

The Principle of Mathematical Induction is equivalent to the Well-Ordering Principle; that is, if we take the Well-Ordering Principle as an axiom, then the Principle of Mathematical Induction can be proven. Without giving a rigorous argument, here is the idea of the equivalence of the two principles. Given an integer a, let W_a be the set $\{n | n \in \mathbb{N}, n \geq a\}$. Let M be a nonempty subset of W such that

(1) $a \in M$ and

(2) if $k \in M$, then $k + 1 \in M$ also.

We claim that $M = W_a$. Suppose not; that is, suppose that $W_a \setminus M$ is not empty. Then by the Well-Ordering Principle, $W_a \setminus M$ has a least element, x. Since $a \in M$ (by (1) above), $x > a$. Thus, $x - 1 \geq a$, in which case $x - 1 \in W_a$. Since x is the smallest element of $W_a \setminus M$, $x - 1 \notin W_a \setminus M$. Therefore, $x - 1 \in M$. But, by (2) above, this means that $x \in M$, which contradicts our assumption that $x \in W_a \setminus M$. We conclude that $W_a \setminus M$ is empty, so $M = W_a$.

Interestingly, utilizing the Principle of Mathematical Induction in a proof is *deductive* in the sense that doing so involves a specific application of a valid, general principle. In the examples that follow, note the two-step nature of each proof: first proving the basis step, then proving the inductive step. When proving the inductive step, it is completely valid to assume that $P(k)$ is true for an arbitrary value of k; this is called the **inductive hypothesis**.

Example 69. Prove that for any $n \in \mathbb{N}$,

$$1^2 + 2^2 + \cdots + n^2 = \frac{n(n + 1)(2n + 1)}{6}.$$

First Thoughts. In order to construct a proof by induction, we must first identify the predicate $P(n)$ that is to be proven for some infinite set of positive integers, as well as the value of a that must be used in demonstrating the basis step. In this case, $P(n)$ is given explicitly in the proof request:[3]

$$P(n) : 1^2 + 2^2 + \cdots + n^2 = \frac{n(n + 1)(2n + 1)}{6}.$$

Since we wish to prove the statement for all $n \in \mathbb{N}$, the basis step is to prove $P(1)$.   ◇

Proof. <u>Basis Step.</u> Letting $n = 1$, we obtain the statement to be proven in the basis step:

$$P(1) : 1^2 = \frac{1(1 + 1)(2 \cdot 1 + 1)}{6}.$$

This is clearly true, since $1^2 = 1$ and $\dfrac{1(1 + 1)(2 \cdot 1 + 1)}{6} = \dfrac{1 \cdot 2 \cdot 3}{6} = 1$.

<u>Inductive Step.</u> Now we *assume* that $P(k)$ is true for some arbitrary $k \geq 1$ and show that $P(k + 1)$ must also be true.

---

[3]Note that, in this case, the predicate $P(n)$ is a label assigned to the given equation. Because $P(n)$ is not *equal* to any part of the equation that follows, we use a colon rather than an equal sign, as we did for propositions and predicates in Chapters 3 and 4.

Remark. $P(k)$ can be obtained by simply replacing each n in $P(n)$ with a k; likewise, $P(k+1)$ results from replacing each n in $P(n)$ with $k+1$:

$$P(k) : 1^2 + 2^2 + \cdots + k^2 = \frac{k(k+1)(2k+1)}{6},$$

$$P(k+1) : 1^2 + 2^2 + \cdots + (k+1)^2 = \frac{(k+1)((k+1)+1)(2(k+1)+1)}{6}.$$

Remember that $P(k)$ is what we are assuming to be true and $P(k+1)$ is what we must prove to be true. One helpful approach for establishing that $P(k+1)$ is true in algebraic examples is to manipulate both sides of the (assumed) equation given in $P(k)$. ◇

In this case, we add $(k+1)^2$ to both sides of

$$1^2 + 2^2 + \cdots + k^2 = \frac{k(k+1)(2k+1)}{6},$$

so that the left-hand side of the equation matches the left-hand side of $P(k+1)$. This gives

$$1^2 + 2^2 + \cdots + k^2 + (k+1)^2 = \frac{k(k+1)(2k+1)}{6} + (k+1)^2$$
$$= \frac{k(k+1)(2k+1)}{6} + \frac{6(k+1)^2}{6}$$
$$= \frac{(k+1)[k(2k+1) + 6(k+1)]}{6}$$
$$= \frac{(k+1)(2k^2 + 7k + 6)}{6}$$
$$= \frac{(k+1)((k+1)+1)(2(k+1)+1)}{6}.$$

Since the final line in the preceding string of equivalent statements equals the right-hand side of $P(k+1)$, we have successfully carried out the inductive step.

Therefore, by the Principle of Mathematical Induction, we have proven that $1^2 + 2^2 + \cdots + n^2 = \frac{n(n+1)(2n+1)}{6}$ for any $n \in \mathbb{N}$. □

Further Thoughts. Notice the role that factoring out $(k+1)$ played in developing the statement for $P(k+1)$. While this may not be an obvious strategy to those who are new to proofs by induction, this type of factorization is frequently helpful. In this situation, the fact that $P(k+1)$ is expressed in terms of $(k+1)$ provides an incentive for the factorization used. ◇

Example 70. For each integer n, $n \geq 1$, prove $P(n)$, where:

> $P(n)$: *Given n points in the plane, no two of which are collinear, there are exactly $\frac{n^2-n}{2}$ segments joining any two of the points.*

Proof. Basis Step.

When $n = 1$, there are no edges present. Since $(1^2 - 1)/2 = 0$, the base case is true.

Inductive Step. Assume that for some arbitrary $k \geq 1$, $P(k)$ is true; i.e., among any set of k points in the plane with no two collinear, there will be exactly $(k^2 - k)/2$ segments joining two points from the set. Now consider a set of $k+1$ points in the plane, no two of which are collinear. We need to prove that there will be $((k+1)^2 - (k+1))/2$, or $(k^2 + k)/2$, segments between these points.

Set one point aside and note that the remaining k points have the property that no two of them are collinear. Therefore, by the inductive hypothesis, there will be $(k^2 - k)/2$ segments joining two points from these k points. Furthermore, there will be k different segments joining the set-aside point to one of these k points. All segments will be formed in one of these two ways, so there will be a total of $(k^2 - k)/2 + k$ segments. This sum simplifies to $\frac{k^2-k}{2} + \frac{2k}{2} = \frac{k^2+k}{2}$, as desired, and completes the proof by induction. $\square$

Further Thoughts. There is a faster and more elegant way to prove this statement than by induction; we will see it in Chapter 8. $\diamond$

Example 71. Prove that $3^{5n} - 5^{3n}$ is divisible by 59 for any $n \in \mathbb{N}$.

Proof. For each integer greater than or equal to 1, we wish to prove $P(n)$: $3^{5n} - 5^{3n}$ is divisible by 59.

Basis Step. $P(1)$ states that $3^{5 \cdot 1} - 5^{3 \cdot 1} = 3^5 - 5^3$ is divisible by 59. Since $3^5 - 5^3 = 243 - 125 = 118 = 2 \cdot 59$, $P(1)$ is a true statement.

Inductive Step. Let $k \geq 1$ and assume $P(k)$ is true:

$$P(k): 3^{5k} - 5^{3k} \text{ is divisible by } 59.$$

That is, $3^{5k} - 5^{3k} = 59m$ for some $m \in \mathbb{N}$.

Remark. To prove $P(k+1)$ is true, we wish to rewrite $3^{5(k+1)} - 5^{3(k+1)}$ in a way that is helpful for ultimately showing that it is divisible by 59. To make use of the inductive hypothesis, we will need to obtain the expression $3^{5k} - 5^{3k}$ as part of the rewritten version. How do we do so? The first step is often very natural. After that, one must use a combination of trial and error, algebra, and experience to proceed. In this case, we see that the useful trick of subtracting the term we need, namely $3^5(5^{3k})$, and then adding it back in again to preserve the equality brings success. $\diamond$

$$3^{5(k+1)} - 5^{3(k+1)} = 3^5 \cdot 3^{5k} - 5^3 \cdot 5^{3k}$$
$$= 3^5(3^{5k}) - 3^5(5^{3k}) + 3^5(5^{3k}) - 5^3(5^{3k})$$
$$= 3^5(3^{5k} - 5^{3k}) + (3^5 - 5^3)(5^{3k}).$$

Since $3^{5k} - 5^{3k}$ is divisible by 59 by the inductive assumption and $3^5 - 5^3$ is divisible by 59 as shown in the Basis Step, we can conclude that $3^{5(k+1)} - 5^{3(k+1)}$ is also divisible by 59. Therefore, by the Principle of Mathematical Induction, we have proven that $P(n)$ holds for any $n \in \mathbb{N}$. $\square$

Example 72. You have a collection of 3^n gold coins that are identical in appearance. However, you have heard a rumor that one of the coins is counterfeit, being made partially of silver rather than of pure gold. If this is true, although the coins all look the same, the counterfeit coin should be lighter, since silver is lighter than gold. Devise an efficient method for finding a counterfeit (lighter) coin among your collection of 3^n coins using a two-pan balance. How many weighings are required? Prove that your answer is correct.

First Thoughts. Since you are asked to devise the method yourself, you should begin by considering how you would accomplish the task for one or two small values of n. For instance, suppose $n = 1$. In this case, you have a set of three identical-looking coins and a two-pan balance to use for determining if one of them is lighter than the others. How will you do it? Try to come up with a scheme before reading further. Once you have an idea, see if the procedure you've designed can be extended to a set of $3^2 = 9$ coins.   ◇

Hopefully your experimentation with real or imagined sets of 3^n coins for small values of n, such as $n = 1$ (3 coins) and $n = 2$ (9 coins), led you to conjecture the following:

> $P(n)$: Isolating one counterfeit (lighter) coin among 3^n coins can always be accomplished in at most n weighings with a two-pan balance.[4]

Solution. We prove $P(n)$ as stated above.

Basis Step. $P(1)$ states that if one of three identical-looking coins is lighter than the other two, the lighter coin can be identified in one weighing using a two-pan balance. To do this, simply choose any two of the three coins and place them in opposite pans of the scales. If one of the two is lighter than the other, it must be the counterfeit coin; if the two coins in the balance are equal in weight, the counterfeit coin must be the third coin, which has not been weighed.

Inductive Step. Suppose that for some arbitrary $k \geq 1$, k weighings are sufficient for finding the counterfeit lighter coin among a set of 3^k coins. Now, consider a set of $3^{k+1} = 3 \cdot 3^k$ coins. Divide the 3^{k+1} coins into three piles of 3^k coins each; call them piles A, B, and C. Place pile A in one of the pans of the balance and pile B in the other. If the two pans do not balance, then the counterfeit coin must be in the pile (A or B) on the lighter side; if the pans do balance, then the counterfeit coin must be in pile C. Either way, after this first weighing, you have determined that the counterfeit coin is in one of pile A, B, or C, each of which contains 3^k coins. By the inductive assumption, the counterfeit coin can be isolated from a set of 3^k coins in at most k weighings. Thus, the total number of weighings needed to find the counterfeit coin in a set of 3^{k+1} coins is at most $k + 1$. By the Principle of Mathematical Induction, we have proven that $P(n)$ holds for any $n \in \mathbb{N}$.   □

Example 73. A "right triomino" is a 2×2 square, minus one of the small (1×1) squares in one of the four corners. (See the left image in Figure 6.1.) Prove that for any positive integer n, any $2^n \times 2^n$ checkerboard with any one square removed can be tiled using right triominoes (i.e., covered completely, with no overlapping triominoes). The middle image in Figure 6.1 is an 8×8 checkerboard with one square removed (denoted by an asterisk, in section C).

First Thoughts. As was the case with Example 72, although this problem does not require arithmetic computations, because the request is to establish a statement as being

---

[4]An additional weighing may be needed to verify that the suspect coin is actually counterfeit if one doesn't know a priori that there is exactly one counterfeit coin.

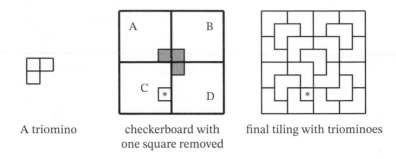

A triomino   checkerboard with   final tiling with triominoes
                        one square removed

Figure 6.1

true for all natural numbers, proving the result with mathematical induction should be explored. We need to prove the following:

$P(n)$: $\forall n \in \mathbb{N}$, a $2^n \times 2^n$ checkerboard with any one square removed can be tiled using right triominoes.   ◇

Proof. We use induction to prove $P(n)$ as stated above.

Basis Step. $P(1)$ says that a 2×2 checkerboard with any square removed can be tiled using right triominoes. Since a 2×2 checkerboard with one square removed is itself a triomino, this statement is obviously true.

Inductive Step. Let $k \geq 1$, and assume that a $2^k \times 2^k$ checkerboard with any one square removed can be tiled using right triominoes. Now, consider any $2^{k+1} \times 2^{k+1}$ checkerboard that has one square removed. This larger checkerboard can be partitioned into four smaller $2^k \times 2^k$ checkerboards; label these smaller regions as A, B, C, D, as shown in Figure 6.1. The square that has been removed is in one of A, B, C, or D; assume it is in C, as shown. Place a right triomino in the center of the $2^{k+1} \times 2^{k+1}$ checkerboard in such a way that the triomino covers one square in each of A, B, and D. This triomino is shown as shaded in the middle image of Figure 6.1.

Each of the $2^k \times 2^k$ checkerboards, A, B, C, and D, have one square that is either missing or already covered. By the inductive assumption, the remaining squares in each of A, B, C, and D can now be filled using right triominoes—resulting in a tiling of the $2^{k+1} \times 2^{k+1}$ checkerboard. The third image in the figure shows the final tiling once the procedure is complete. By the Principle of Mathematical Induction, we have proven that $P(n)$ holds for any $n \in \mathbb{N}$.   □

For our final example, we prove an inequality. Our solutions thus far have been very formal, and $P(n)$ was explicitly stated in each proof. Our solution below is streamlined and is written the way a proof would be written by someone who has mastered the method of induction and is writing for an experienced audience.

Example 74. Prove that for $n \geq 5$, $2^n > n^2$.

Solution. The base case is true, since $2^5 = 32 > 25 = 5^2$. Now assume that $2^n > n^2$ for some $n \geq 5$. Then,

$$2^{n+1} = 2(2^n) > 2(n^2),$$

which is greater than $(n + 1)^2$ when $n \geq 5$. This proves the result.   □

Further Thoughts. At the end of the proof above, we simply stated without proof that $2n^2 > (n+1)^2$ when $n \geq 5$. Some audiences would accept this, while others may not. Here are more details:

$$2n^2 = (n^2 + 2n + 1) + (n^2 - 2n + 1) - 2 = (n+1)^2 + (n-1)^2 - 2.$$

The final expression exceeds $(n+1)^2$ as long as $(n-1)^2 - 2 > 0$. This is true when $n \geq 5$ ..., but is this clear? Again, it depends on the audience. We (the authors) would say it is obvious that $(n-1)^2 > 2$ when $n \geq 5$ and leave it at that. But, if pushed for a reason, we point out that $n - 1 > \sqrt{2}$ when $n \geq 5$. Why? Because in that case $n \geq 5 > \sqrt{2}+1 \approx 2.4$. Ultimately, it comes down to basic arithmetic. One should show enough steps to convince, but not insult, the audience reading the proof.   ◇

Exercises and Problems.

(1) Find a formula for the sum of the first n positive integers, $1 + 2 + \cdots + n$.

    (a) Use induction to prove that your answer is correct by carrying out the following steps.

        (i) Write your hypothesized formula as a predicate, $P(n)$.

        (ii) Carry out the basis step by showing that $P(1)$ is true.

        (iii) State $P(k+1)$.

        (iv) Carry out the inductive step by proving that $P(k+1)$ is true if $P(k)$ is true.

    (b) Find a proof that does not use induction.

(2) Recall the Fibonacci sequence from Example 64, defined as $f_1 = 1$, $f_2 = 1$, and $f_{n+1} = f_n + f_{n-1}$ for $n \geq 2$. In this problem we will prove that the sum of the first m odd-indexed Fibonacci numbers is a Fibonacci number.

    (a) Consider $P(m) : f_1 + f_3 + f_5 + \cdots + f_{2m-1} = f_{2m}$. Show that $P(1)$ and $P(2)$ are true.

    (b) State $P(m+1)$.

    (c) Use the Fibonacci recursion to simplify the right-hand side of the statement of $P(m+1)$.

    (d) Use induction to prove that $P(m)$ is true for all $m \geq 1$.

(3) Use induction to prove $P(n) : 1 + 2 + 4 + \cdots + 2^n = 2^{n+1} - 1$ for $n \geq 1$.

(4) Use induction to prove that the sum of the cubes of the first n positive integers, $1^3 + 2^3 + \cdots + n^3$, is equal to $\dfrac{n^2(n+1)^2}{4}$.

(5) Prove that the sum of the measures of the interior angles of a convex[5] polygon with $n \geq 3$ sides is $180(n-2)°$.

(6) Prove that for $n \geq 0$, $3^n > (n-1)^3$.

(7) Find a positive integer n_0 for which you believe $2^n > n^3$ for all $n \geq n_0$. Prove the statement is true using induction.

(8) Determine the positive integers n for which $n! > n2^n$, and prove your result.

(9) Determine the positive integers n for which $3^n > 10n^2$, and prove your result.

---

[5]The statement is true for simple nonconvex polygons as well, but the proof is more difficult.

(10) (a) Prove that for $n \geq 2$, $(1 + 1/3)^n > 1 + n/3$.

     (b) (Bernoulli's Inequality) Prove that for any nonzero real number x, $x > -1$, and every integer $n \geq 2$,

$$(1 + x)^n > 1 + nx.$$

(11) Use induction to prove that $x^n - 1$ is divisible (as a polynomial) by $x - 1$ for $n \in \mathbb{N}$. What is the quotient? (Do not simply use the well-known theorem from algebra, which states that a polynomial $f(x)$ is divisible by $x - a$ if and only if a is a root of $f(x)$.)

(12) Find a closed form for the finite geometric sum $1 + r + r^2 + \cdots + r^n$, $n \geq 0$, and prove it is correct. See Problems 1 and 11.

(13) Use induction to prove that an n-element set has 2^n subsets, $n \geq 1$.

(14) State and prove an extension to De Morgan's Law as it pertains to the union of sets $A_1 \cup A_2 \cup \cdots \cup A_n$, $n \geq 2$.

(15) When A and B are square matrices of the same size, $(AB)^T = B^T A^T$. State and prove an extension of this result for matrices $A_1, A_2, \dots, A_n$, $n \geq 1$.

(16) Use induction to prove the following statements about natural numbers. (This problem also appears as Problem 6 in Section 7.4, where congruences are used.)

    (a) $6 \mid n(n^2 + 5)$.

    (b) $5 \mid (n^5 - n)$.

    (c) $7 \mid (n^7 - n)$.

---

(17) Let f_n be the Fibonacci sequence and consider the following two statements:

$$P(n) : f_{2n-1} = f_{2n}^2 + f_{2n+1}^2 \quad \text{and} \quad Q(n) : f_{2n} = f_n(f_n + 2f_{n-1}).$$

    (a) Show that $P(1)$ and $Q(1)$ are true.

    (b) State $P(k + 1)$ and $Q(k + 1)$.

    (c) Write each of f_{2k+1} and f_{2k+2} in terms of f_{2k} and f_{2k-1}.

    (d) Prove $P(n)$ and $Q(n)$ are true for $n \geq 1$. (Prove them simultaneously by assuming both $P(k)$ and $Q(k)$ are true for some arbitrary $k \geq 1$ and then showing $P(k + 1)$ and $Q(k + 1)$ are true.)

(18) Find the error in the "proof" of the following statement: $P(n)$: In any collection of n crayons ($n \geq 1$), the crayons will all be the same color.

"Proof". $P(1)$ is clearly true since any crayon is the same color as itself.

Now assume that any collection of $k \geq 2$ crayons will be the same color, and consider a collection of $k + 1$ crayons numbered from 1 to $k + 1$. On one hand, crayons 1 through k form a collection of k crayons, so they must all be the same color, by the inductive hypothesis. Similarly, crayons 2 through $k + 1$ also form a collection of k crayons, so they will all be the same color as well. Since crayon 2 lies in both collections, all crayons are the same color as crayon 2. $\square$

(19) Use induction to prove the following statements for natural numbers. (This problem also appears as Problem 7 in Section 7.4, where congruences are used.)

(a) $9|(4^n + 15n - 1)$.

(b) $64|(3^{2n+3} + 40n - 27)$.

(c) 15^n has remainder 1 when divided by 7.

(20) For $n \geq 2$, prove that

$$\frac{3}{4} \cdot \frac{8}{9} \cdot \frac{15}{16} \cdots (1 - 1/n^2) = \prod_{k=2}^{n} (1 - 1/k^2) = \frac{n+1}{2n}.$$

(21) The Tower of Hanoi puzzle was invented by the French mathematician François Édouard Anatole Lucas (1842–1891) in 1883. It is well known in both the computer sciences and mathematics communities for being an excellent application of algorithms, stacks, recursion, and induction.

Here is the problem. Given a tower of n disks, initially stacked in increasing size on one of three posts, transfer the entire tower to one of the other two posts, given the following two rules: (i) each move consists of moving exactly one disk from one post to another post and (ii) a disk may never be placed on top of a disk smaller than itself. Prove that this task can be accomplished in $2^n - 1$ moves.

(22) Consider the following variation of the Tower of Hanoi puzzle. Suppose the three posts are lined up in a row and an additional rule is imposed: (iii) one may only move a disk to an adjacent post. Let $f(n)$ be the number of moves it takes to transfer a tower of n disks from the first post to the third post.

(a) Find a recursive solution for $f(n)$, thereby proving that the problem can be solved.

(b) Prove that $f(n) = 3^n - 1$.

(23) Here is yet another variation of the Tower of Hanoi puzzle. Suppose that n disks are arranged on the three posts (not necessarily the same post) in such a way that a larger disk is never on top of a smaller disk. Assuming that the largest disk must be moved at least once, prove that one can transfer all n disks to one post using the two rules given in Problem 21. Show that this can always be done in at most $2^n - 1$ moves.

Remark. It seems that the above property might be useful in determining the minimum number of moves to solve the four-post Hanoi puzzle. (The four-post puzzle is identical to the original Tower of Hanoi problem, except there are four posts instead of three.) However, the authors know of no solution to the four-post version and believe it is an unsolved problem.   ◇

(24) Letting f_n denote the terms of the Fibonacci sequence, as in Problem 2, prove that

$$\sum_{i=1}^{n} f_i^2 = f_n \cdot f_{n+1}, \qquad n \geq 1.$$

(25) The following problem is a generalization of the opening hook problem for this chapter. Each of two bright math students, Terence and Srinivasa, has a number on his forehead. Each can see the number on the other's forehead, but not the number on his own. They are told that the two numbers are consecutive positive integers. They start the conversation with Terence saying, "I don't know my number." Srinivasa thinks for a moment and replies, "I don't know my number either."

They then repeat this exact same conversation, for a total of n times, at which point Terence exclaims, "Hey, I just figured out my number!" Srinivasa immediately follows, "Yeah? Me too!" Prove that Terence's number is $2n + 1$ if Srinivasa's number is even and it is $2n + 2$ if Srinivasa's number is odd.

(26) (De Moivre's Theorem) Prove that for any real number θ and natural number n,
$(\cos\theta + i\sin\theta)^n = \cos(n\theta) + i\sin(n\theta)$.

(27) Show that
$$\sqrt{6 + \sqrt{6 + \sqrt{6 + \cdots + \sqrt{6}}}} < 3,$$
where there are n occurrences of "6" in the expression on the left.

Remark. It turns out that the expression on the left side of the given inequality actually converges to 3. (This means that it gets arbitrarily close to 3 as n tends to ∞.) One way to prove this is to first show that indeed the expressions even converge to some positive number, L. From that point, it is easy to see that $\sqrt{6 + L} = L$, so $L = 3$. However, proving convergence is not trivial. Typically, one uses a theorem from calculus that says that an increasing, bounded sequence of numbers must have a limit. Proving the inequality in this problem shows that the terms are bounded above by 3. Is it obvious that their values increase as the number of occurrences of "6" increases?   ◇

---

(28) Show that
$$\sqrt{2 + \sqrt{2 + \sqrt{2 + \cdots + \sqrt{2}}}} = 2\cos\left(\frac{\pi}{2^{n+1}}\right),$$
where there are n occurrences of "2" in the expression on the left.

(29) The **harmonic numbers**, h_n, are defined to be the sum of the reciprocals of the first n natural numbers. That is,
$$h_n = 1 + \frac{1}{2} + \frac{1}{3} + \cdots + \frac{1}{n}.$$
Prove that $h_n \geq 1 + \frac{\log_2(n)}{2} = 1 + m/2$ when $n = 2^m$ for some $m \geq 0$.

Note: This proves that the sequence of harmonic numbers is unbounded. In fact, one can show (using integral calculus) that $\ln(n) < h_n < 1 + \ln(n)$.

(30) Refer to Problem 13 in Section 5.4 for the following problem.

   (a) Prove that for all natural numbers
   $$n, \quad (1 + \sqrt{2})^n + (1 - \sqrt{2})^n, \quad \text{and} \quad \left(\frac{1}{\sqrt{2}}\right)\left((1 + \sqrt{2})^n - (1 - \sqrt{2})^n\right)$$
   are integers.

   (b) Prove that for all natural numbers n, a, and b, $(a + \sqrt{b})^n + (a - \sqrt{b})^n$ and $\left(\frac{1}{\sqrt{b}}\right)\left((a + \sqrt{b})^n - (a - \sqrt{b})^n\right)$ are even integers.

(31) Design a method, using $n + 2$ weighings in a two-pan balance, for finding a counterfeit coin and whether it is heavier or lighter than the other coins in a collection of $4 \cdot 3^n$ coins. The coins are identical in appearance to each other but one of them is a different weight than the others. Prove that your answer is correct. (When $n = 0$, in addition to the four coins, one of which is counterfeit, you may also assume that you have a few other authentic coins you may use for comparison.)

(32) Use induction on n to prove that for any natural number k, there is a natural number M (dependent on k) such that $n! > k^n$ for all $n \geq M$. I.e., prove the following:

$$\forall k \in \mathbb{N}, \exists M \in \mathbb{N}, ((n \geq M) \Rightarrow (n! > k^n)).$$

6.4 Strong Induction

The following problem is similar in spirit to Problem 5 in Chapter 1.

Suppose that you have an unlimited supply of 4-cent and 7-cent stamps. Describe completely the (exact) amounts of postage that can be formed using only these denominations.

Let us think about this problem in light of induction.

By experimentation, one will discover that the postage values that can be created appear sporadically (for example, we can represent 7-cent and 8-cent amounts, but not 9-cent and 10-cent amounts) until we reach $n = 18$ cents. When we get to 18 cents, however, we seem to be able to create all subsequent integer values of postage.

- $4 + 7 + 7 = 18$ cents,
- $4 + 4 + 4 + 7 = 19$ cents,
- $4 + 4 + 4 + 4 + 4 = 20$ cents,
- $7 + 7 + 7 = 21$ cents,
- $4 + 4 + 7 + 7 = 22$ cents.

In fact, once we know that we can find a representation for 18, 19, 20, and 21, it is clear that we can find a representation for any n larger than 21 as well. Simply add 4 to the representation for $n - 4$. This is an example of **strong induction**; rather than relying on just $P(k)$ to prove $P(k + 1)$, it was necessary to know that at least the four preceding values held. We state this principle formally below.

Principle 6.4 (Principle of Strong Mathematical Induction). *Let $P(n)$ be a statement that is defined for all $n \in \mathbb{N}$. Suppose the following two statements are true.*

(1) *$P(a)$ is true for some natural number a. (Often $a = 1$, but not necessarily.)*

(2) *For all $k \geq a$, if $P(a), P(a + 1), P(a + 2), \ldots, P(k)$ are true, then $P(k + 1)$ is true.*

Then $P(n)$ is true for all $n \geq a$.

Observe that the Principle of Induction (sometimes referred to as "weak induction") and the Principle of Strong Induction are both equivalent to the method of proof by smallest counterexample. If we assume that n is the smallest positive integer for which the given statement is not true, we are implicitly assuming that the statement holds for every positive integer less than n (i.e., the assumption of strong induction). The word *strong* is used because the assumptions of the inductive hypothesis are stronger than in regular induction: $P(n)$ is assumed to be true for $k = 1 \leq n \leq k$ for strong induction while only $P(k)$ is assumed to be true with weak induction.

Recall the statement of Theorem 6.2:

Every integer $n \geq 2$ is either prime or the product of primes.

The proof of this statement may be viewed through the lens of strong induction.

Proof. Basis Step. An integer p is prime if its only positive divisors are 1 and p. Since 1 and 2 are the only positive divisors of 2 and since 2 is prime, $P(2)$ holds.

Inductive Step. Assume that for some integer $k \geq 2$, all integers i for which $2 \leq i \leq k$ are either prime or the product of primes. Now consider $k + 1$. If $k + 1$ is prime, we are done. If $k + 1$ is not prime, then $k + 1 = ab$, where integers a and b are each less than $k + 1$ but greater than or equal to 2. By the inductive hypothesis, each of them is either prime or a product of primes, say $a = p_1 p_2 \dots p_m$ and $b = q_1 q_2 \dots q_r$. Hence, $k + 1 = (p_1 p_2 \dots p_m)(q_1 q_2 \dots q_r)$, which shows it is a product of primes.

By the Principle of Strong Mathematical Induction, we have proven that $P(n)$ holds for any $n \geq 2$.   □

Further Thoughts. Note that strong induction was very important here. Using regular induction, if we had simply assumed that k was either prime or the product of primes for some k, we would not have been able to make any assumptions about the factors of $k + 1$, because those factors certainly would not have included k.   ◇

Recall the Fibonacci sequence, first introduced in Example 64. There are many interesting properties about the Fibonacci sequence and about integers as they relate to elements of the Fibonacci sequence. Several appeared in the previous problem set, and we prove another in the following example.

Example 75. Prove that every positive integer can be represented as the sum of one or more distinct Fibonacci numbers in such a way that the sum does not include any two consecutive Fibonacci numbers.[6]

Proof. We prove $P(n)$: n is the sum of one or more Fibonacci numbers, no two of which are consecutive.

Basis Step. Certainly, $P(1)$, $P(2)$, and $P(3)$ are true, since 1, 2, and 3 are themselves Fibonacci numbers. $P(4)$ is also true, since $4 = 1 + 3$.

Inductive Step. Assume that $P(i)$ is true for all $1 \leq i \leq k$, for some integer $k \geq 4$. Now consider $k + 1$. If $k + 1$ is itself a Fibonacci number, then $P(k + 1)$ is true. If not, then there exist successive Fibonacci numbers, f_j and f_{j+1} such that $f_j < k + 1 < f_{j+1}$. Let $m = k + 1 - f_j$. Since $m + f_j = k + 1 < f_{j+1} = f_j + f_{j-1}$, we conclude $m < f_{j-1}$.

Since $m < k + 1$, $P(m)$ is true; that is, m can be represented as the sum of one or more distinct Fibonacci numbers in such a way that the sum does not include any two consecutive Fibonacci numbers. Moreover, because $m < f_{j-1}$, this sum does not contain f_{j-1} as one of its terms. Therefore, $k + 1$ can be represented as the sum of f_j and the distinct Fibonacci numbers whose sum is m.[7] Thus, $P(k + 1)$ is true.

By the Principle of Strong Mathematical Induction, we have proven that $P(n)$ holds for any $n \in \mathbb{N}$.   □

---

[6]This is known as Zeckendorf's Theorem.

[7]In fact, it turns out that the representation is unique, but we will not prove that here.

Example 76. Consider the recursively-defined sequence $a_0 = 1, a_1 = 1, a_n = 2a_{n-1} + 3a_{n-2}$ for $n \geq 2$. Determine a closed (i.e., explicit) formula for a_n and prove that your answer is correct.

First Thoughts. A first step in solving this problem is to start with small values of n and look for a pattern. The pattern is not immediately obvious.

$$a_0 = 1 = 2/2,$$
$$a_1 = 1 = 2/2,$$
$$a_2 = 2a_1 + 3a_0 = 5 = 10/2,$$
$$a_3 = 2a_2 + 3a_1 = 13 = 26/2,$$
$$a_4 = 2a_3 + 3a_2 = 41 = 82/2,$$
$$a_5 = 2a_4 + 3a_3 = 121 = 242/2.$$

We notice that the numerators are very close to the successive powers of 3, alternating from being one above and one below the integers $1, 3, 9, 27$, etc. Thus, the closed form seems to be $a_n = \frac{1}{2}(3^n + (-1)^n)$. One would typically prove this form is correct by substituting into the recurrence relation, but we will prove it by induction.   ◇

Proof. We seek to prove

$$P(n) : a_n = \frac{1}{2}(3^n + (-1)^n) \text{ for each } n \geq 0.$$

Basis Step. We check that the closed formula gives the correct value for the smallest values of n:

$$a_0 = \frac{1}{2} \cdot 3^0 + \frac{1}{2} \cdot (-1)^0 = \frac{1}{2} + \frac{1}{2} = 1,$$
$$a_1 = \frac{1}{2} \cdot 3^1 + \frac{1}{2} \cdot (-1)^1 = \frac{3}{2} - \frac{1}{2} = 1.$$

Inductive Step. Assume that $a_i = \frac{1}{2} \cdot 3^i + \frac{1}{2} \cdot (-1)^i$ for $1 \leq i \leq k-1$, for some $k \geq 2$. Using the recursive definition along with the inductive assumption, we have

$$a_k = 2a_{k-1} + 3a_{k-2}$$
$$= 2\left(\frac{1}{2} \cdot 3^{k-1} + \frac{1}{2} \cdot (-1)^{k-1}\right) + 3\left(\frac{1}{2} \cdot 3^{k-2} + \frac{1}{2} \cdot (-1)^{k-2}\right)$$
$$= 3^{k-1} + \frac{3}{2} \cdot 3^{k-2} + (-1)^{k-1} + \frac{3}{2} \cdot (-1)^{k-2}$$
$$= \frac{3}{2} \cdot 3^{k-1} + \frac{1}{2} \cdot (-1)^k$$
$$= \frac{1}{2}(3^k + (-1)^k).$$

Thus, $P(k)$ holds. (Note our slight variation in the usual technique; it is valid to assume that $P(k-1)$ and previous statements are true and then proceed in showing that $P(k)$ was true.)

By the Principle of Strong Mathematical Induction, we have proven that $P(n)$ holds for any $n \geq 0$.   □

Exercises and Problems.

(1) Explain in your own words why the method of smallest counterexample is equivalent to the method of strong induction.

(2) Consider the following two-person game. There are two piles of matches, containing $n > 1$ and $m > 1$ matches, respectively. The players alternate turns, each player removing at least one match from a pile of choice. The player who removes the last match wins. Prove that if $m = n$, the second player can play in a way that will guarantee that he wins the game, and if $m \neq n$, the first player can always play in a way that guarantees a win. (Can you extend the problem to a game with k stacks of n matches each?)

(3) Suppose that you have an unlimited supply of 4-cent and 9-cent stamps. Let $P(n)$ be the statement that a postage of n cents can be formed using stamps of just these denominations. For which n does $P(n)$ hold? Use strong induction to prove that $P(n)$ is true for all $n \geq 24$.

Note: In general, if one has an unlimited supply of stamps with values a and b, the largest value that cannot be formed using stamps of these types is $ab - a - b$, called the **Frobenius number** of $\{a, b\}$.

---

(4) Use strong induction on n to prove that $\dfrac{n^2}{b^2} \neq 2$ for any positive integers n and b. (Note that this proves that $\sqrt{2}$ is irrational.)

(5) Consider the sequence (a_n), where $a_1 = 2$, $a_2 = 3$, $a_3 = 7$, and for $n \geq 4$, $a_n = a_{n-1} + a_{n-2} + a_{n-3}$. Prove that $a_n \leq 2^n$ for all n.

---

(6) Use an approach similar to the one in Problem 11 of Section 6.3 to prove by induction that $a^n - b^n$ is divisible (as polynomials) by $a - b$ for all real numbers a and b, $b \neq 0$.

(7) Given a positive integer s_1, let s_2 be the sum of the squares of the digits of s_1. Let s_3 be the sum of the digits of s_2, and so on. Prove that for any choice of s_1, the sequence (s_n) will eventually reach either 1 or 42.

Note: A number s_1 whose resulting sequence eventually reaches 1 is called a **happy number**, while one that reaches 42 is called a **hitchhiker number** (not really).

(8) (Pick's Theorem) A lattice point in a coordinate plane is one whose coordinate values are both integers. Given a simple polygon (a polygon whose edges intersect only at the vertices) in the coordinate plane whose vertices are all lattice points, show that the area of the polygon is given by $I + B/2 - 1$, where I is the number of lattice points lying in the interior of the polygon and B is the number of lattice points that lie on the boundary of the polygon.

(9) Consider $[n] = \{1, 2, \ldots, n\}$. From this set, choose a subset that contains no consecutive integers; multiply the elements of the subset together and square the result. Let S_n be the sum of the result of carrying out this process on all such subsets of $[n]$. (This problem comes from `artofproblemsolving.com`.)

(a) Find the values of S_1, S_2, and S_3.

(b) Express S_{k+1} recursively in terms of S_k and S_{k-1}.

(c) Calculate a few more terms using the recursion and make a conjecture on a closed expression for S_n. Use strong induction to prove that your formula is correct.

6.5 Proof of the Division Theorem

In Section 2.2, we stated but did not prove the Division Theorem (Theorem 2.6):

> *For any two integers a and b, b $\neq$ 0, there exists a unique pair of integers q and r, $0 \leq r < |b|$, such that a = qb + r.*

We will use the Well-Ordering Principle to prove this theorem.

Proof. We break the proof into two parts: first, the existence of q and r and, second, their uniqueness.

Existence. Let us first assume that $a \geq 0$ and $b > 0$; all other cases will be easily reduced to this one. Intuitively, the theorem states that by "walking" along the x-axis with a step of length b (as shown in Figure 6.2), one can reach a or stop before reaching a at distance r from a, where $r < b$.

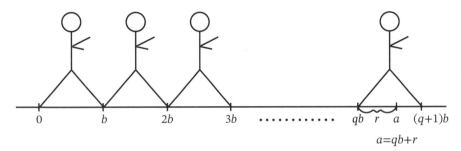

$$a = qb + r$$

Figure 6.2

We form a set $A = \{a - xb : a \geq xb, x \in \mathbb{Z}\}$ by subtracting from a all multiples of b that do not exceed a. Clearly, $A \neq \emptyset$, since $a = a - 0 \cdot b \in A$ and all elements of A are nonnegative integers. By the Well-Ordering Principle, A contains a smallest element; call this element r. Since r is in A, it is of the form $r = a - qb$, for some $q \in \mathbb{Z}$. We are going to show that q and r satisfy the statement of the theorem.

Note that $a = qb + r$, as needed. To show that $0 \leq r < |b| = b$, we assume the contrary, namely $r \geq b$, and arrive at a contradiction. Under the assumption that $r \geq b$, let $r' = r - b$; clearly, $r' \geq 0$. Then $a = qb + r = (qb + b) + (r - b) = (q + 1)b + r'$, and therefore $r' = a - (q + 1)b$. Since $r' \geq 0$ and r' is of the form $a - xb$, $r' \in A$. Since $0 \leq r' < r$, we have found an element in A that is smaller than r. This contradicts the designation of r as being the smallest member of A. The source of the contradiction is our assumption that $r \geq b$. Therefore, $0 \leq r < b$, and the proof of the existence is completed (in this case).

The case when $a \geq 0$ but $b < 0$ can be reduced to the previous one. Indeed, since $-b > 0$, there are integers q' and r such that $a = q'(-b) + r = (-q')b + r$ and

$0 \le r < -b = |b|$ (by the preceding case). Setting $q = -q'$, we get $a = qb + r$ with $0 \le r < |b|$, and the proof is complete.

The case when $a < 0$ and $b > 0$ can also be reduced to the first case. Since $-a > 0$, there are integers q' and r' such that $-a = q'b + r'$, with $0 \le r' < b$. If $r' = 0$, then $a = (-q')b + 0$, and the proof is complete. If $0 < r' < b$, then $a = (-q')b - r' = (-q' - 1)b + (b - r')$. Setting $q = -q' - 1$ and $r = b - r'$, we get $a = qb + r$, where $0 < r < b = |b|$, which completes the proof in this case.

The case when $a < 0$ and $b < 0$ is left to the reader.

Uniqueness. We must show that if $a = qb + r = q_1 b + r_1$, with $0 \le r < |b|$ and $0 \le r_1 < |b|$, then $q = q_1$ and $r = r_1$. Indeed, $qb + r = q_1 b + r_1$ if and only if $r - r_1 = (q_1 - q)b$, so

$$|r - r_1| = |q_1 - q||b|. \tag{6.1}$$

If $q = q_1$, then $|r - r_1| = 0 \iff r = r_1$, and the statement is proven. If $q \ne q_1$, then $|q - q_1| \ge 1$, and the right-hand side of (6.1) is *at least* $|b|$. But the left-hand side of (6.1) represents the distance between two integer points in the set $\{0, 1, 2, \ldots, |b| - 1\}$, and therefore is *at most* $|b| - 1$. The obtained contradiction shows that the case $q \ne q_1$ is not possible. This proves that the representation is unique and completes the proof of the Division Theorem.   □

Suppose you have $n = 18$ equally spaced points around the perimeter of a circle, numbered 0 through 17. Start at 0 and move clockwise around the circle, connecting every 8th point, until you eventually complete drawing a star-shaped figure upon revisiting the point 0. How many of the eighteen original points will be visited in creating this star; i.e., what is the value of p in this p-pointed star?

Repeat the above problem if every kth point is visited for $k = 4$, 7, and 15. For what values of k, $1 \leq k \leq 17$, will you visit all n points? Can you find a formula for p in terms of general values of n and k?

7

Number Theory

God may not play dice with the universe,
but something strange is going on with the prime numbers.
—Paul Erdős (1913–1996)

Along with geometry, number theory is one of the oldest areas of mathematics. Indeed, the theoretical study of numbers and their properties dates back almost 4,000 years to the early Babylonians, who explored systems of equations, Pythagorean triples, and solutions of certain cubic equations. Unlike plane and solid geometry, however, which are not active subjects of current research, number theory has grown in prominence over the years, weaving its way into many other areas of both continuous and discrete mathematics. In the last fifty years, coinciding with increased technological capabilities and requirements, new applications have been found for the transmission, retrieval, and security of information.

Our brief venture into number theory will focus on the integers. The reader is invited to review Section 2.2, especially the Division Theorem, and the accompanying problem set, where we introduced integers and gave some elementary properties related to divisibility.

7.1 Primes

Chebyshev said it and I'll say it again.
There is always a prime between n and 2n.[1]

Recall that for integers a and b, $b \neq 0$, a is divisible by b, denoted by $b|a$, if there is an integer q such that $a = bq$. In Chapter 2, we defined a positive integer $p \neq 1$ as a **prime number**, or **prime**, if it is divisible only by ± 1 and $\pm p$, while a positive integer with more than two positive divisors is called a **composite number**, or **composite**. The

---

[1]This rhyme is often attributed to Paul Erdös, since he found his own proof as a teenager. The Wolfram MathWorld website claims this is an incorrect attribution and states that the rhyme was first uttered by N. J. Fine.

importance of primes in number theory is mainly due to the following theorem, which simply claims that the primes are the building blocks for a multiplicative construction of the integers.

Theorem 7.1 (Prime Factorization Theorem). *Every integer $n \geq 2$ is prime or a product of primes. If n is represented as a product of primes in two ways, then these representations differ only in the order of their factors.*

Equivalently, every integer $n \geq 2$ is prime or a product of powers of distinct primes with positive integer exponents:

$$n = p_1{}^{e_1} p_2{}^{e_2} \cdots p_k{}^{e_k}. \tag{7.1}$$

Assuming $p_1 < p_2 < \cdots < p_k$, this representation is unique.

For example, $48 = 2^4 \cdot 3^1 = 3^1 \cdot 2^4$, and $90 = 2 \cdot 3 \cdot 5 \cdot 3 = 3 \cdot 5 \cdot 2 \cdot 3 = 2^1 \cdot 3^2 \cdot 5^1$.

Clearly, the two statements of the theorem are equivalent. The proof has two parts, existence and uniqueness. We proved the existence of the prime factorization in Theorem 6.2 and we delay the proof of the uniqueness of that factorization until Section 7.2.

When m divides n, the divisors of m are a subset of the divisors of n, as we see in Corollary 7.2.

Corollary 7.2. *Let $n = p_1{}^{e_1} p_2{}^{e_2} \cdots p_k{}^{e_k}$, $e_i \geq 1$, $i = 1, 2, \ldots, k$. Then*

$$m|n \iff m = p_1{}^{l_1} p_2{}^{l_2} \cdots p_k{}^{l_k},$$

where $0 \leq l_i \leq e_i$, $i = 1, 2, \ldots, k$.

Proof. If $m|n$, then $n = p_1{}^{e_1} p_2{}^{e_2} \cdots p_k{}^{e_k} = qm$ for some integer q. If either one of the prime factorizations of q or m contains a power of a prime distinct from each p_i or if it contains a power of p_i with exponent greater than e_i, then it will violate the uniqueness of the prime factorization of n. This proves the implication. To prove the converse, note that

$$n = (p_1{}^{e_1-l_1} p_2{}^{e_2-l_2} \cdots p_k{}^{e_k-l_k}) \cdot (p_1{}^{l_1} p_2{}^{l_2} \cdots p_k{}^{l_k})$$
$$= (p_1{}^{e_1-l_1} p_2{}^{e_2-l_2} \cdots p_k{}^{e_k-l_k}) \cdot m.$$

Since $e_i - l_i \geq 0$ for $1 \leq i \leq k$, it follows that $m|n$. $\qquad\qquad\square$

For example, $(3^2 \cdot 17)|(2 \cdot 3^3 \cdot 5 \cdot 17^2)$, and when we say that neither 11 nor 3^4 divides $2 \cdot 3^3 \cdot 5 \cdot 17^2$, we actually use the contrapositive of Corollary 7.2.

Prime numbers have fascinated mathematicians for centuries, and much is still not known about them, as we will see. Here are several natural questions that we will attempt to answer, with varying degrees of success.

- How does one determine whether a number is prime?
- How many prime numbers are there?
- Is there any pattern in the distribution of primes?
- Is there any formula for producing prime numbers?

How does one determine whether a number is prime? Suppose you are given a large integer n and you have to determine whether it is a prime. An obvious approach—namely trying to divide n by all positive integers less than $n/2$—will work, but it is rather slow. A refinement of this idea which is slightly more efficient is given below.

Theorem 7.3. *Let $n \geq 2$. If no prime number $p \leq \sqrt{n}$ divides n, then n is prime.*

Proof. We prove the contrapositive statement: if n is a composite number, then it is divisible by a prime p, $2 \leq p \leq \sqrt{n}$. Suppose $n = ab$, for some integers a and b, $1 < a \leq b < n$. Then, $a \cdot a \leq a \cdot b = n$, so $a \leq \sqrt{n}$. From Theorem 7.1, a has a prime factor p, and $2 \leq p \leq a \leq \sqrt{n}$. Moreover, since $p|a$ and $a|n$, it follows that $p|n$. $\square$

For example, to check whether 167 is prime, it is sufficient to try to check for divisibility by 2, 3, 5, 7, and 11, since 11 is the largest prime not exceeding $\sqrt{167}$. Since none of these primes divide 167, 167 is prime.

For large numbers the above method is slow, even for a computer. Much better methods for testing primality have been developed, but their theory is considerably more involved. Such tests are important in modern cryptography. Interestingly, while these methods are efficient in determining with high probability whether a given number is prime, they are *not* effective for factoring the number.

How many prime numbers are there? If we try to list the terms in the sequence of primes, we will notice that the frequency of their appearance decreases with growth. For example, there are 25 primes among the integers 1 to 100, but only 16 primes among those from 1,001 to 1,100 and only 6 primes among the integers from 100,001 to 100,100. One may start wondering whether we will eventually exhaust all of them. This will never happen! The following proof goes back to Euclid's ($\sim$ 300 BC) *Elements* (Book IX, Prop. 20). It is often used as an example of the incredible power of mathematical thinking, in particular of the method of a proof by contradiction, and it should be studied carefully.

Theorem 7.4. *There are infinitely many prime numbers.*

Proof. We prove the statement by contradiction. To this end, suppose there are only finitely many primes. If there are n primes, we can order them in the following finite sequence:
$$p_1 = 2 < p_2 = 3 < p_3 < \cdots < p_{n-1} < p_n.$$
Now consider the number $N = p_1 p_2 \cdots p_n + 1$. Since $N > p_n$, N is not in the sequence, and therefore it is composite. However, any composite number is divisible by a prime due to Theorem 7.1. Since all primes are listed in the sequence, there exists some prime p_i, $1 \leq i \leq n$, such that $p_i|N$. On the other hand, $N = qp_i + 1$, where q is the product of all primes except p_i. Thus, $N \bmod p_i = 1$, which contradicts that $p_i|N$. The source of the obtained contradiction is our assumption that there are finitely many primes. Therefore the set of primes is infinite. $\square$

We postpone answers to the latter two questions (regarding patterns and formulas for primes) until Chapter 12.

We conclude the section with the following classical demonstration of the irrationality of $\sqrt{2}$. In other words, there are no integers m and n satisfying $(n/m)^2 = 2$. Students should study this example enough that they could prove it in their sleep.

Example 77. Prove there are no nonzero integers satisfying the equation $n^2 = 2m^2$.

Proof. Suppose there is a solution (m, n), with $m \neq 0$ and $n \neq 0$. Every prime power in the prime factorization of n^2 (and m^2) has an even exponent, since each such exponent is twice that of the corresponding exponent in n. Therefore, the exponents of 2 in the prime factorizations of n^2 and $2m^2$ are even and odd, respectively. Since $m^2 = 2n^2$, this contradicts the uniqueness of the prime factorization of integers, which completes the proof.   □

Exercises and Problems.

(1) Are the following integers prime? Prove your answers.

   (a) 127   (b) 667   (c) 1987
   (d) $2^{50} - 3^{20}$ (e) $2^{50} + 10^{50}$ (f) $2^{50} + 15^{50}$
   (g) 111...111 (126 ones)   (h) $2^{2,018} - 1$ (i) $2^{2,018} + 1$

(2) Prove that the numbers given in each part cannot be prime simultaneously.

   (a) $n + 5$ and $n + 10$, $n \geq 0$
   (b) n, $n + 2$, and $n + 5$, $n \geq 2$
   (c) $2^n - 1$ and $2^n + 1$, $n \geq 3$

(3) Prove that if a cube of a number is divisible by 17, then the number is divisible by 17.

(4) Prove that the sum of four consecutive positive integers is never a prime number.

(5) Find the greatest integer n such that 3^n divides 30!.

---

(6) (a) Prove that if $2^n - 1$ is prime, then n is prime. Does the converse hold?

   (b) Prove that if $2^n + 1$ is prime, then n is a power of 2. Does the converse hold?

(7) Let p_n denote the nth prime. Thus $p_1 = 2$, $p_2 = 3$, $p_3 = 5$, etc. Prove that for $n \geq 5$, $p_n > 2n$.

(8) Find all $n \in \mathbb{N}$ such that all three numbers n, $n + 10$, and $n + 14$ are prime.

(9) Prove that if n is a perfect square, then n has an odd number of distinct positive divisors, and that if n is not a perfect square, then n has an even number of distinct positive divisors.

(10) Prove that the only solution to each of the following equations is $(0, 0)$.

   (a) $n^2 = 5m^2$.
   (b) $n^3 = 40m^3$.

(11) Show that $n^4 + 4$ is a composite number for all integers $n \geq 2$.

(12) Let p be a prime integer, $p \geq 5$. Prove that $p^2 - 1$ is divisible by 24.

(13) What is the greatest integer n such that 2^n divides $\binom{1,000}{500} = 1{,}000! / (500!)^2$? (It can be shown that this number is an integer, but you do not have to do it.)

(14) The 1,000 students at a school arrived and found that their 1,000 lockers were open. The first student entered and closed all 1,000 lockers. Then the second student entered and changed the state of every second locker; i.e., he opened lockers numbered $2, 4, 6, \ldots, 1{,}000$. Then the third student entered and changed the condition of every third locker; i.e., he opened locker 3, closed locker 6, opened locker 9, and so on. Eventually the 1,000th student changed the state of the 1,000th locker. Which lockers were closed after this process?

---

(15) Let p be a prime and let k be the greatest integer such that p^k divides n! Prove that
$$k = \lfloor n/p \rfloor + \lfloor n/p^2 \rfloor + \lfloor n/p^3 \rfloor + \cdots.$$

(16) Prove that there are infinitely many prime numbers of the form $4n + 3$, $n \geq 1$; that is, there are infinitely many primes p for which $p \bmod 4 = 3$. One proof is similar to Euler's proof that there are infinitely many primes.

(17) Prove by induction with respect to a that for any prime p and any integer a, $a^p - a$ is divisible by p. (This statement is known as "Fermat's Little Theorem".)

(18) For fixed positive integer k, assume that the probability that a random positive integer is divisible by k is $1/k$.[2]

   (a) Given positive integers a and b, what is the probability that (i) both are divisible by 3? (ii) Neither is divisible by 3? (iii) 3 is not a common divisor of a and b, i.e., 3 is not a divisor of both a and b?

   (b) Given positive integers a and b and prime p, what is the probability that p is not a common divisor of a and b?

   (c) Given integers a and b and distinct primes p and q, what is the probability that a and b have neither p nor q as a common divisor?

   (d) Given positive integers a and b, what is the probability that a and b are relatively prime? In other words, what is the probability that they do not share 2, 3, 5, 7, or any other prime number as a common divisor? Express your answer as an infinite product.

   (e) Estimate the product found in (d) by finding the product of the first ten factors. Compare your estimate with the known actual probability of $6/\pi^2$.

   Note: An equivalent formulation of the question in (d) is, "If one is at the origin, O, what is the proportion of lattice points in the first quadrant that are viewable?" (A point $A : (a, b)$ is "viewable" if no other lattice points lie on the segment $\overline{OA}$.)

7.2 The Euclidean Algorithm

7.2.1 Greatest Common Divisors. Given two integers a and b, any integer d that divides both of them is called a **common divisor** of a and b. If $a = b = 0$, then any nonzero number d is a common divisor, and the set of all common divisors is infinite. If at least one of a or b is not 0, say $b \neq 0$, then the set of common divisors is finite, since by Theorem 2.5(6), for every divisor d of b, $|d| \leq |b|$. In what follows we

---

[2]Technically, this is not well stated. It would be more accurate to say that the probability that a number in $[n]$ is divisible by k is $1/k$ in the limit, as $n \to \infty$.

will always assume $b \neq 0$. For example, here are the sets of common divisors of the given integer pairs:

- 12 and 30: $\{\pm 1, \pm 2, \pm 3, \pm 6\}$,
- 0 and -8: $\{\pm 1, \pm 2, \pm 4, \pm 8\}$,
- 14 and 15: $\{1, -1\}$.

Two integers that have ± 1 as their only common divisors, as is the case with 14 and 15 above, are said to be **relatively prime**.

It turns out, and it will be proven below, that among common divisors of a and b there will always be two (one positive and one negative) that are divisible by each of the common divisors of a and b. In the examples above they are ± 6, ± 8, and ± 1, respectively. Concentrating on positive ones we provide the following definition: given a and b in $\mathbb{Z}$, $b \neq 0$, the **greatest common divisor** of a and b, denoted $\gcd(a, b)$, is a (the!) positive common divisor of a and b that is divisible by each of their common divisors. Notice that when a and b are relatively prime, $\gcd(a, b) = 1$. The concept of the gcd is very useful, and it will allow us to discover many deep properties of integers.

We have a few comments before proceeding. First, it actually must be proven that $\gcd(a, b)$ exists and is unique, and we will do so in Theorem 7.7 (the Euclidean Algorithm). Second, based on previous experience, the reader may wonder why we did not define $\gcd(a, b)$ to simply be the greatest (in terms of magnitude) common divisor of a and b. It can be shown that this definition would be equivalent to ours, but a proof of the equivalence would be comparable in difficulty to the one of Theorem 7.7. The primary reason for choosing the definition we gave lies outside the scope of this course: it is easier to generalize the gcd to many algebraic systems other than $\mathbb{Z}$, e.g., to polynomials, where the concept of the gcd plays as important a role as it does in $\mathbb{Z}$, but where the notion of "greatest" is not clear.

Assuming for now that the greatest common divisor of two integers always exists, we present some useful facts that collectively will help explain why the Euclidean Algorithm finds it.

Theorem 7.5. *Let a and b be positive integers. Then:*

(1) $\gcd(a, b) = \gcd(b, a)$.

(2) $\gcd(a, a) = a$.

(3) $\gcd(a, b) = \gcd(a - b, b)$.

(4) $\gcd(a, 0) = a$.

Proof. Part (1) is obvious based on the symmetry of a and b in the definition of $\gcd(a, b)$.

If $d = \gcd(a, a)$, then d is a positive divisor of a. By definition, each divisor of a must divide d, and since a is a divisor of itself, $a \mid d$. Thus, since $d \mid a$ and $a \mid d$ and $a > 0$, it must be the case that $d = a$, which proves (2).

To prove part (3), let $d = \gcd(a, b)$. Then d is a common divisor of a and b, divisible by all other common divisors; we must show that $d = \gcd(a - b, b)$. Since $d \mid a$ and $d \mid b$, d divides their difference, which means that d is a common divisor of $a - b$ and b. Now let c be another common divisor of $a - b$ and b. Then c divides their sum, $(a - b) + b = a$. But if $c \mid a$ and $c \mid b$, then $c \mid d$, since $d = \gcd(a, b)$.

Part (4) follows immediately from parts (2) and (3), which completes the proof. □

Example 78. We use Theorem 7.5 to find the greatest common divisor of 44 and 8:

$$\gcd(44, 8) = \gcd(36, 8) = \gcd(28, 8) = \gcd(20, 8) = \gcd(12, 8)$$
$$= \gcd(4, 8) = \gcd(8, 4) = \gcd(4, 4) = 4.$$

Notice that, in the above example, each of the first five steps involved subtracting 8 (the smaller value) from the larger value. We could therefore have skipped right from finding $\gcd(44, 8)$ to finding $\gcd(4, 8)$ if we had seen to subtract five 8's from the initial value $a = 44$. Why five of them? Based on the Division Theorem! (Note that $44 = 8 \cdot \mathbf{5} + 4$.)

The above observation gives us a corollary to Theorem 7.5 and also generalizes into a procedure that both proves the existence of $\gcd(a, b)$ and gives an effective method to compute it.

Corollary 7.6. *Let a and b be positive integers, and suppose $a = bq + r$, $0 \leq r < b$. Then $\gcd(a, b) = \gcd(a, r)$.*

7.2.2 The Algorithm. The method itself goes back to Euclid's *Elements* (Book VII, Prop. 2) and is often called the **Euclidean Algorithm**. With no exaggeration, it is considered to be one of the most fruitful ideas in mathematics. Here is a "pseudo code" description of the algorithm.

- Use the Division Theorem with integers a and b to write $a = q_1 b + r_1$. If $r_1 = 0$, i.e., if $b|a$, then set $d = |b|$, and stop. Else
- Use the Division Theorem with integers b and r_1 to write $b = q_2 r_1 + r_2$. If $r_2 = 0$, i.e., $r_1|b$, then set $d = r_1$, and stop. Else
- Use the Division Theorem with integers r_1 and r_2 to write $r_1 = q_3 r_2 + r_3$. If $r_3 = 0$, i.e., $r_2|r_1$, then set $d = r_2$, and stop. Else, and so on.

Since we have $|b| > r_1 > r_2 > \cdots \geq 0$, the algorithm has to terminate; otherwise, there would be infinitely many integers between b and 0. Let the algorithm take n divisions to terminate; i.e., $r_n = 0$ for some $n \geq 1$. Then we have

$$a = q_1 b + r_1,$$
$$b = q_2 r_1 + r_2,$$
$$\cdots$$
$$r_{n-4} = q_{n-2} r_{n-3} + r_{n-2},$$
$$r_{n-3} = q_{n-1} r_{n-2} + r_{n-1},$$
$$r_{n-2} = q_n r_{n-1}.$$

(7.2)

Using Corollary 7.6 repeatedly and then Theorem 7.5(4), we see that

$$\gcd(a, b) = \gcd(a, r_1) = \gcd(r_1, r_2) = \gcd(r_2, r_3)$$
$$= \cdots$$
$$= \gcd(r_{n-2}, r_{n-1}) = \gcd(r_{n-1}, 0) = r_{n-1}.$$

A formal proof of the statement in the preceding line would use induction, giving us the following important theorem.

Theorem 7.7. *For $a, b \in \mathbb{Z}$, $b \neq 0$, the greatest common divisor of a and b exists and equals the last nonzero remainder found by the Euclidean Algorithm.*

Example 79. Use the Euclidean Algorithm to find $\gcd(78, 32)$.

Solution. Using the Division Theorem at each step of the algorithm, we obtain

$$78 = 2 \cdot 32 + 14,$$
$$32 = 2 \cdot 14 + 4,$$
$$14 = 3 \cdot 4 + \mathbf{2},$$
$$4 = 2 \cdot \mathbf{2} + 0.$$

(7.3)

Therefore, $\gcd(78, 32) = 2$, the last nonzero remainder.   □

7.2.3 Linear Combinations of two integers. The Euclidean Algorithm gives us a mechanism for writing $\gcd(a, b)$ in the form $xa + yb$ for some integers x and y. Such an expression is called a **linear combination** of a and b, and it will prove quite useful. For example:

- 388 is a linear combination of 44 and 20, since $388 = 7 \cdot 44 + 4 \cdot 20$.
- -2 is a linear combination of 7 and -3 (and of 4 and 10), since $-2 = 4 \cdot 7 + 10 \cdot (-3) = 7 \cdot 4 + (-3) \cdot 10$.
- 5 is a linear combination of 5 and -18, since $5 = 1 \cdot 5 + 0 \cdot (-18)$.
- $0 = 0 \cdot a + 0 \cdot b$ is a linear combination of any two integers a and b.

As we saw above, linear combinations are not unique. Let us denote the set of all linear combinations of a and b by $L_{a,b}$; i.e.,

$$L_{a,b} = \{xa + yb : x, y \in \mathbb{Z}\}.$$

For example, $-2 \in L_{4,10}$ (with $x = 2$, $y = -1$, or with $x = 12$, $y = -5$), $7 \in L_{23,-8}$ (with $x = 1$, $y = 2$), and $5 \in L_{5,18}$ (with $x = 1$, $y = 0$).

Let $n\mathbb{Z} = \{nt : t \in \mathbb{Z}\}$ denote the set of all multiples of n. The following theorem gives two important relationships between $\gcd(a, b)$ and the set $L_{a,b}$.

Theorem 7.8. *Let $a, b \in \mathbb{Z}$ and $b \neq 0$, and let $d = \gcd(a, b)$. Then*

(1) *d is the smallest positive linear combination of a and b, and*

(2) *$L_{a,b} = d\mathbb{Z} = \{\dots, -2d, -d, 0, d, 2d, \dots\}$.*

Before proceeding with a proof, we illustrate (1) with an example.

Example 80. Show that $2 = \gcd(78, 32)$ is a linear combination of 78 and 32.

Solution. From the third equality of (7.3),

$$2 = 14(1) + 4 \cdot (-3).$$

So 2 is a linear combination of 14 and 4. From the second equality of (7.3), $4 = 32 + 14 \cdot (-2)$, and substituting this result into the equation $2 = 14 + 4 \cdot (-3)$, we obtain

$$2 = 14(1) + (32 + 14 \cdot (-2)) \cdot (-3) = 32 \cdot (-3) + 14 \cdot 7.$$

So 2 is a linear combination of 32 and 14. Solving the first equality of (7.3) for 14 and substituting the result into the equation $2 = 32 \cdot (-3) + 14 \cdot 7$, we obtain

$$2 = 32 \cdot (-3) + (78 + 32 \cdot (-2)) \cdot 7 = 78 \cdot 7 + 32 \cdot (-17).$$

So $2 = \gcd(78, 32)$ is a linear combination of 78 and 32. Furthermore, it is the smallest positive element of $L_{78,32}$, since each element of $L_{78,32}$ is divisible by 2.   □

Proof of Theorem 7.8. (1) First we show that $d \in L_{a,b}$. By Theorem 7.7, $d = r_{n-1}$. From the $(n-1)$st division with remainder in the Euclidean Algorithm (7.2), we obtain

$$d = r_{n-1} = r_{n-3} - q_{n-1}r_{n-2} = r_{n-3} \cdot 1 + r_{n-2}(-q_{n-1}),$$

so d is a linear combination of r_{n-3} and r_{n-2}. Expressing r_{n-2} from the $(n-2)$nd division of (7.2) and substituting it above, we get

$$d = r_{n-1} = r_{n-3} \cdot 1 + (r_{n-4} - q_{n-2}r_{n-3})(-q_{n-1})$$
$$= r_{n-3}(1 + q_{n-1}q_{n-2}) + r_{n-4}(-q_{n-1}),$$

so d is a linear combination of r_{n-4} and r_{n-3}. Expressing r_{n-3} from the $(n-3)$rd division of (7.2) and substituting it above, we get that d is a linear combination of r_{n-5} and r_{n-4}, and so on. "Moving up" this way we eventually obtain that d is a linear combination of a and b. It must be the smallest positive one, since d divides each element of $L_{a,b}$ (Theorem 2.5(4)).

(2) Since d divides both a and b, it divides every element of $L_{a,b}$, and $L_{a,b} \subseteq d\mathbb{Z}$. Since $d \in L_{a,b}$ (by (1)), $d = ua + vb$ for some u and v. Therefore any multiple of d is again a linear combination of a and b: $dt = (ut)a + (vt)b \in L_{a,b}$. Hence $d\mathbb{Z} \subseteq L_{a,b}$. Having both inclusions, we conclude that $L_{a,b} = d\mathbb{Z}$. □

As we can see from Example 80 and the proof above, integers u and v in a representation $\gcd(a, b) = ua + vb$ can be computed via subsequent "backward" substitutions of the remainders appearing in the Euclidean Algorithm.

As an immediate corollary of Theorem 7.8, we have the following very useful statement.

Corollary 7.9. *Given integers a and b, the following three statements are equivalent.*

(1) *a and b are relatively prime.*

(2) *The number 1 is a linear combination of a and b.*

(3) *$L_{a,b} = \mathbb{Z}$; i.e., every integer is a linear combination of a and b.*

For example, 6 and 25 are relatively prime, as are 12 and 19, so 1 can be written as a linear combination of the integers in each pair. Indeed, $1 = 6(-4) + 25(1)$ and $1 = 12(8) + 19(-5)$.

The following statements further illustrate the importance of the notion of relative primeness. The proof heavily relies on Corollary 7.9.

Theorem 7.10.

(1) *If $c|ab$ and $\gcd(a, c) = 1$, then $c|b$.*

(2) *If $\gcd(a, c) = \gcd(b, c) = 1$, then $\gcd(ab, c) = 1$.*

(3) *If $a|c$, $b|c$, and $\gcd(a, b) = 1$, then $ab|c$.*

(4) *$\gcd(a, b) = d \iff \gcd(a/d, b/d) = 1$.*

Proof. (1) Since $\gcd(a, c) = 1$, $1 = ax + cy$ for some integers x, y. Then $b = abx + cby$. But since $c|ab$, $ab = qc$ for some integer q. Hence $b = (qc)x + cby = c(qx + by)$, which implies that $c|b$.

(2) Since $\gcd(a, c) = \gcd(b, c) = 1$, $1 = ax + cy = bu + cv$ for some integers x, y, u, v. Then $1 = 1 \cdot 1 = (ax + cy)(bu + cv) = (ab)(xu) + c(ybu + axv + cyv)$. Therefore $1 \in L_{ab,c}$, and $\gcd(ab, c) = 1$.

(3) Since $a|c$, $c = qa$ for some $q \in \mathbb{Z}$. Therefore $b|qa$. Since $\gcd(a, b) = 1$, by (1), $b|q$. Therefore $q = q_1 b$ and $c = (q_1 b)a = q_1(ab)$, which implies that $ab|c$.

(4) Left to the reader. $\square$

Here are a few typical applications of Theorem 7.10.

- If $8|(25n)$, then $8|n$ (by (1)).
- A number is divisible by 15 if and only if it is divisible by both 3 and 5 (by (3) and Theorem 2.5(2)).
- To prove that for all $n \in \mathbb{N}$, $30|(n^5 - n)$, it is sufficient to show that $n^5 - n$ is divisible by 5 and 6, or by 2, 3, and 5 (by (3)).

7.2.4 Proof of Uniqueness in the Prime Factorization Theorem. We now prove the uniqueness portion of the Prime Factorization Theorem. The following lemma, a corollary of Theorem 7.10, will be helpful in that endeavor.

Lemma 7.11. *If p is prime and $p|a_1 a_2 \cdots a_s$, then $p|a_i$ for some i, $1 \le i \le s$.*

Proof. If $\gcd(p, a_i) = p$ for at least one i, then p divides the corresponding a_i, and the proof is finished. If for each i, $\gcd(p, a_i) = 1$, then $\gcd(p, a_1 a_2 \cdots a_s) = 1$ (just generalize Theorem 7.10(2) by induction on s). This contradicts that $p|a_1 a_2 \cdots a_s$. Therefore p divides some a_i, which completes the proof. $\square$

Theorem (Theorem 7.1 (Prime Factorization Theorem)). *Every integer $n \ge 2$ is prime or a product of powers of distinct primes with positive integer exponents:*

$$n = p_1^{e_1} p_2^{e_2} \cdots p_k^{e_k}. \tag{7.4}$$

Assuming $p_1 < p_2 < \cdots < p_k$, such a representation is unique.

Proof of uniqueness. Our proof proceeds by strong induction on n. If $n = 2$, a prime, the statement is clearly true. Suppose the theorem is proven for all integers m such that $2 \le m < n$. We want to show that it is true for $m = n$. If n is prime, there is nothing to prove. If n is not prime, let

$$n = p_1 \cdots p_k = q_1 \cdots q_t$$

be two representations of n as product of primes (not necessarily distinct).

We want to show that $k = t$ and, after a proper rearrangement, if necessary, $p_i = q_i$ for all $i = 1, \dots, k$. Since p_1 is prime and divides $q_1 \cdots q_t$, by Lemma 7.11, $p_1|q_i$ for some i, $1 \le i \le k$. Since q_i is also prime, $p_1 = q_i$. Relabeling the q_i if necessary, we may assume that $p_1 = q_1$. Thus we have

$$n = p_1 p_2 \cdots p_k = p_1 q_2 \cdots q_t.$$

Dividing by p_1, we obtain $n/p_1 = p_2 \cdot p_3 \cdots p_k = q_2 \cdot q_3 \cdots q_t < n$. By the inductive hypothesis, n/p_1 is prime or is the product of primes, and such representation is unique up to the order of primes. So $k = t$, and after rearrangement, if necessary, $p_i = q_i$ for $i = 2, \dots, k$. Multiplying both sides of $p_2 \cdots p_k = q_2 \cdots q_k$ by p_1 proves the uniqueness of the factorization of n. $\square$

Further Thoughts. The definitions of factors and primes involve solely the operation of multiplication and have no references to that of addition. The same is true about our proof of the existence of prime factorization. At first glance it may seem that the proof of the uniqueness is also independent of the addition operation. This is not the case, since we used Lemma 7.11, whose proof depended on the Euclidean Algorithm. The latter is clearly inseparable from the addition of integers. There are proofs of Theorem 7.1 that do not use the Euclidean Algorithm, but all of them use the additive properties of $\mathbb{Z}$.

Therefore one can ask a question: is it possible to prove the Prime Factorization Theorem by using the multiplicative properties of integers *only*? It turns out that no such proof can ever be found; i.e., it is not a matter of our cleverness but is rather an intrinsic property of integers! How can one prove a statement like this? The following ingenious argument is due to American mathematician David Hilbert (1862–1943). Consider the set S of all positive integers of the form $4k + 1, k \in \mathbb{N}$:

$$S = \{1, 5, 9, 13, 17, 21, 25, 29, ...\}.$$

Multiplying any two numbers from S, we get another number from S (see Theorem 7.15(2)). Call a number from S **primish** if it is different from 1 and is not a product of two smaller numbers from S. For example, numbers $5, 9, 13, 17, 21, 29, 33, 49$ are all primish, but not 25 ($= 5 \cdot 5$) or 45 ($= 5 \cdot 9$) or 117 ($= 9 \cdot 13$). It is true that every number from S is either primish or can be factored as a product of primish integers, and this can be proven in just the same way as in Theorem 7.1. But it is not true that the factorizations are unique! For example, number $441 = 21 \cdot 21 = 9 \cdot 49$, and these two factorizations are the product of different primish numbers.

On the other hand, the axioms of the multiplication operations (see Property 2.1 in Section 2.1) on S and $\mathbb{Z}$ are the same; namely, each set is closed under multiplication (i.e., the products of any two elements of a set is an element of the set), both multiplications are commutative and associative and each set contains a multiplicative identity (the number 1). At the same time the prime factorization is unique in $\mathbb{N}$ (or $\mathbb{Z}$), but not in S. This argument shows that a proof of uniqueness cannot be based solely on multiplicative axioms of integers.   ◇

7.3 Linear Diophantine Equations

The reader may wonder why so much attention was given in the previous section to finding the greatest common divisor of two integers. Is there any use for the gcd besides using it to simplify fractions? Additionally, why was it helpful to write $\gcd(a, b)$ as a linear combination of a and b? The following problem, while perhaps not realistic, is representative of the type of problem that utilizes these concepts. You need \$3.88 postage to mail a small package. Unfortunately, all you have is a collection of old stamps with denominations of 44 cents and 20 cents. Is it possible to obtain the exact amount of postage? Mathematically, the question can be phrased as, "Are there nonnegative integers x and y for which $44x + 20y = 388$?" or as "Is 388 a linear combination of nonnegative multiples of 44 and 20?" In this section, we explore solutions to equations like these.

An equation of the form $ax + by = c$, where a, b, and c are given integers and x and y are unknown integers, is called a **linear Diophantine equation** with two unknowns. The term "Diophantine" commemorates an ancient Greek mathematician, Diophantus (~ 250 years A.D.), who investigated integer solutions of different equations. When we refer to an equation as "Diophantine" it usually means that the constants in such an equation are integers and that we are interested in integer solutions only. Examples include $3x - 4y = 10$, $x^2 + 2y^2 = z^2$, and $3^x - 2^y = 1$. For many classes of Diophantine equations it is extremely hard to find all their solutions. One of the few successes in this regard is the class of linear Diophantine equations. The statement of the following theorem should be studied carefully.

Theorem 7.12. *Let $a, b, c \in \mathbb{Z}$, $b \neq 0$, and $d = \gcd(a, b)$. The equation*

$$ax + by = c \qquad (7.5)$$

has a solution if and only if $d | c$. If $d | c$, then a particular solution (x_0, y_0) to (7.5) can be found by means of the Euclidean Algorithm. The solution set of (7.5) is

$$\left\{ (x, y) : x = x_0 - \frac{b}{d}t, \ y = y_0 + \frac{a}{d}t, \ t \in \mathbb{Z} \right\}. \qquad (7.6)$$

Proof. Let $d = \gcd(a, b)$. Equation (7.5) has a solution if and only if c is a linear combination of a and b. By Theorem 7.8(2), this happens if and only if $c = c'd$ for some integer c'. Dividing both sides of (7.5) by d, we obtain an equivalent equation

$$a'x + b'y = c', \qquad (7.7)$$

where $a' = \frac{a}{d}$, $b' = \frac{b}{d}$, and, by Theorem 7.10(4), $\gcd(a', b') = 1$. By using "backward" substitutions in the Euclidean Algorithm applied to a' and b', one can represent 1 as their linear combination, say $a'u + b'v = 1$. This gives us a particular solution, $(x_0, y_0) = (c'u, c'v)$ of (7.5), and therefore of (7.5) as well, since $a'x_0 + b'y_0 = a'(c'u) + b'(c'v) = c'$.

We now describe the set of all solutions of (7.7) (or (7.5)). Let (x, y) represent one of them. Then $a'x + b'y = c'$. Since $a'x_0 + b'y_0 = c'$, subtracting these equalities, we get

$$a'(x - x_0) = b'(y_0 - y).$$

Then $a' | b'(y_0 - y)$. Since $\gcd(a', b') = 1$, by Theorem 7.10(1), $a' | (y_0 - y)$. Let $y_0 - y = a't$. Then $x - x_0 = b't$, and we obtain that every solution of (7.5) is contained in the set (7.6).

The only point left to check is that set (7.6) does not contain any "extraneous" pairs, i.e., that every element of (7.6) is a solution of (7.5). Indeed, let $x = x_0 - \frac{b}{d}s = x_0 - b's$ and $y = y_0 + \frac{a}{d}s = y_0 + a's$ for some integer s. Then

$$\begin{aligned} a'x + b'y &= a'(x_0 - b's) + b'(y_0 + a's) \\ &= (a'x_0 + b'y_0) + (-a'b' + b'a')s \\ &= c' + 0 \\ &= c'. \end{aligned}$$

Thus every element of (7.6) is a solution of (7.7), and therefore (7.6) is the solution set of (7.5). $\square$

Example 81. Solve the Diophantine equation $858x + 253y = 33$.

Solution. First we find $\gcd(858, 253)$ by using the Euclidean Algorithm:

$$858 = 3 \cdot 253 + 99,$$
$$253 = 2 \cdot 99 + 55,$$
$$99 = 1 \cdot 55 + 44, \tag{7.8}$$
$$55 = 1 \cdot 44 + \mathbf{11},$$
$$44 = 4 \cdot \mathbf{11} + 0.$$

Therefore $\gcd(858, 253) = 11$, and since 33 is divisible by 11, our equation has solutions. By repeatedly using backward substitutions in (7.8), we obtain

$$11 = 55 + (-1) \cdot 44$$
$$= 55 + (-1) \cdot (99 + (-1) \cdot 55) = 2 \cdot 55 + (-1) \cdot 99$$
$$= 2 \cdot (253 + (-2) \cdot 99) + (-1) \cdot 99 = 2 \cdot 253 + (-5) \cdot 99$$
$$= 2 \cdot 253 + (-5) \cdot (858 + (-3) \cdot 253) = (-5) \cdot 858 + 17 \cdot 253.$$

Thus $858 \cdot (-5) + 253 \cdot 17 = 11$. We can then multiply both sides of this equation by 3 to find that

$$(x_0, y_0) = (3 \cdot (-5), 3 \cdot 17) = (-15, 51)$$

is a particular solution of $858x + 253y = 33$. Using (7.6), we see that the general solution of the equation is

$$x = -15 - \left(\frac{253}{11}\right)t = -15 - 23t, \tag{7.9}$$
$$y = 51 + \left(\frac{858}{11}\right)t = 51 + 78t,$$

where t ranges over all integer values. $\qquad\square$

Further Thoughts. (1) When t takes on the values 0, 1, and -1, we get the particular solutions $(-15, 51)$, $(-38, 129)$, and $(8, -27)$, respectively. Note that another choice of a particular solution would change only the *form* in which the general solution was written. For example, replacing $(-15, 51)$ with $(8, -27)$, we get the solution set

$$\{(x, y) : x = 8 - 23s, \ y = -27 + 78s, \ s \in \mathbb{Z}\}. \tag{7.10}$$

It is important to understand that the set of (x, y) satisfying (7.9) and the set in (7.10) are equal: a pair $(-15 - 23t, 51 + 78t)$ from the first set appears in the second set when s takes the value $t + 1$, and a pair $(8 - 23s, -27 + 78s)$ appears in the first set when $t = s - 1$.

(2) The logic of our solution of $858x + 253y = 33$ did not follow the precise path of our solution of the general equation $ax + by = c$ in the proof of Theorem 7.12. Here, a particular solution of the equation was found from the Euclidean Algorithm applied to the original numbers a and b rather than to the reduced numbers a' and b'. We did this because we started our solution with finding $\gcd(858, 253)$, and it would be extra work to perform a new Euclidean Algorithm for the reduced numbers 78 and 23 (even though the latter could be obtained by dividing each term in the equations of (7.8) by a factor of 11). On the other hand, if we see immediately that the original equation can be reduced, it would save time to do so. With smaller

numbers, we can often find a particular solution simply by inspection, without invoking the Euclidian Algorithm. For example, consider the Diophantine equation $100x - 40y = 260$. Reducing by 20, we get $5x - 2y = 13$. Now it is easy to notice that $(1, -4)$ is a particular solution. Since $\gcd(5, -2) = 1$, using (7.9) we see that the general solution is $\{(x, y) : x = 1 + 2t, \ y = -4 + 5t, \ t \in \mathbb{Z}\}$.   ◇

Exercises and Problems.

(1) Use the Euclidean Algorithm to find $\gcd(112, 356)$.

(2) Use the Euclidean Algorithm to find $\gcd(5{,}005, 33)$.

(3) Explain how you know that the equation $33x + 5{,}005y = 22$ has an integer solution, without actually finding a solution.

(4) Describe the set of all integers c such that the equation $112x + 356y = c$ has an integer solution (x, y).

(5) Find the general solution for each of the following linear Diophantine equations.

    (a) $17x + 10y = 3$.

    (b) $540x - 300y = 3{,}540$.

    (c) $315x + 66y = 94$.

(6) Prove each of the following. Assume each term is a positive integer.

    (a) $\gcd(n, n + 1) = 1$.

    (b) $\gcd(n, n + 2) = 1$ or 2.

    (c) $\gcd(a, b) = \gcd(a, a + b)$.

    (d) $\gcd(3n + 1, 10n + 3) = 1$.

(7) The difference of two odd integers x and y is 4. Prove that x and y are relatively prime.

---

(8) Form the converse statements for (2) and (3) of Theorem 7.10. Prove or disprove them.

(9) Prove part (4) of Theorem 7.10.

(10) Find the smallest positive integer $N \geq 2$ that gives remainder 1 upon division by each of the integers 3, 4, 5, and 7.

(11) Let r, s, t be three integers, no two of which are zero.

    (a) Suppose $d = \gcd(\gcd(r, s), t)$. Prove that $d > 0$ and divides each of the integers r, s, t.

    (b) Prove that d (from (a)) is divisible by any common divisor of the integers r, s, t.

    (c) Prove that $\gcd(\gcd(r, s), t) = \gcd(\gcd(s, t), r) = \gcd(\gcd(r, t), s)$.

Properties (a) and (b) suggest we can call d the greatest common divisor of integers r, s, and t, and denote it by $\gcd(r, s, t)$. Property (c) shows that this definition does not depend on the order of the integers, as before.

(12) (Opening hook problem for the chapter.) Suppose you have $n = 18$ equally spaced points around the perimeter of a circle, numbered 0 through 17. Start at 0 and move clockwise around the circle, connecting every 8th point, until you eventually complete drawing a star-shaped figure upon revisiting the point 0. How many of the 18 original points will be visited in creating this star; i.e., what is the value of p in this p-pointed star?

Repeat the above question if every kth point is visited for $k = 4$, 7, and 15. For what values of k, $1 \leq k \leq 17$, will you visit all n points? Can you find a formula for p in terms of general values of n and k?

---

(13) Prove that for any positive integer n there are two integers a and b such that the Euclidean Algorithm applied to a and b consists of exactly n divisions.

(14) Let a and b be two nonzero integers. Then a common multiple of a and b is an integer divisible by each of them. We call a positive integer m a (the!) **least common multiple** of a and b, denoted $\text{lcm}(a, b)$, if

  (a) m is a common multiple of a and b and

  (b) if n is any other common multiple of a and b, then $m | n$.

This definition of $\text{lcm}(a, b)$ is probably not the one taught in middle school. There, the definition is likely to be the smallest positive integer that is divisible by both a and b. Prove that the two definitions are equivalent.

(15) Prove that for integers a and b, $ab = \gcd(a, b) \cdot \text{lcm}(a, b)$. For example $12 \cdot 28 = 336$ and $\gcd(12, 28) \cdot \text{lcm}(12, 28) = 4 \cdot 84 = 336$ as well.

(16) Try to find an example of an infinite sequence of integers with the property that every two of its members are relatively prime. Of course, a sequence of all prime numbers would do, but try to find a sequence that does not rely on the fact that there are infinitely many prime numbers.

Show that the sequence of Fermat numbers, $F_n = 2^{2^n} + 1$, $n \geq 0$, provides such an example; i.e., prove that
$$\gcd(F_n, F_m) = 1$$
for each pair of distinct nonnegative integers m, n.

(17) Prove that there are infinitely many primes using the sequence in Problem 16. A proof based on this idea was suggested by G. Pólya (1887–1984), as cited in G. H. Hardy and E. M. Wright, *An Introduction to the Theory of Numbers*, Fifth Edition, Oxford Science Publications, 1996.

7.4 Congruences

When studying the divisibility of integers, important information is often obtained not just from the actual integers involved, but from the remainders they produce when divided by a given number. This phenomenon is captured well through the definition of a congruence, introduced by the German mathematician Carl Friedrich Gauss (1777–1855). The language of congruences has proven to be very convenient in number theory and its applications.

The concept of congruence itself is not new, even to those who are unfamiliar with the terminology. If it is 10:00 a.m. and one embarks on a 4-hour trip, the expected arrival time is 2:00 p.m. In this sense, then, one could take the view that 10+4=2. Similarly, if one leaves on a Friday evening on a 22-night cruise, it is not hard to calculate that the return day will be a Saturday, because 22 days equals three weeks plus one day. In calculating the return day, one could treat 22 and 1 as being equal.

Of course, the above statements "14 = 2" and "22 = 1" violate the sacredness with which mathematicians use the = sign. Therefore, we use a different symbol, $\equiv$, to denote this "equivalence" of two numbers. It is also helpful to clarify what number takes on the role of "zero", as 12 and 7 do in the two examples above, respectively. We do that with the word "mod", short for modulus, which was first introduced in Chapter 2. Here, we would write $14 \equiv 2 \pmod{12}$ and $22 \equiv 1 \pmod 7$.

Given a positive integer m, we define two integers a and b to be **congruent modulo** m, written $a \equiv b \pmod m$, if $a - b$ is divisible by m. As examples, notice that $16 \equiv 6 \pmod 5$, since $5|10$; $24 \equiv -12 \pmod 4$, since $4|36$; $35 \equiv 0 \pmod 7$, since $7|35$; and any two integers are congruent modulo 1. It is not too difficult to see that $a \equiv b \pmod m$ if and only if a and b give equal remainders when divided by m—that is, if $a \bmod m = b \bmod m$. See Problem 5 in Section 4.5.

The next three theorems hold for any modulus $m \in \mathbb{N}$ and all integers a, b, c, and d. Theorem 7.13 clarifies the alternative ways of viewing congruence. In each case, the proof follows from the definition of congruence, namely that $a \equiv b \pmod m$ if and only if $b - a$ is a divisor of m.

Theorem 7.13. (1) $a \equiv b \pmod m \iff a = mt + b$ for some $t \in \mathbb{Z}$.
(2) If $a \equiv r \pmod m$ and $0 \le r < m$, then r is the remainder upon division of a by m.

Example 82. Give the smallest positive integer b that satisfies each congruence below.
(1) $34 \equiv b \pmod 6$.
(2) $-8 \equiv b \pmod{12}$.

Solution. (1) $34 \equiv 4 \pmod 6$.
  (2) $-8 \equiv 4 \pmod{12}$.   □

The next theorem proves that $\equiv$ is an equivalence relation, since it is reflexive, symmetric, and transitive. We observed this in Example 53(5) in the special case where $m = 7$.

Theorem 7.14. (1) *Reflexive property:* $a \equiv a \pmod m$.
(2) *Symmetric property:* $a \equiv b \pmod m \iff b \equiv a \pmod m$.
(3) *Transitive property:* If $a \equiv b \pmod m$ and $b \equiv c \pmod m$, then $a \equiv c \pmod m$.

Proof. The proofs of the first two properties follow immediately from the definition of congruence. For the transitive property, suppose $a \equiv b \pmod m$ and $b \equiv c \pmod m$. Then from Theorem 7.13, $a = mt_1 + b$ and $b = mt_2 + c$ for some integers t_1 and t_2. It follows that $a = mt_1 + (mt_2 + c) = m(t_1 + t_2) + c$. Hence, $a - c$ is divisible by m, so $a \equiv c \pmod m$. This proves that $\equiv$ is transitive.   □

Congruences have many properties that remind us of the corresponding properties of equalities. For example, congruences by the same modulus can be added, subtracted, and multiplied, and both sides of a congruence can be raised to the same power; in all cases the congruence will be preserved. Theorem 7.15 collects the properties we will need.

Theorem 7.15. (1) *If* $a \equiv b \pmod{m}$ *and* $c \equiv d \pmod{m}$, *then* $a + c \equiv b + d$ (mod m) *and* $a - c \equiv b - d \pmod{m}$.

(2) *If* $a \equiv b \pmod{m}$ *and* $c \equiv d \pmod{m}$, *then* $ac \equiv bd \pmod{m}$. *In particular,* $ac \equiv bc \pmod{m}$.

(3) *Cancellation property: If* $ac \equiv bc \pmod{n}$ *and* $\gcd(c, n) = 1$, *then* $a \equiv b \pmod{n}$.

(4) *If* $a \equiv b \pmod{m}$, *then* $a^n \equiv b^n \pmod{m}$, *for* $n \in \mathbb{N}$.

Proof. (1) By definition of congruence, $a - b = q_1 m$ and $c - d = q_2 m$, for some integers q_1 and q_2. Then $(a + c) - (b + d) = (a - b) + (c - d) = q_1 m + q_2 m = (q_1 + q_2)m$, and therefore $m | [(a + c) - (b + d)]$. Thus, $a + c \equiv b + d \pmod{m}$, and the additive proof is complete. The proof that $a - c \equiv b - d \pmod{m}$ is very similar and is left to the reader. □

(2) We need to show that $m | (ac - bd)$. We use the trick of rewriting $ac - bd$ as $(a - b)c + b(c - d)$. By definition, $m | (a - b)$ and $m | (c - d)$, so by Theorem 2.5(4), $m | [(a - b)c + b(c - d)]$. Hence, $m | (ac - bd)$, and the proof is complete.

(3) Note that $ac \equiv bc \pmod{n} \iff ac - bc = (a - b)c \equiv 0 \pmod{n} \iff n | (a - b)c$. From Theorem 7.10(1), since n and c are relatively prime, if $n | (a - b)c$, then $n | (a - b)$. This proves that $a \equiv b \pmod{n}$.

(4) One way to proceed is by induction. For $n = 2$ the statement follows from (2), using $a = c$ and $b = d$. Suppose the statement is proven for $n = k \geq 2$, so $a^k \equiv b^k \pmod{m}$. We want to show that it is correct for $n = k + 1$, i.e., that $a^{k+1} \equiv b^{k+1} \pmod{m}$. This follows immediately from (2) if we multiply two congruences: $a^k \equiv b^k \pmod{m}$ (which is correct by the inductive hypothesis) and $a \equiv b \pmod{m}$ (given).

Another proof could be obtained by using the formula

$$a^n - b^n = (a - b)(a^{n-1} + a^{n-2}b + \cdots + ab^{n-2} + b^{n-1}).$$

If $m | (a - b)$, then $m | (a^n - b^n)$. Then, by definition, $a^n \equiv b^n \pmod{m}$. □

The following examples give several immediate illustrations of the theorem.

Example 83. (1) To find $32{,}517 \cdot 5{,}328$ mod 14, we can first divide each factor by 14 and note the remainders. We then multiply the obtained remainders and computer their product modulo 14. Using parts (1) and (2) of Theorem 7.15, this can be written as

$$32{,}517 \equiv 9 \pmod{14},$$

$$5{,}328 \equiv 8 \pmod{14},$$

$$32{,}517 \cdot 5{,}328 \equiv 9 \cdot 8 \equiv 72 \equiv 2 \pmod{14}.$$

(2) Since $2^4 \equiv 16 \equiv 1 \pmod{15}$, it follows that

$$2^{1000} \equiv (2^4)^{250} \equiv 1^{250} \equiv 1 \pmod{15}.$$

Hence, $2^{1{,}000}$ mod $15 \equiv 1$. (Here we mainly used part (4) of Theorem 7.15.)

(3) $10x \equiv 35 \pmod{27}$ implies $2x \equiv 7 \pmod{27}$, where cancelation is valid by part (3). □

Example 84. Let $n = 2^{100} \cdot 375 - 35^{87}$. Find n mod 6.

Solution. Here we stop writing the references to the various parts of Theorem 7.15, though we do use them constantly. Notice that we use "=" when the numbers are actually equal and we use "≡" when they are congruent modulo 6:

$$2^{100} = (2^5)^{20} = 32^{20} \equiv 2^{20} = (2^5)^4 = 32^4 \equiv 2^4 = 16 \equiv 4 \quad \text{and} \quad 375 \equiv 3.$$

Since $35 \equiv -1$, we get that $35^{87} \equiv (-1)^{87} = -1$.

Therefore, $n = 2^{100} \cdot 375 - 35^{87} \equiv 4 \cdot 3 - (-1) = 13 \equiv 1$. Thus n mod 6 = 1. □

Please look again at Examples 5 and 6 in Section 2.2, where we used exhaustive casework to show the given expressions were divisible by 8 and 6, respectively. Here we rewrite the solutions of the examples using the language of congruences—the advantages are readily apparent.

Example (Example 5). *When n is divided by 8, the remainder is 5. What is the remainder of the division of $n^3 + 5n$ by 8?*

Solution. Since $n \equiv 5 \pmod 8$,

$$n^3 \equiv 5^3 = 25 \cdot 5 \equiv 1 \cdot 5 = 5 \pmod 8 \quad \text{and} \quad 5n \equiv 5 \cdot 5 = 25 \equiv 1 \pmod 8.$$

Adding, we obtain

$$n^3 + 5n \equiv 5 + 1 = 6 \pmod 8.$$

Therefore the remainder is 6. A shorter presentation could be just the following line:

$$n^3 + 5n \equiv 5^3 + 5 \cdot 5 = 5^2(5+1) = 25 \cdot 6 \equiv 1 \cdot 6 = 6 \pmod 8.$$ □

Example (Example 6). *Prove that $M = m(m+1)(2m+1)$ is divisible by 6 for all integers m.*

Solution. Any given integer m is congruent modulo 6 to one and only one of the numbers in $\{0, 1, 2, 3, 4, 5\}$. All congruences below are done modulo 6.

(1) If $m \equiv 0$, then $M \equiv 0 \cdot 1 \cdot 1 = 0$.

(2) If $m \equiv 1$, then $M \equiv 1 \cdot 2 \cdot 3 = 6 \equiv 0$.

(3) If $m \equiv 2$, then $M \equiv 2 \cdot 3 \cdot 5 = 6 \cdot 5 \equiv 0$.

(4) If $m \equiv 3$, then $M \equiv 3 \cdot 4 \cdot 7 = 12 \cdot 7 \equiv 0$.

(5) If $m \equiv 4$, then $M \equiv 4 \cdot 5 \cdot 9 = 36 \cdot 5 \equiv 0$.

(6) If $m \equiv 5$, then $M \equiv 5 \cdot 6 \cdot 11 \equiv 0$.

As we see, in each case $M \equiv 0 \pmod 6$, and therefore $6|M$. □

Further Thoughts. Recall Example 69, where we used induction to prove that for any $m \in \mathbb{N}$,

$$6(1^2 + 2^2 + \cdots + m^2) = m(m + 1)(2m + 1).$$

That fact gives another proof of our result here. ◇

7.4.1 Divisibility Rules. The reader may be familiar with rules for determining whether a given integer is divisible by certain other integers, such as 3, 4, 8, 9, or 11. The statements of the following theorem on congruences are actually stronger than these divisibility rules, which follow from the theorem immediately.

Theorem 7.16. *Let $N = (a_{n-1} \ldots a_1 a_0)$ be an n-digit positive integer, where a_0 is the number of units, a_1 is the number of tens, and so on. Then:*

(1) $N \equiv a_{n-1} + \cdots + a_1 + a_0 \pmod 3$.

(2) $N \equiv a_{n-1} + \cdots + a_1 + a_0 \pmod 9$.

(3) $N \equiv (-1)^{n-1}a_{n-1} + (-1)^{n-2}a_{n-2} + \cdots - a_1 + a_0 \pmod{11}$.

(4) $N \equiv (a_1 a_0) \pmod 4$, *where $(a_1 a_0)$ is the number formed by the last two digits of N.*

(5) $N \equiv (a_2 a_1 a_0) \pmod 8$, *where $(a_2 a_1 a_0)$ is the number formed by the last three digits of N.*

So, we see that a number is divisible by 3 (or 9) if and only if the sum of its digits is divisible by 3 (or 9), and a number is divisible by 11 if and only if the alternating sum of its digits is divisible by 11. A number is divisible by 4 (or 8) if and only if the number formed by its last two (three) digits is divisible by 4 (or 8). For example,

$$254{,}361 \equiv 2 + 5 + 4 + 3 + 6 + 1 \equiv 2 + 2 + 1 + 0 + 0 + 1 = 6$$
$$\equiv 0 \pmod 3, \text{ so } 3 | 254{,}361;$$
$$254{,}361 \equiv 2 + 5 + 4 + 3 + 6 + 1 \equiv 3 \pmod 9,$$
$$\text{so } 254{,}361 \bmod 9 = 3;$$
$$254{,}361 \equiv -2 + 5 - 4 + 3 - 6 + 1 \equiv -3 \equiv 8 \pmod{11},$$
$$\text{so } 254{,}361 \bmod 11 = 8;$$
$$123{,}356 \equiv 56 \equiv 0 \pmod 4, \text{ so } 4 | 123{,}356; \text{ and}$$
$$123{,}356 \equiv 356 \equiv 4 \pmod 8,$$
$$\text{so } 123{,}356 \bmod 8 = 4.$$

Proof. (1) If $a_0, a_1, \ldots, a_{n-1}$ are digits of N, then

$$N = a_{n-1}10^{n-1} + a_{n-2}10^{n-2} + \cdots + 10a_1 + a_0.$$

But $1 \equiv 10 \equiv 10^2 \equiv \cdots \equiv 10^{n-1} \pmod 3$. Hence $N \equiv a_0 + a_1 + \cdots + a_{n-1} \pmod 3$.

(2) The proof is virtually identical to the one for (1).

(3) $10 \equiv -1 \pmod{11}$. So $-1 \equiv 10 \equiv 10^3 \equiv 10^5 \equiv \cdots \pmod{11}$ and $1 \equiv 10^2 \equiv 10^4 \equiv 10^6 \equiv \cdots \pmod{11}$. Therefore

$$N = a_{n-1}10^{n-1} + a_{n-2}10^{n-2} + \cdots + 10a_1 + a_0$$
$$\equiv (-1)^{n-1}a_{n-1} + \cdots - a_1 + a_0 \pmod{11}.$$

(4) One can prove this by simply using the Division Theorem. (See Problem 8.) Here is a congruence solution:

$$N = a_{n-1}10^{n-1} + a_{n-2}10^{n-2} + \cdots + 10a_1 + a_0$$
$$= 100(a_{n-1}10^{n-3} + a_{n-2}10^{n-4} + \cdots + a_2) + (10a_1 + a_0)$$
$$\equiv 10a_1 + a_0 \equiv (a_1 a_0) \pmod 4.$$

(5) Left to the reader. See Problem 11 for an extension. $\qquad\square$

Exercises and Problems.

(1) Below, use the fact that the last digit of a number n equals n mod 10.

   (a) What is the last digit of the number 7^{40}?

   (b) What is the last digit of the number 3^{100}?

   (c) By finding the last digit of the number $9^{1,972} - 7^{1,972}$, prove that this number is divisible by 10.

(2) Find the remainder of the division of $37^{2,018}$ by 17.

(3) Give the smallest positive integer b that satisfies each congruence below. Do not use a calculator.

   (a) $183 \equiv b \pmod 4$.

   (b) $-18 \equiv b \pmod 7$.

   (c) $185^{18} \equiv b \pmod 6$.

   (d) $106^3 \cdot 203^5 \equiv b \pmod{100}$.

(4) Find the last two digits of the number $37^{2,018}$.

(5) Form the converse statements for (2)–(4) of Theorem 7.15. By giving counterexamples, show that all of them are false.

---

(6) Use congruences to prove the following. They were proven by induction in Problem 16 of Section 6.3.

   (a) $6 | n(n^2 + 5)$.

   (b) $5 | (n^5 - n)$.

   (c) $7 | (n^7 - n)$.

(7) Use congruences to prove the following. They were proven by induction in Problem 19 in Section 6.3.

   (a) $9 | (4^n + 15n - 1)$.

   (b) $64 | (3^{2n+3} + 40n - 27)$.

   (c) $15^n \bmod 7 = 1$.

(8) Use the Division Theorem to give an alternate proof of Theorem 7.16(4), the rule for divisibility by 4.

(9) Use induction to prove that if each of k integers is congruent to 1 modulo n, then their product is congruent to 1 modulo n.

(10) Prove that the sum of the squares of three integers cannot give remainder 7 when divided by 8. Are there three integers x, y, and z such that

$$x^2 + y^2 + z^2 = 23{,}654{,}009{,}839?$$

(11) Extend Theorem 7.16 by stating and proving a rule for determining whether a number is divisible by 2^k for an integer $k \geq 1$.

(12) Prove that among any $n+1$ natural numbers, there are at least two whose difference is divisible by n.

(13) Prove that a 6-digit number of the form $(abcabc)$, where a, b, and c are the digits, is always divisible by 7, 11, and 13.

(14) Prove the following rule for divisibility by 8. An even integer that ends in the digits (abc) is divisible by 8 if and only if $(ab) + c/2$ is divisible by 4. For example, the number 3,768 is divisible by 8 because $76 + 8/2 = 80$ is divisible by 4.

(15) **(Exploratory Problem)** Find a "mental arithmetic" rule for determining whether a number n that ends in the digits $(abcd)$ is divisible by 16. Your rule should be different than the one from Problem 11, though it should still depend on the values of a, b, c, and d only. Several answers are possible.

(16) Prove that for all $n \in \mathbb{N}$, $133 | (11^{n+2} + 12^{2n+1})$.

---

(17) Use modular arithmetic to solve this problem from Chapter 1. A group of 100 prisoners is told they will be given a collective chance at a pardon if they can work together to solve the following challenge. At the appointed time all 100 prisoners will come into a large room to gather in a circle. Each prisoner will have a hat placed on his head as he enters, and it will be one of seven colors (red, orange, yellow, green, blue, indigo, or violet). Each prisoner will be able to see all hats except his own. One at a time, each prisoner will conjecture out loud the color of his or her own hat. If at most one of the 100 prisoners makes an incorrect conjecture, they will all go free; otherwise they will all be executed. The prisoners can plan a strategy ahead of time, but once they are brought into the room, there will be no communication, other than that each prisoner can hear all of the other guesses. No clues can be given based on eye contact, positioning of the prisoners, timing of responses, etc. What strategy could the prisoners adopt to best guarantee survival? Under this strategy, what is the probability of survival?

(18) Consider any positive integer N whose (decimal) digits read from left to right are in nondecreasing order, except the last two digits (tens and ones), which are in strictly increasing order. For example,

$$N_1 = 1,778, \qquad N_2 = 2,344,459, \qquad N_3 = 12,225,557,779.$$

Note that when each of these numbers is multiplied by 9, the sum of digits in the result is 9:

$$9N_1 = 16,002, \qquad 9N_2 = 21,100,131, \qquad 9N_3 = 110,030,020,011.$$

Prove that the sum of the digits of $9N$ is 9 for any N that meets the given requirements.

(19) Is there a moment of time when the hour, the minute, and the second hands of a clock form $120°$ angles with each other? (Assume that the hands move continuously.)

(20) Are there integers x, y, and z such that $x^3 + y^3 + z^3 = 1,234,567,894$?

(21) Let A be an arbitrary 1,972-digit number divisible by 9. Let a be the sum of digits of A, let b be the sum of digits of a, and let c be the sum of digits of b. Prove that the value of c is the same for every A and find it.

7.5 Applications

7.5.1 Check Digits. One of the simple but extremely useful applications of divisibility lies in the area of identification numbers. Identification numbers are ubiquitous, as we see them on credit cards, as International Standard Book Numbers (ISBN), as Universal Product Codes (UPC), on airline tickets, as routing numbers on personal checks, etc. What is not generally known, however, is that a *check digit* is embedded within many of these numbers. Such a digit is usually the last digit of the identification number, and it functions as an error-detecting mechanism. There are a variety of schemes involving division with remainder, but all of them use the check digit as verification that the rest of the identification number was properly received. In this section, we explore several such schemes.

Example 85. Several types of traveler's checks use divisibility by 9 for the encoding of the check digit. Specifically, the check digit is chosen so that the entire identification number (and therefore the sum of its digits) will be divisible by 9. Therefore, the identification number 254536902 has check digit 2, because $2+5+4+5+3+6+9+0 = 34$, and $34 + 2$ is divisible by 9.   □

Some schemes use a weighted sum of the digits to determine the check digit in order to increase error detection capabilities. In particular, the check digit method used for the 10-digit ISBN will correct all single-digit and transposition errors (errors where two digits are transposed). In this scheme, the ISBN $a_1a_2a_3a_4a_5a_6a_7a_8a_9a_{10}$ includes the check digit a_{10}, chosen so that $10a_1+9a_2+8a_3+7a_4+6a_5+5a_6+4a_7+3a_8+2a_9+a_{10}$ is divisible by 11.[3]

Example 86. Verify that the check digit is correct for Vera Pless's book, *The Theory of Error Correcting Codes*, which has ISBN 0-471-61884-5.

Solution. Notice that

$$10(0) + 9(4) + 8(7) + 7(1) + 6(6) + 5(1) + 4(8) + 3(8) + 2(4) + 1(5) = 209,$$

and $209 = 11(19)$ is divisible by 11, so 5 is the correct check digit.   □

Further Thoughts. In the scheme from Example 85, not all single-digit errors will be detected. In particular, if a digit x is mistakenly entered as a y, then the new digit sum will differ from the original sum by $x - y$. Thus, if $x - y$ is divisible by 9 (quick—for what values of x and y could this happen?), the error will go undetected. However, for the ISBN scheme in Example 86, if the ith digit, a_i (which is multiplied by $11 - i$), is entered incorrectly as a y, then the weighted sum will change by $(11 - i)(a_i - x)$. Since this product cannot be divisible by 11 (11 is prime and neither factor can equal 0 or 11), we see that all single-digit errors will be detected. A similar argument may be used to show that all transposition errors are detected (see Problem 9).   ◇

Example 87. The 12-digit UPC number $a_1a_2a_3a_4a_5a_6a_7a_8a_9a_{10}a_{11}a_{12}$ found on items in retail stores has its check digit a_{12} selected so that

$$3(a_1 + a_3 + a_5 + a_7 + a_9 + a_{11}) + a_2 + a_4 + a_6 + a_8 + a_{10} + a_{12}$$

---

[3]If it turns out that a_{10} needs to be 10, then publishers will use the symbol "X" as the check digit. This is a minor drawback with the scheme.

is divisible by 10. This method will detect all single-digit errors, but not all transposition errors (see Problem 10).

7.5.2 Cryptography. One of the most useful applications of congruences is in the area of **cryptography**, the study of encrypting and decrypting secret messages. The **shift cipher**, also known as the Caesar Cipher, is an example that has been around since antiquity, utilized by Julius Caesar more than 2,000 years ago. In this type of cipher, letters of the alphabet are shifted forward by a prescribed number, k, of positions to a different letter of the alphabet, wrapping around the end of the alphabet appropriately. Caesar reportedly shifted the alphabet by $k = 3$ positions, so $A \rightarrow D$, $J \rightarrow M$, and $Y \rightarrow B$.

To represent this scheme mathematically with our current alphabet (it would work similarly with others), we associate each letter from A to Z, in order, with the appropriate integer from $(0, 1, 2, \dots, 24, 25)$. The function $f(m) = (m + k) \bmod 26$ will then provide the appropriate shift of the letters.

Example 88. To send the message "HARD WORK PAYS OFF" using a shift of $k = 7$, one would first transform the letters into the numeric sequence (7, 0, 17, 3; 22, 14, 17, 10; 15, 0, 24, 18; 14, 5, 5). Applying the function f to each letter in the sequence yields the encrypted sequence (14, 7, 24, 10; 3, 21, 24, 17; 22, 7, 5, 25; 21, 12, 12). In turn, using the same correspondence as before, this sequence yields the message that will be delivered: "OHYK CVYR WHFZ VMM".

Of course, when the above message is received, in order to encrypt it, one would need to use the inverse function, f^{-1}, to the numeric sequence in order to decrypt. In this example, that is tantamount to applying f using $k = -7$, which corresponds to a left shift in the alphabet by 7 positions. The general form would be $f^{-1}(m) = (m - k) \bmod n$, where n is the number of characters in the alphabet. $\qquad\square$

The shift cipher is not very secure, in that someone who intercepted the message could fairly quickly decipher it, especially if it was suspected that a shift cipher were used. The scheme could be made slightly more difficult by using the encryption function $f(m) = g(m) \bmod 26$, where $g : \mathbb{Z} \rightarrow \mathbb{Z}$ is any function that is one-to-one on the set $\{0, 1, 2, \dots, 25\}$. A couple of examples are given in the problem set. However, even if the function g were complicated enough that one could not determine it explicitly, the encryption scheme could still be cracked by observing the frequency of characters used in the message. This is true of any method in which a single letter is always replaced by the same single letter. More sophisticated schemes will change the encryption function much more often or will replace each block of letters of a certain length with a different block of letters of the same length.

Exercises and Problems.

(1) Determine the check digit, x, for the traveler's check with identification number $52390847x$.

(2) Determine the check digit, x, for the traveler's check with identification number $47265235x$.

(3) Verify the check digit for the UPC identification number 7-18103-00921-8, taken off of a spiral notebook.

(4) Determine the check digit, c, for the 10-digit ISBN 0-07-322972-c.

(5) What should be the second digit, x, of the 12-digit UPC code number 3-x2903-09012-3 found on a plastic tool box?

(6) Use the Caesar Cipher (with $k = 3$) to encrypt the text "The quick brown fox jumps over the lazy dog."

(7) Use the encryption function $f(m) = (5m + 2) \bmod 26$ to encrypt the text "Math is cool."

(8) Suppose the encryption function $f(m) = (5m + 2) \bmod 26$ was used to send a message, and the text

"WPMLSZTQUP BGPMTQUPO KYOT GW QPDWLTQGFW"

was received. Decrypt the message.

---

(9) Prove that in the ISBN check-digit scheme presented in Example 86, all transposition errors will be detected.

(10) Prove that the UPC check-digit scheme presented in Example 87 will detect all single-digit errors. Give an example of a transposition error that will not be detected.

(11) Search the Internet and find a check-digit scheme that is different than any presented in this section. Explain the method and give an example of its implementation. Note whether all single-digit errors and transposition errors are detected.

7.6 Additional Problems

This section consists of a variety of challenging problems that can be solved using techniques found in the chapter. The problems are not arranged in any particular order.

(1) How many zeros are at the end of 2,018! ?

(2) (a) Prove that for all $n \in \mathbb{N}$, the fraction $\frac{12n+1}{30n+2}$ is irreducible.

(b) Find all integers n such that $\frac{19n+17}{7n+11}$ is an integer.

(3) Let n and k be nonnegative integers.

(a) Prove that $0 \leq k \leq n$, $k!$ divides $n(n-1)(n-2)\cdots(n-k+1)$.

(b) Prove that the binomial coefficient $\binom{2n}{n} = \frac{(2n)!}{n!n!}$ is divisible by 2 for all $n \geq 1$.

(c) Prove that the binomial coefficient $\binom{2n}{n} = \frac{(2n)!}{n!n!}$ is divisible by $n+1$ for all $n \geq 0$.

(d) Prove that the multinomial coefficient $\binom{kn}{n,n,...,n} = \frac{(kn)!}{(n!)^k}$ is divisible by $k!$ for all $k \geq 1$. (Multinomial coefficients will be studied in Chapter 11.)

(4) Prove that for all integers a, b, and c, if $6|(a+b+c)$, then $6|(a^3+b^3+c^3)$.

(5) Prove that $n^2 + 3n + 5$ is not divisible by 121 for any integer n.

(6) Suppose that integers a and b are each a sum of squares of two integers. Then ab is the sum of squares of two integers. Prove it.

(7) Is the number 111...11, with 300 ones, a perfect square?

(8) Can the sum of the digits of a perfect square be 2,018?

(9) Prove that for every $n \in \mathbb{N}$, there exists $x \in \mathbb{N}$, such that the number $nx + 1$ is composite.

(10) Prove that if positive integers m and n are relatively prime, then the same is true for $2^m - 1$ and $2^n - 1$.

(11) Prove that for any integer $n \geq 2$, the sum $1 + 1/2 + 1/3 + \cdots + 1/n$ is never an integer.

(12) Let a and b be integers. If $a^2 + b^2$ is divisible by 21, then it is divisible by 441. Prove it.

(13) Let $a, b, x_0 \in \mathbb{N}$. Prove that some terms of the sequence $(x_0, x_1 = ax_0 + b, x_2 = ax_1 + b, \ldots, x_{n+1} = ax_n + b, \ldots)$ are composite numbers.

(14) Consider the sequence of Fibonacci numbers: $f_1 = f_2 = 1$ and $f_n = f_{n-1} + f_{n-2}$ for $n \geq 3$. Prove that $5 | f_{5k}$ for each $k \in \mathbb{N}$. (More generally, it can be shown that for $n > 2$, $n | m$ if and only if $f_n | f_m$.)

(15) When the process of long division is used to find the decimal form for the quotient of two positive integers, say m and n, the resulting decimal fraction either terminates or repeats; i.e.,

$$\frac{m}{n} = a_n a_{n-1} \cdots a_1 a_0 \, . \, b_1 \cdots b_k c_1 \cdots c_p c_1 \cdots c_p c_1 \cdots c_p \cdots$$

$$= a_n a_{n-1} \cdots a_1 a_0 \, . \, b_1 \cdots b_k \overline{c_1 \cdots c_p}.$$

Here each a_i represents a digit of the integer part of the fraction, each b_j represents a digit which appears after the decimal point and which precedes the repeating string of digits, and each c_k is one of the digits which form the shortest repeating string of digits $c_1 \ldots c_p$, called a *period* of the decimal fraction. Here the terminating-decimal fractions may also be viewed as infinite with 0 repeated. For example,

$$20/7 = 2.857142857142\ldots = 2.\overline{857142} \quad \text{(the period is 857142, so } p = 6\text{)},$$

$$74/8 = 9.25000\ldots = 9.25\overline{0} = 9.25 \quad \text{(the period is 0, so } p = 1\text{)},$$

$$1127/90 = 12.5222\ldots = 12.5\overline{2} \quad \text{(the period is 2, so } p = 1\text{)}.$$

Prove that the number of digits, p, in a period is never greater than $n - 1$, for $n \geq 2$.

Remark. In fact, it is true that $p | \phi(n)$, where $\phi(n)$ is the Euler totient function. This function denotes the number of integers from 1 to n that are relatively prime to n; in particular, $\phi(n) = n - 1$ when n is prime. For more discussion on $\phi(n)$, see Section 12.3, particularly Problem 10.   ◇

(16) For each irreducible fraction $\frac{a}{b} \in (0, 1)$, consider an open interval $\left(\frac{a}{b} - \frac{1}{4b^2}, \frac{a}{b} + \frac{1}{4b^2}\right)$. Does the union of all these intervals cover the interval $(0, 1)$; i.e., is

$$(0, 1) \subseteq \bigcup_{\frac{a}{b} \in (0,1)} \left(\frac{a}{b} - \frac{1}{4b^2}, \frac{a}{b} + \frac{1}{4b^2}\right)?$$

(17) Given a finite set of numbers $a_1, a_2, \ldots, a_n$, the arithmetic mean of the set is $(a_1 + a_2 + \cdots + a_n)/n$ and the harmonic mean is $n/(1/a_1 + 1/a_2 + \cdots + 1/a_n)$.

   (a) Given a set $A = \{a, b\}$ of two nonnegative integers, prove that the product of the arithmetic and harmonic means of A is ab.

   (b) Given a positive integer N, define A_N and H_N to be the arithmetic and harmonic means, respectively, of the divisors of N. Prove that $N = A_N H_N$.

(18) Consider a sequence of n integers $a_1, a_2, \ldots, a_n$. Prove that there are integers i and j such $1 \le i \le j \le n$ and $n|(a_i + a_{i+1} + \cdots + a_j)$.

(19) Let a, b, k be positive integers, let $\gcd(b, 10) = 1$, and let $N = aaa \ldots aa$ be the concatenation of k copies of a. (For example, if $a = 2{,}446$ and $k = 4$, then $N=2{,}446{,}244{,}624{,}462{,}446$.) Prove that given a and b, k can be chosen in such a way that $b|N$.

You and 49 of your friends are considering entering the daily lottery. To win this lottery, one must correctly pick all six integers that appear on the six ping pong balls selected from a group of 40 balls. Suppose you and your friends each decide to play the lottery every day until one of you wins. If each day all 50 of you select a different set of six numbers, how many consecutive days would you need to play until the probability that one of you has won reaches 0.01? What if you had 99 friends? 999 friends? How many friends would you need to have play the lottery with you if you wanted to win within a year with a probability of 0.5?

8

Counting

Not everything that can be counted counts,
and not everything that counts can be counted.
—Albert Einstein (1879–1955)

8.1 What Is Counting?

In discrete math, the term *counting* decidedly does not refer to the naming of the ordinal numbers in sequence: $1, 2, 3, \ldots$. Rather, counting is the term we use for the collection of techniques utilized to answer "how many" questions. A classic joke among mathematicians is that there are three types of people in the world, those who can count and those who cannot. However, we suspect that even those in the latter category would be able to determine that a room with six rows of chairs and five chairs per row contains 30 chairs, without counting them. Counting questions abound in both real life and in mathematics. For example:

(1) In a lottery, how many different tickets can be played?

(2) In how many ways can teams be seeded in a tournament?

(3) How many outcomes are possible in a race with eight horses?

(4) How many six-element subsets does a ten-element set have?

(5) How many passwords can be formed that meet certain requirements?

(6) Given ten points on a circle, in how many ways can one construct five chords that use all ten points?

(7) How many 5-digit positive integers have the property that the digits of the integer are arranged in strictly decreasing order?

(8) Into how many regions can five lines divide a plane?

(9) How many one-to-one functions are there from a six-element set to a ten-element set?

For some people, answering "How many?" questions without actually counting is quite natural and intuitive, while for others the questions are difficult to answer, even if given a collection of applicable methods. Our desire is for both types of people to benefit from the chapter. We hope that those in the former group will gain additional understanding through learning the formal mathematics behind their intuitive approach, while those in the latter group will build up a collection of counting techniques and learn to match techniques with classes of problems.

At this point we offer a general strategy for studying discrete mathematics: when you do not know how to approach a problem, experiment with a similar problem that uses smaller numbers. The answer can often be obtained by merely listing all of the possibilities and counting them. Then try the problem again, continuing until a general pattern emerges.

8.2 Counting Techniques

The first counting technique we mention is the **Addition Rule**. Because the technique applies to the partition of a set, it is also sometimes called the **Partition Rule**. The concept is simple—if two or more pairwise disjoint (nonoverlapping) groups are combined, the size of the combined group equals the sum of the individual group sizes. For example, to obtain the number of students in a high school, one could add together the numbers of freshmen, sophomores, juniors, and seniors. Alternatively, the students could be partitioned according to many other demographic factors (political persuasion, race, gender, academic track, etc.), as long as the factor was well-defined. The formal statement of the rule is as follows.

Theorem 8.1 (Addition Rule). *The number of elements in the union of a finite number of pairwise disjoint sets is the sum of the cardinalities of those sets.*

Example 89. Consider ten points lying on a circle. How many chords can be drawn whose endpoints are among the ten points?

Solution. If the points are labeled 1 through 10, the chords can be divided into disjoint sets. See Figure 8.1; there will be 9 chords containing the first point (shown in the first drawing of the circle), 8 chords containing the second point that do not contain the first point (second drawing), 7 chords containing the third point that do not contain the first two points (third drawing), and so on. The total number of chords is therefore $9 + 8 + 7 + \cdots + 1 = 45$. □

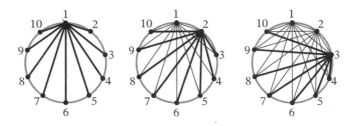

Figure 8.1. Example 89

Further Thoughts. Clearly, the solution above can be generalized to n points lying on a circle forming $(n-1) + (n-2) + \cdots + 2 + 1 = \sum_{i=1}^{n-1} i$ chords. While a summation representation may be correct, it is always desirable to simplify as far as possible. Since $\sum_{i=1}^{k} i = (k+1)k/2$, as shown in Exercise 1 of Section 6.3, we see that in the general problem with n points there are $n(n-1)/2$ chords.   ◇

Notice that the given Addition Rule is only valid when the given sets have no elements in common. For example, consider the sets $A = \{1, 3, 5\}$ and $B = \{2, 5\}$. The union of A and B is $A \cup B = \{1, 2, 3, 5\}$, a set with four elements. However $|A| + |B| = 3 + 2 = 5$. One cannot simply add the number of elements of A to the number of elements of B to obtain the number of elements in $A \cup B$, because the element(s) in their intersection ($A \cap B = \{5\}$ in this case) would be counted twice. By subtracting one from five, we do arrive at four, the size of $|A \cup B|$. This is a specific instance of the general statement we formulated in Section 3.4.

Proposition (Proposition 3.1). *Given finite sets A and B,*

$$|A \cup B| = |A| + |B| - |A \cap B|.$$

For a more interesting application of the proposition, we find the number of integers in $\Omega = \{1, 2, \ldots, 99, 100\}$ that are divisible by 4 or 10 (or both). Let A_4 be the integers in Ω that are divisible by 4 and let A_{10} be the integers in Ω that are divisible by 10. Clearly, $|A_4| = 25$ and $|A_{10}| = 10$. The intersection of A_4 and A_{10} consists of the integers in Ω that are both multiples of 4 and multiples of 10, i.e., multiples of 20; so $A_4 \cap A_{10} = \{20, 40, 60, 80, 100\}$. Hence, $|A_4 \cup A_{10}| = 25 + 10 - 5 = 30$.

Proposition 3.1 is itself a special case of the Principle of Inclusion-Exclusion, which gives a formula for finding the number of elements in the union of two or more sets. This principle will be discussed in Section 11.1, where a general proof is given.

Recall that if A is a subset of a universal set Ω, the complement of A is $\overline{A} = \{x \in \Omega : x \notin A\}$. Since A and $\overline{A}$ are disjoint, when we implement the Addition Rule, we get the following trivial (but useful!) result.

Theorem 8.2 (Complement Rule). *For a finite set Ω, if $A \subseteq \Omega$,*

$$|A| + |\overline{A}| = |\Omega|.$$

Example 90. Given ten points lying on a circle, how many chords can be drawn between nonadjacent points?

Solution. This question is similar to the one posed in Example 89, except we do not want to include the ten chords connecting adjacent points around the circle. Therefore, there are $45 - 10 = 35$ chords connecting nonadjacent points. (Formally, to apply the Complement Rule, we would let Ω be the set of all 45 chords between the ten points and A be the set of ten chords between adjacent points, in which case $|\overline{A}| = 35$.)   □

The Addition Rule is not a tool that can or should be used for all problems. Figure 8.2 depicts three cities, A, B, and C, connected by an assortment of roads. Roads d, e, and f join A to B, while roads g and h connect B to C. Consider these two questions.

(1) Altogether, how many roads are present?

(2) How many different routes are there from A to C, assuming one never backtracks?

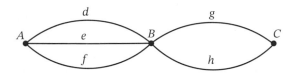

Figure 8.2

The Addition Rule (though it applies, by adding 3 and 2) is not needed to answer the first question: there are five different roads. To answer the second question, we simply list the routes and see that there are six of them:

dg	dh
eg	eh
fg	fh

What if a fourth city, D, were added to the map, with four routes from C to D (call them i, j, k, and l), as shown in Figure 8.3? Then how many routes would there be from A to D if no backtracking occurs?

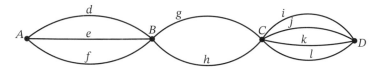

Figure 8.3

For each of the six routes from A to C shown in the table above, we simply append one of the roads i, j, k, or l to reach D, giving $6 \cdot 4 = 24$ different routes from A to D, as shown in the table below.

dgi	dhi	dgj	dhj	dgk	dhk	dgl	dhl
egi	ehi	egj	ehj	egk	ehk	egl	ehl
fgi	fhi	fgj	fhj	fgk	fhk	fgl	fhl

Alternatively, one can envision expanding the first 3×2 table into a $3 \times 2 \times 4$ "table" (box) in three dimensions with 24 entries (dgi, dgj, etc.).

Multiplication clearly was at work here—there were three ways to go from A to B, then two ways to go from B to C, and then four ways to go from C to D. The product of the numbers of routes to take at each juncture gave us the total number of routes. The **Multiplication Rule** (also known as the **Fundamental Theorem of Counting** or the **Fundamental Counting Principle**) is simply a generalization of this approach.

Theorem 8.3 (Multiplication Rule). *If a sequence of k tasks needs to be completed in order, with n_1 ways to complete the first task, n_2 ways to complete the second task, regardless of how the first is completed, n_3 ways to complete the third task, regardless of how the first two are completed, etc., and n_k ways to complete the last task, regardless of how the first $k - 1$ tasks are completed, then there are $n_1 n_2 \cdots n_k$ ways to complete the tasks, in order.*

Proof. Our proof is by induction on k. The first $k - 1$ tasks can be grouped together and considered as one task, to be completed in $n_1 n_2 \cdots n_{k-1}$ ways (by the inductive hypothesis). There are then n_k ways to complete the last task. We omit the formal details of this proof, including the base case with $k = 2$, which can be shown by using repeated applications of the Addition Rule.   □

Example 91. Given six (or n) people, how many different ways can all six of them form a line?

Solution. By "a line", we assume an order is chosen and the people are arranged, say, from left to right. The person selected for the first position can be chosen in 6 ways. No matter whom we chose, the person for the second position can be chosen in 5 ways. No matter whom we chose for the first and the second positions, there are 4 candidates for the third position, and so on. Hence, by the Multiplication Rule, the people can form $6 \cdot 5 \cdot 4 \cdot 3 \cdot 2 \cdot 1 = 720$ different lines.

A similar argument for n people gives $n! = n \cdot (n-1) \cdots 2 \cdot 1$ lines.   □

Further Thoughts. Note that in Example 91, unlike in the roads and city example, the available choices at the various stages are not the same, depending on previous choices made. However, the number of choices at a given stage will always be the same, which is all that matters.   ◇

The following result was actually proven in Chapter 3, but it is equivalent to the Multiplication Rule.

Corollary 8.4. *If A_i is a finite set for $i = 1, 2, \dots, k$, then*
$$|A_1 \times A_2 \times \cdots \times A_k| = |A_1| \cdot |A_2| \cdots |A_k|.$$

The Sum and Multiplication Rules, on the surface, are quite simple. However, these two rules may be used in conjunction to handle much more complex counting problems. To that end, the remainder of this chapter will be devoted to building an assortment of methods for addressing such problems.

One natural question to consider is, "How many subsets does a finite set A have?" Equivalently, "What is $|\mathcal{P}(A)|$, the size of the power set of A?" We stated the result in Chapter 3 and it appeared as Problem 13 of Section 6.3, but let us take time to explore this problem again carefully. Ultimately the result is simple to state and fairly easy to prove, but for now we simply begin with small values of n and look for a pattern. Let $a_n = |\mathcal{P}([n])|$, the number of subsets of $[n] = \{1, 2, \dots, n\}$. Below is a table listing the elements of $\mathcal{P}(n)$ for $n \le 3$.

$\mathcal{P}(\varnothing)$	$\mathcal{P}([1])$	$\mathcal{P}([2])$	$\mathcal{P}([3])$
$\varnothing$	$\varnothing$	$\varnothing$	$\varnothing$
			$\{3\}$
		$\{2\}$	$\{2\}$
			$\{2, 3\}$
	$\{1\}$	$\{1\}$	$\{1\}$
			$\{1, 3\}$
		$\{1, 2\}$	$\{1, 2\}$
			$\{1, 2, 3\}$

The empty set has only one subset (the empty set is a subset of every set!), so $a_0 = 1$. We also see that $a_1 = 2$, $a_2 = 4$, and $a_3 = 8$. The pattern seems to be emerging that the number of subsets doubles each time n increases by one; that is, $a_n = 2a_{n-1}$ for $n \geq 1$, and $a_n = 2^n$. Of course, we have not proven that this is the case, and showing that the recurrence holds for several more (or even several thousand more) values of n would not prove it either. Therefore, we will deduce it logically.

In each column after the first one, notice how we constructed the subsets of $[n]$ from the subsets of $[n-1]$ in the previous column. First, each element of $\mathcal{P}([n-1])$ became an element of $\mathcal{P}([n])$; second, n was inserted into each element of $\mathcal{P}([n-1])$ to create an element of $\mathcal{P}([n])$. All elements of $\mathcal{P}([n])$ were formed in one of these two ways, which is why the number of elements doubled in moving from $\mathcal{P}([n-1])$ to $\mathcal{P}([n])$. This idea is used to formally prove the general case below.

Theorem 8.5. *An n-element set has 2^n subsets.*

Proof. The subsets of $[n] = \{1, 2, \ldots, n\}$ can be divided into two groups—those that contain n and those that do not. There is a one-to-one correspondence between these two groups, namely the function which maps a set B containing n to the set $B \setminus \{n\}$. Since the members of $\mathcal{P}([n])$ that do not contain n form $\mathcal{P}([n-1])$, we have shown that the recursive relationship $a_n = 2a_{n-1}$ holds for $n \geq 1$. Since the empty set has only one subset, $a_0 = 1$, and the solution to the recurrence is $a_n = 2^n$. (We omit the straightforward inductive proof that 2^n is the solution to this recurrence relation.) □

The name *power set* likely comes from what we just proved, that $|\mathcal{P}(A)| = 2^{|A|}$. The succinct nature of this expression suggests we could have found it in some other way. In defining a subset of $\mathcal{P}([n])$, one needs to know n pieces of information, namely whether 1 is in the subset, whether 2 is in the subset, whether 3 is in the subset, etc., up to and including whether n is in the subset. Hence, by the Multiplication Rule, since there are two possibilities (in or out) for each of the n elements, the total number of ways to construct a subset of $[n]$ is 2^n.

The previous discussion leads to the following representation of subsets of $[n]$ as 0-1 sequences of length n. Let the ith number of the sequence be a 1 if i is in the subset and let it be 0 if it is not. See the following table for a complete listing of the $n = 3$ case.

Binary Sequence	Subset
$(0, 0, 0)$	$\varnothing$
$(1, 0, 0)$	$\{1\}$
$(0, 1, 0)$	$\{2\}$
$(0, 0, 1)$	$\{3\}$
$(1, 1, 0)$	$\{1, 2\}$
$(1, 0, 1)$	$\{1, 3\}$
$(0, 1, 1)$	$\{2, 3\}$
$(1, 1, 1)$	$\{1, 2, 3\}$

For another example, if $n = 5$, the sequence 01100 corresponds to the subset $\{2, 3\}$ and the sequence 10011 corresponds to the subset $\{1, 4, 5\}$.

Example 92. How many 3-digit numbers can be made out of digits from $\{1, 4, 5, 8, 9\}$ if

(1) the digits in a number can repeat?

(2) no digit can repeat?

(3) the number is odd and digits do not repeat?

(4) at least one of the digits is a 5?

Solution. (1) We apply the Multiplication Rule, where the tasks are to choose a digit for the first, second, and third position, in order. There are five choices for each of the three tasks, giving $5 \cdot 5 \cdot 5 = 125$ numbers where digits can repeat.

(2) Once the first digit is chosen, there are four choices for the second digit, and then three choices for the third digit. Again using the Multiplication Rule, we obtain $5 \cdot 4 \cdot 3 = 60$ numbers.

(3) If we proceed as in part (2), we will not know how many choices are available for the third digit, because we would not know how many odd digits are available. However, if we build the number from right to left, there are three (odd) choices for the third digit, then four remaining choices for the second digit, and three choices for the first digit. Thus there are $3 \cdot 4 \cdot 3 = 36$ odd numbers with distinct digits from the given set.

(4) *First Thoughts.* While it would be possible to enumerate the separate cases where 5 is in each of the first three positions, these cases would overlap, and one would need to be careful not to double-count when combining them. Fortunately, there is a much faster way, and it is often used in "at least" questions like this one. The set of numbers with at least one 5 is the complement of the set of numbers without a 5, so we can use the Complement Rule. ◇

Solution. There are $4^3 = 64$ numbers with three digits that can be formed from the set $\{1, 4, 8, 9\}$, not using the digit 5. Using part (1), by the Complement Rule, there are $125 - 64 = 61$ numbers that have at least one 5. □

Example 93. What is the sum of all numbers that meet the requirements in part (3) of Example 92?

Solution. We first calculate the sum of the digits that appear in each position of the 36 odd numbers formed in part (3); we will handle the place values later. This will take some casework.

Of the 36 numbers, by symmetry there will be 12 each that end in the digits 1, 5, and 9. The sum of the third digits of the 36 numbers will therefore be $12(1+5+9) = 180$.

Of the 12 numbers that end in a 1, an equal number (namely 3) of them will contain 4, 5, 8, and 9 as the second digit. The sum of these digits will therefore be $3(4+5+8+9) = 78$. Similarly, the respective sums of the numbers that end in a 5 and 9 are $3(1+4+8+9) = 66$ and $3(1+4+5+8) = 54$. Hence, the sum of the middle digits of the 3-digit numbers in question is $78 + 66 + 54 = 198$.

The distribution of digits occupying the first position of our 3-digit numbers will be identical to the distribution of those occupying the second position. Therefore, the sum of the first digits will also be 198.

Now that we know the sums of the digits in the units, tens, and hundreds positions of the 3-digit numbers, we can calculate the sum of the 36 numbers. It will be $198(100) + 198(10) + 180(1) = 21,960$. $\qquad\square$

Further Thoughts. Is the answer reasonable? The average of the five digits used is $27/5 = 5.4$, which will be the approximate average value of each position. Therefore, the average of the 36 numbers will be approximately $100(5.4) + 10(5.4) + 5 = 599$, which would give an approximate sum of $36(599) = 21,564$. This is close; the reader is invited to determine why the actual sum was slightly higher. $\qquad\diamond$

Example 94. Let A and B be finite sets of orders m and n, respectively. How many relations are there from A to B? How many functions are there from A to B?

Solution. By definition, a relation from A to B is a subset of $A \times B$, a set with mn elements. By Theorem 8.5, $|A \times B|$ has 2^{mn} subsets, so there are 2^{mn} relations from A to B.

Each function $f : A \to B$ is defined by choosing a unique value $f(a)$ for each $a \in A$. For fixed $a \in A$, this can be done in $|B| = n$ ways, no matter how the values of f for other elements of A are assigned. The number of such functions is $n \cdot n \cdots n$, where there are $m = |A|$ factors of n. Thus, there are n^m functions from A to B. $\qquad\square$

In Problem 5 of Section 8.2 and Problem 6 of Section 8.5, the reader will be asked to determine the number of one-to-one functions and the number of onto functions from a set A to a set B under various scenarios.

Exercises and Problems.

(1) Out of 35 students, 20 excel in mathematics, 11 excel in physics, and 10 excel in neither.

   (a) How many students excel in both mathematics and physics?

   (b) How many students excel in mathematics only?

(2) There are six roads which lead to the top of the mountain. In how many ways can one walk to the top and then walk back down under the assumption that different roads are used for going up and down.

(3) In the discussion following Example 89, we noted that $n(n-1)/2$ different chords could be drawn between n points on a circle, by simplifying a summation. Explain how to arrive at this value using a counting argument that includes the Multiplication Rule.

(4) Consider a roll of six dice of different colors.

   (a) How many different rolls are possible? (Assume the colors of the dice are relevant for distinguishing between rolls.)

   (b) How many different rolls have at least one 1 appear among the six dice?

   (c) How many different rolls have at least one 1 or at least one 5 among the six dice?

(5) Consider sets A and B with $|A| = s$ and $|B| = t$, $s \geq t \geq 1$.

   (a) How many onto functions are there from A to B if $t = 1$?

   (b) How many onto functions are there from A to B if $t = s$?

(c) How many onto functions are there from A to B if $t = 2$?

(6) There are 20 teams in a tournament. In how many ways can the gold, silver, and bronze medals be distributed at the end of the tournament?

(7) A single-elimination tournament consists of 64 teams, broken into four regions of sixteen teams each. In each region, the teams are seeded from 1 to 16. In the process of playing the tournament, one semifinalist will emerge from each region to create a "Final Four". How many Final Four configurations are possible in each of the following cases?

   (a) Every game is an upset (i.e., the team with the larger seed number always wins).
   (b) No number 1 seeds reach the Final Four.
   (c) Exactly three number 1 seeds reach the Final Four.
   (d) At least one number 5 seed reaches the Final Four.
   (e) Each team that reaches the Final Four has a seed number that is a power of 2.

---

(8) How many integers in the set $[1{,}000] = \{1, 2, \ldots, 1{,}000\}$ are

   (a) divisible by 5? divisible by 3?
   (b) divisible by 5 and divisible by 3?
   (c) divisible by 5 or by 3?
   (d) divisible by neither 3 nor 5?
   (e) divisible by exactly one of the numbers 3 or 5?

 (9) How many integers in the set $[3^{100}]$ are relatively prime to 3^{100}?

(10) How many integers in $[2^{1,776} \cdot 5^{2,017}]$ are relatively prime to 10?

(11) How many integers in the set $A = [3^{10} \cdot 5^{14}]$ are relatively prime to $3^{10} \cdot 5^{14}$?

(12) In a standard game of Tic-Tac-Toe, on a two-dimensional 3×3 grid, there are eight different winning rows (three vertical, three horizontal, and two diagonal). In each part below, find the number of winning rows for a Tic-Tac-Toe game with the given grid size and dimension. In each case, a winning row of squares (or cubes, in the three-dimensional case) is a row whose length is equal to the length of one side of the grid.

   (a) On a 4×4 grid.
   (b) On an $n \times n$ grid.
   (c) On a $3 \times 3 \times 3$ grid.
   (d) On an $8 \times 8 \times 8$ grid.
   (e) On an $n \times n \times n$ grid.

(13) A 4-digit number is a number which is represented in the decimal system by four digits with the first (left-most) digit not equal to zero. How many 4-digit numbers can be made out of digits $0, 1, 2, 3, 8, 9$ if

   (a) the digits in a number may repeat?
   (b) no digit can repeat?
   (c) the number is even and digits may repeat?

(d) the number is odd and digits may not repeat?

(e) at least one of the digits is odd and digits may repeat?

(14) How many distinct positive divisors does a number N have in each case?

(a) $N = p^2$, where p is a prime.

(b) $N = p^e$, where p is a prime and $e \in \mathbb{N}$.

(c) $N = pq$, where p and q are distinct primes.

(d) $N = p^k q^l$, where p and q are distinct primes and $k, l \in \mathbb{N}$.

(e) $N = p_1^{e_1} p_2^{e_2} \cdots p_k^{e_k}$, where all p_i are distinct primes and all e_i are positive integers.

(f) Use the above results to find the number of positive divisors of the integer 1,188.

(15) Let Noga and Peter be among twelve people forming a line at the author's book signing. In how many such lines

(a) does Noga stand immediately in front of Peter?

(b) do Noga and Peter stand next to each other?

(c) are there exactly three people between Noga and Peter?

(d) does Noga stand somewhere ahead of Peter?

(16) A sequence $a_1, a_2, \ldots, a_n$ is called a **palindrome** if for every $i \in [n]$, $a_i = a_{n+1-i}$. For example, sequences, $(1, 2, 2, 1)$ and $(3, 3, 4, 5, 4, 3, 3)$ are palindromes. Suppose all possible sequences of length seven are formed using letters from the set $\{R, A, C, E\}$.

(a) How many of the sequences are palindromes? Which one of them is an English word?

(b) How many of the sequences are not palindromes?

(17) Endre forgot the combination for his bike lock. He only remembers that it has five digits, that there is a 2 followed immediately by a 3, and that there is a 3 immediately followed by a 7. What is the maximum number of 5-digit sequences he would have to try in order to open the lock?

(18) Among the set of all 8-digit positive integers are there more with or without 1 as a digit?

---

(19) How many positive integers have the property that their digits are in strictly increasing order? For example, the integer 258 has the property but the integers 91 and 667 do not.

(20) (a) Consider all $m \times n$ matrices whose entries are equal to 1 or -1 only. In how many of these matrices is the product of entries in each row and each column equal to 1?

(b) Consider all $m \times n$ matrices whose entries are equal to 0 or 1 only. In how many of these matrices is the sum of the entries in each row and each column even?

(c) Consider all $m \times n$ matrices whose entries are from the set $\{0, 1, 2, \ldots, k - 1\}$. In how many of these matrices is the sum of the entries in each row and each column divisible by k?

8.3 Permutations and Combinations

In this section we develop two fundamental tools in combinatorics. They are actually both extensions of the Multiplication Rule, but many other counting problems can and will be reduced to them, so they deserve their own special notation and treatment.

Given a finite set A and a positive integer k, a k-**permutation of** A is defined as a sequence of k distinct elements of A. Note that if k is not mentioned, then we assumed that $k = |A|$ and simply call the sequence a **permutation of** A. Clearly, if $k > |A|$, there are no k-permutations of A. Since k-permutations are sequences, not sets, they are typically denoted with parentheses rather than braces. For example, there are exactly six 2-permutations of $A = \{1, 2, 3\}$, namely $(1, 2), (1, 3), (2, 1), (2, 3), (3, 1)$, and $(3, 2)$.

On the other hand, a k-**combination of** A is defined as a subset of k (distinct) elements of A. Notice that for both k-permutations and k-combinations the terms/elements are distinct. The difference between k-permutations and k-combinations, then, is simply that order is important for k-permutations and not for k-combinations. While there are six 2-permutations of the set A above, it has only three 2-combinations, namely $\{1, 2\}, \{1, 3\}$, and $\{2, 3\}$.

The natural questions to ask are generalizations of the example considered above: How many k-permutations and how many k-combinations can be formed from an n-element set? We will see that the answers to these two questions are quite related. The number of k-permutations can be determined using the Multiplication Rule. When $A = \{1, 2, 3\}$, there were three choices for the first element in a 2-permutation. Once that selection was made, there remained two choices for the second element, because of the requirement that it be different from the first. Using identical reasoning for larger values of n and k gives the following generalization.

Theorem 8.6. *Let k and n be integers with $1 \le k \le n$. The number of k-permutations that can be formed from an n-element set is*

$$P(n, k) = n(n - 1)(n - 2) \cdots (n - k + 1).$$

In the case where $k = n$, the permutation involves all members of A, and

$$P(n, n) = n(n - 1)(n - 2) \cdots (3)(2)(1) = n! \, .$$

This was the case in Example 91, where $P(6, 6) = 6 \cdot 5 \cdot 4 \cdot 3 \cdot 2 \cdot 1 = 6!$ By convention we define 0! to be 1, in order that many of the formulas involving factorials will still be true when 0! occurs. Using factorial notation, we have the following alternate representation for $P(n, k)$:

$$
\begin{aligned}
P(n, k) &= n(n - 1)(n - 2) \cdots (n - k + 1) \\
&= n(n - 1) \cdots (n - k + 1) \cdot \frac{(n - k)(n - k - 1) \cdots (2)(1)}{(n - k)(n - k - 1) \cdots (2)(1)} \\
&= \frac{n!}{(n - k)!}.
\end{aligned}
$$

The formula $P(n, k) = n! / (n - k)!$ suggests that we should define $P(n, 0) = 1$ for integers $n \ge 0$, and this is a common convention, even if the combinatorial interpretation is somewhat ambiguous.

Here are several examples which illustrate the notation:

- $P(5,3) = 5 \cdot 4 \cdot 3 = 60$.
- $P(4,0) = 1$.
- $P(5,5) = 5! = 5 \cdot 4 \cdot 3 \cdot 2 \cdot 1 = 120$.
- $P(11,2) = 11 \cdot 10 = 110$.

Notice that $P(n,k)$ is the product of k factors, with first factor n and each factor decreasing by one. Also, each value is the simplified quotient of two factorials.

Example 95. How many 4-digit numerical passwords can be formed if the digits must be distinct?

Solution. Each password will be a 4-permutation of the set of ten digits 0 through 9. There are $P(10,4) = 10 \cdot 9 \cdot 8 \cdot 7 = 5{,}040$ of them. □

Example 96. In an eleven-team conference, the top eight teams qualify for the post-season, where they are seeded for the tournament. Prior to the season (i.e., before it is known which teams will qualify for the postseason), how many different seedings are possible?

Solution. The set consists of eleven teams and each seeding of the top eight teams corresponds to exactly one 8-permutation. Therefore, the number of seedings is $P(11,8) = 11 \cdot 10 \cdots 5 \cdot 4 = 6{,}652{,}800$. □

Example 97. A committee of fourteen people must select a president, a vice president, a treasurer, and a secretary. In how many ways can this be done?

First Thoughts. Although the problem does not specify that the four offices must be held by distinct persons, that is a reasonable assumption, and we will make it here. You will sometimes find that counting problems require one to make such assumptions, though clearly worded problems are certainly preferable!   ◇

Solution. We have four distinct offices being filled, which means we need the number of 4-permutations of a fourteen-element set. There are $P(14,4) = 14 \cdot 13 \cdot 12 \cdot 11 = 24{,}024$ ways to fill the four offices. □

We now turn to counting the number of k-combinations of an n-element set. This number is often denoted $C(n,k)$, where the C stands for "combination". Ultimately we seek a formula for $C(n,k)$ but let us begin by exploring some special cases, using just the definition of a k-combination.

(1) $C(n,0) = 1$, since there exists only one subset of size 0, the empty set.

(2) $C(n,n) = 1$, because only one subset contains all n elements. This is the counterpart of $C(n,0) = 1$.

(3) $C(n,1) = n$, because an n-element set has n different one-element subsets.

(4) The counterpart of $C(n,1)$ is $C(n,n-1)$. This is the number of subsets that contain all but one element. There are n choices for the element that gets left out, so there are n different subsets containing $n-1$ elements. Thus, $C(n,n-1) = n$.

(5) $C(n, k) = C(n, n - k)$ for $0 \leq k \leq n$. This is a generalization of the previous case. There are $C(n, k)$ ways to choose k elements from an n-element set, and for each such set there is a unique complement with $n - k$ elements. For a more formal proof, consider the bijection $A \mapsto \overline{A}$ on the power set of $[n]$. It maps k-combinations to $(n - k)$-combinations, and vice versa. Hence, the cardinalities of A and $\overline{A}$ are equal.

(6) Determining $C(n, 2)$, the number of subsets with two elements, is a little harder. Recall the example of $n = 3$ given at the beginning of the section. In this case, there were six *ordered* pairs formed by distinct elements of $\{1, 2, 3\}$, but only three 2-combinations of $\{1, 2, 3\}$, as shown:

2-permutations	2-combinations
$(1, 2), (2, 1)$	$\{1, 2\}$
$(1, 3), (3, 1)$	$\{1, 3\}$
$(2, 3), (3, 2)$	$\{2, 3\}$

Because order matters in a permutation, counting the number of 2-permutations literally double-counts the number of 2-combinations, and we must therefore divide their number by two to get the number of 2-combinations.

The same principle holds for general n. We can use the Multiplication Rule to find the number of 2-permutations: there are n choices for the first element in our ordered pair (call it a), and after it is selected, there are $n - 1$ choices for the second element of the ordered pair (call it b). This creates $n(n-1)$ different ordered pairs (a, b). However, for any given pair (a, b), (b, a) is a different pair with the same elements, and together these two ordered pairs correspond naturally with the (unordered) 2-combination $\{a, b\}$. As in the case for $n = 3$, we have double-counted the number of 2-permutations, and we must therefore divide by two to get the number of 2-combinations. Accordingly, $C(n, 2) = \frac{n(n-1)}{2}$.

In Example 96, suppose we want to know how many different combinations of eight teams could make the postseason, without regard for the seeding of the teams. Since such a group is an 8-combination, the number of combinations is denoted $C(11, 8)$. Clearly, there are more 8-permutations than 8-combinations, because for any set of eight teams that make the postseason, there are many possible seedings of those eight teams. In fact, we know exactly how many! There are precisely $8! = 43{,}320$ of them. Since there is a 43,320-to-1 correspondence between the 8-permutations and the 8-combinations, we see that the number of distinct sets of eight postseason teams that can be formed from the initial eleven teams is $\frac{P(11,8)}{8!} = \frac{6{,}652{,}800}{43{,}320} = 165$. The generalization is clear.

Theorem 8.7. *For $0 \leq k \leq n$, the number of k-combinations of an n-element set is*

$$C(n, k) = \frac{P(n, k)}{k!} = \frac{n!}{(n - k)! \; k!}.$$

Proof. Each k-permutation can be constructed uniquely in the following two steps: first choose a k-combination of the n-element set, and then create a sequence of length k using each element from the chosen subset once. The first step can be accomplished in $C(n, k)$ ways, and the second in $k!$ ways. Hence, $P(n, k) = C(n, k) \cdot k!$. The result now follows by dividing by $k!$ □

Here are several examples which illustrate the notation and how the expressions may be simplified:

- $C(5, 3) = \frac{5!}{3! \cdot 2!} = \frac{5 \cdot 4 \cdot 3!}{3! \cdot 2!} = \frac{5 \cdot 4}{2 \cdot 1} = 10$.
- $C(6, 6) = 1$.
- $C(8, 2) = \frac{8!}{6! \cdot 2!} = \frac{8 \cdot 7 \cdot 6!}{6! \cdot 2!} = \frac{8 \cdot 7}{2} = 28$.
- $C(n, 3) = n(n - 1)(n - 2)/3!$.

Example 98. A committee of fourteen people must choose four of its members to form a subcommittee. In how many ways can this be done?

Solution. This is similar to Example 97 except that the assignment of labels to the four members is not important. Therefore we seek the number of 4-combinations of a fourteen-element set, which is

$$C(14, 4) = \frac{14!}{10! \, 4!} = \frac{14 \cdot 13 \cdot 12 \cdot 11}{4 \cdot 3 \cdot 2 \cdot 1} = 7 \cdot 11 \cdot 13 = 1001.$$

It is worth noting that this value is 24 times the value found in Example 97.   □

Notice the use of the word "choose" in Example 98. Even when it may not seem like a "choice" is actually being made, the word is so common in these types of questions that the number of k-combinations of an n-element set is often referred to as "n choose k". The following representations are also commonly used:

$$C(n, k) = {}_nC_k = \binom{n}{k} = \frac{n!}{(n - k)! \, k!}.$$

The symmetry of k and $n - k$ in the final expression gives us another proof of an identity that we established earlier by a counting argument—one that is important enough to highlight as a theorem.

Theorem 8.8. *For $0 \le k \le n$, $C(n, k) = C(n, n - k)$.*

Our understanding of how to count k-permutations and k-combinations of an n-element set can be used together with the Multiplication Rule to answer many different counting questions. We give several examples here, and a variety of others can be found in the problem set.

Example 99. How many different 7-character license plates can be formed in which the first three characters are letters and the last four characters are distinct numerals?

Solution. There are 26 choices for each of the first three characters, resulting in $26^3 = 1,756$ possible 3-letter sequences. Since the four numerals are distinct digits from 0 to 9, there are $P(10, 4) = 5,040$ ways to order the last four characters. Multiplying the number of ways of placing the letters and then the numerals, we arrive at $1,756 \cdot 5,040 = 88,583,040$ license plates.   □

Example 100. In one form of a lottery, each player must select six numbers from $[40] = \{1, 2, \ldots, 40\}$ and then specify one of those numbers as the "red" number. (The red number is used to help break ties if more than one person correctly picks the six numbers.) In how many different ways can one complete such a lottery ticket?

Solution. There are $C(40, 6) = \frac{40!}{6!\,34!} = 3{,}838{,}380$ ways to choose six numbers from the 40 possibilities. For each such choice of a set of six numbers, there are six ways to choose the red number. By the Multiplication Rule, we conclude there are $3{,}838{,}380 \cdot 6 = 23{,}030{,}280$ possible lottery tickets. $\qquad\square$

The previous example provides us with a timely opportunity to answer the probability question in the opening hook problem of the chapter. Though we do not treat probability in a formal way in this book, in lottery questions like this one, it should be apparent how counting the number of possible combinations (tickets) leads to finding the likelihood (probability) of a single ticket being a winner.

Example 101. You and 49 of your friends are considering entering the daily lottery. To win this lottery, one must correctly pick all six integers that appear on the six ping pong balls selected from a group of 40 balls. Suppose you and your friends each decide to play the lottery every day until one of you wins. If each day all 50 of you select a different set of six numbers, how many consecutive days would you need to play until the probability that one of you has won reaches 0.01?

First Thoughts. We could solve this problem by adding up the probabilities of winning the first day, of winning the second day but not the first, of winning the third day but not the first two, etc. However, it is simpler to use the Complement Rule in a similar way to how it was used in the "at least one" question in Example 92, by calculating the probability of a group of friends not winning each day for a set of consecutive days and subtracting from 1. Try this approach before reading further. $\qquad\diamond$

Solution. From the solution to Example 100, the probability of any one player winning on a given day is $p = \frac{1}{3{,}838{,}380}$. If 50 players play on a given day, each choosing a different set of six numbers, the probability of one of them winning is $50p$, and the probability that none of them wins is $1 - 50p$. Using the probability form of the Multiplication Rule, the probability of none of them winning on k consecutive days is $(1 - 50p)^k$. We need to find the value of k for which this probability equals $1 - 0.01 = 0.99$. Taking the natural logarithm of both sides of the equation $(1 - 50p)^k = 0.99$ and solving for k, we obtain

$$k = \frac{\ln 0.99}{\ln(1 - 50p)} \approx 772 \text{ days.}$$

With 99 friends and 999 friends, the number of days needed until the probability of one of you winning equals 0.01 drops to 386 days and 39 days, respectively. For the probability of winning within a year to be 0.5 when n persons play daily, we can solve the equation $(1 - np)^{365} = 0.5$ for n, yielding $n \approx 7{,}282$ players. $\qquad\square$

A **rearrangement** of a word is any ordering of the letters of the word, including the word itself. For example, the word *PROOF* has, among others, *ROOFP*, *FORPO*, *POROF*, and *RPOOF* as rearrangements.

Example 102. How many rearrangements does the word *MISSISSIPPI* have?

Solution. While the word has eleven letters, it has only four distinct letters, M, I, S, and P. Let us view the rearrangements as a sequence of 11 spaces to fill with these letters. There are $C(11, 4)$ ways to choose the positions for the the four I's. Once this is done, there are 7 spaces left and therefore $C(7, 4)$ ways to place the four S's. Then, there are just three spaces left and $C(3, 2)$ ways to place the two P's. The M must go in the remaining spot, and there is just $C(1, 1) = 1$ way to do this. By the Multiplication Rule, there are

$$C(11, 4) \cdot C(7, 4) \cdot C(3, 2) \cdot C(1, 1) = 330 \cdot 35 \cdot 3 \cdot 1 = 34{,}650$$

rearrangements of the word MISSISSIPPI.   □

Rearrangements may be interpreted in such a way to help solve other types of counting problems.

Example 103. How many paths are there in the Cartesian coordinate plane from the origin to the point $(4, 6)$, where each step on the path has length one and moves either right or up?

First Thoughts. With problems like these, it is often helpful to clarify the form of instructions that would completely define the given task. In this case the task is to create a path of length ten from the origin to the point $(4, 6)$ moving a total of four steps to the right and six steps up.   ◇

Solution. Each step of the path may be represented by either an R or a U (for right and up, respectively). Then, each path may be represented by a sequence of length ten containing four R's and six U's. The path corresponding to the sequence $(R, U, U, U, R, R, U, U, R, U)$ is shown in bold in Figure 8.4. There are $C(10, 4) = 210$ ways to choose the four R-steps, and these choices uniquely determine the remaining six U-steps. Thus there are 210 distinct paths.   □

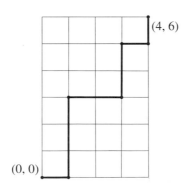

Figure 8.4

The following proposition is a special case of Example 102 with only two characters, as illustrated in Example 103.

Proposition 8.9. *There are $C(m + n, m) = C(m + n, n)$ binary (two characters, say A and B) sequences of length $m + n$, with m terms being equal to A and n terms equal to B.*

Example 104. Given ten labeled points on a circle, how many sets of five chords can be constructed in which each of the ten points is an endpoint of one of the chords?

First Thoughts. In many counting problems, we are asked to find the number of un-labeled objects. This is the case here since the five chords are not put in any specified order. In such situations, it is often easier to count the number of labeled objects, according to some ordering we impose. We then compensate for the overcounting by dividing the number of labeled objects corresponding to each unlabeled object. In the present example, there will be 5! ways to order the chords in each set of five chords. ◇

Solution. Let us label the five chords from one to five as we construct them. There will be $\binom{10}{2} = 10 \cdot 9/2$ ways to form the first chord, since any two points will uniquely determine the chord. At that point, there will then be $\binom{8}{2} = 8 \cdot 7/2$ ways to form the second chord, $\binom{6}{2} = 6 \cdot 5/2$ ways to form the third chord, $\binom{4}{2} = 4 \cdot 3/2$ ways to form the fourth chord. The fifth chord will then be determined by the two remaining points. Multiplying these values together gives $\frac{10!}{2^5}$ ways to form a sequence of five distinct chords. However, the chords were originally unlabeled, and there will be 5! chord-orderings for each set of five chords formed. We thus must divide the above number by 5!. The result simplifies to $\frac{10 \cdot 9 \cdot 8 \cdot 7 \cdot 6}{2^5} = 9 \cdot 7 \cdot 5 \cdot 3 \cdot 1 = 945$ different sets of five chords. □

Further Thoughts. The expression $\frac{10 \cdot 9 \cdot 8 \cdot 7 \cdot 6}{2^5}$ simplifies to $9 \cdot 7 \cdot 5 \cdot 3 \cdot 1$. Upon encountering such a tidy expression in a counting problem, one should try to find a counting argument to match the expression. We do that next—try to find one before reading on.

Denote the "first point" (according to the given labeling) on the circle as A_1. There are then nine ways to form the chord containing A_1. Once this chord is formed, there are eight unused points. Moving clockwise from A_1 around the circle, let A_2 be the first unused point. There are seven ways to form the chord containing A_2, and after this chord is formed, there are then six unused points. Letting A_3 be the first unused point moving clockwise around the circle from A_2, there are five ways to form the chord containing A_3. Repeating this argument until all ten points are used, it is easy to see that there are $9 \cdot 7 \cdot 5 \cdot 3$ different sets of five chords. ◇

Example 105. A father has thirteen identical coins that he plans to distribute to his five children. In how many different ways can he allocate the coins in each scenario?

(1) Each child must receive at least one coin.

(2) Not every child needs to receive a coin.

First Thoughts. One can of course start with a smaller example and look for a pattern. The pattern may be elusive here, though. Another approach is to lay the the coins in a row and then divide them up with four markers, giving the resulting five groups of coins to the five children according to some order (say, by age).

For example, the top configuration in Figure 8.5 depicts a distribution that would give 4, 2, 3, 3, and 1 coins to the children, by age from oldest to youngest. The lower configuration corresponds to a distribution where the children receive 1, 0, 4, 8, and 0 coins, respectively. These figures give us a new way to think about counting the number of distributions of coins; since there is a one-to-one correspondence between

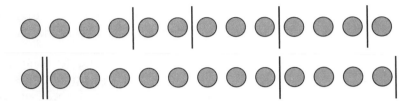

Figure 8.5

the distributions of coins to the children and the configurations of coins and markers, we just need to count the number of possible configurations in each case.   ◇

Solution. (1) By placing the thirteen coins in a row, there will be twelve locations between the coins where the four markers can be placed to divide the coins into five distinct groups of at least one coin each. There will be $C(12, 4) = 495$ ways to place the markers and thus 495 different ways to distribute the coins in such a way that each child receives at least one coin.

(2) If each child does not need to receive a coin, we must permit the markers to be adjacent. So, we simply need to count the number of arrangements of 13 identical coins and 4 identical markers. This is a straightforward application of Proposition 8.9, and there are $C(17, 4) = 2,380$ ways to accomplish the task.   □

Exercises and Problems.

(1) Given twenty labeled points on the circumference of a circle, how many chords can be drawn in which both endpoints are among these twenty points?

(2) A college student is taking five different courses. She has 7 notebooks of different colors. In how many ways can she assign notebooks to the courses, assuming exactly one notebook per course, with two going unused?

(3) The 112th United States Senate consisted of 57 Democrats, 41 Republicans, and 2 Independents. How many different 5-person committees could be formed if the committee were required to consist of

   (a) three Democrats and two Republicans?

   (b) two Democrats, two Republicans, and one Independent?

   (c) at most two Republicans?

   (d) no Independents?

   (e) at least three Democrats?

(4) How many sets of 12 homework exercises containing 4 easy, 5 medium level, and 3 difficult exercises can be selected from a set of 70 exercises, of which 22 are easy, 20 are medium, and 28 are difficult?

(5) Solve for $n \in \mathbb{N}$.

   (a) $\binom{n}{3} = \frac{1}{5}\binom{n+2}{4}$.

   (b) $\binom{n}{5} > \binom{n}{4}$.

(6) Suppose we have $n \geq 3$ points in the space. Choose any three of them and consider a plane containing the chosen points. What is the greatest (smallest) number of distinct planes that can be obtained this way?

(7) How many rearrangements does the word *MATHEMATICS* have?

(8) How many rearrangements does the word *ANTIDISESTABLISHMENTARIANISM* have?

(9) Given n distinct items, in how many ways can one choose an even number of them and in how many ways can one choose an odd number of them for

   (a) $n = 8$?

   (b) $n = 9$?

   (c) general n?

(10) A woman has six friends. Each evening, for five consecutive days, she invites three of them over for dinner in such a way that the same group is never invited twice. How many ways are there to do this? (Assume that the order in which groups are invited matters.)

(11) Consider n red and n blue points on a plane. In how many different ways can one draw n segments such that each segment has one endpoint red and the other endpoint blue and every point is an endpoint of exactly one segment?

---

(12) (a) Suppose we have $n \geq 2$ points lying in a plane. Consider the set A of all lines that pass through two points of the set. What are the smallest and largest possible values of $|A|$?

   (b) If $n = 5$, what are the possible numbers of distinct lines that can be formed (give the smallest and largest numbers possible, as well as all possible values in between).

(13) Ten players get together to play basketball. They need to form two teams of five players each. One player argues that the number of ways of dividing the players into two teams is clearly $\binom{10}{5}$, since five players out of ten must be chosen to create a team, and then the other five players are on the other team. Another player argues that the answer is $\binom{9}{4}$ because that is the number of ways that she can choose her teammates from the other nine players. Who is right, and why?

(14) Fifteen players get together to play basketball. They need to form three teams of five players each.

   (a) In how many ways can this be done?

   (b) In how many ways can the team selection be done if the players line up and count off by threes, with the 1's forming the team that gets the ball first, the 2's forming the team that plays defense first, and the 3's forming the team that sits out the first game?

(15) Ten people are sitting around a round table. How many different sittings are possible in the following scenarios?

   (a) One of the chairs is identified as being the head of the table.

   (b) All chair positions are identical, so sittings are identified only according to the relative positions with each other. (Here it only matters who sits on one's right and who sits on one's left.)

    (c) Orientation does not matter, so two sittings are considered identical if each person sits beside the same two people in each sitting.

(16) Suppose a piece of music is composed having 4 beats per measure, where an eighth note is half a beat, a quarter note is one beat, a half note is two beats, and a whole note is four beats. How many ways are there to compose one measure using these four types of notes? One such example is $\left(\frac{1}{2}, 1, \frac{1}{2}, 2\right)$ and another is $(2, 1, 1)$.

(17) Each of seven computer users tries to establish a wireless connection with five networks in the local vicinity. Not all attempts are necessarily successful. Call the complete set of connections that is actually established a configuration.

    (a) How many distinct configurations between the users and networks can be established?

    (b) In how many configurations does each computer user establish a wireless connection with

        (i) exactly one wireless network?

        (ii) at least one wireless network?

        (iii) at most one wireless network?

(18) Find $|S|$, where $S = \{(a, b, c) | a, b, c \in [100], a < b, a < c\}$.

(19) Let $A = \{1, 2, \ldots, n\}$, $n \in \mathbb{N}$.

    (a) Given $k \in A$, show that the number of subsets of A in which k is the maximum element is 2^{k-1}.

    (b) Apply (a) to show that $\sum_{i=0}^{n} 2^i = 2^n - 1$.

(20) In how many ways can ten identical coins be distributed to three distinct wallets?

(21) Given p white identical balls and q black identical balls, in how many ways can they be arranged in linear order such that no two black balls are next to each other?

(22) Suppose that all streets in a city form a rectangular grid with five horizontal streets and eight vertical streets.

    (a) In how many ways can one walk from the South-West corner to the North-East corner if the only directions one can walk in are north and east?

    (b) Repeat the first part of the problem, but assume that the street AB in one block is closed (see Figure 8.6).

    (c) Answer questions (a) and (b) with "five" replaced by k and "eight" replaced by n.

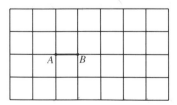

Figure 8.6

(23) How many 4-digit numbers ($abcd$), $a \neq 0$, are there with

    (a) $a < b < c$?

    (b) $a < b < c < d$?

    (c) $a > b > c > d$?

---

(24) In one form of the lottery, each participant must select a set of six integers from 1 to 40.

    (a) In how many ways can this be done where none of the integers chosen are consecutive?

    (b) In how many ways can this be done where exactly two of the integers chosen are consecutive?

(25) The integers from 1 to 40 are numbered consecutively around a circle. In how many ways can one select six of the integers if none of the integers are adjacent to each other?

(26) In how many arrangements of the 26 letters of the alphabet will no two of the five vowels be adjacent to each other?

(27) Let A, B, and C be three different persons in a group of $n \geq 3$ people. Suppose these n people form a line to the cashier's office. In how many of these lines does A stand somewhere before B while C stands somewhere after B?

(28) In how many ways can 40 identical candies can be distributed among six people such that every person gets at least three candies?

(29) How many 6-digit numbers have three even and three odd digits? (The first digit cannot be zero.)

(30) Fifteen points are taken on the circumference of a circle. Through any two of them a chord is drawn. If no three of the chords intersect at a point inside the circle, how many points of intersection of these chords are there (including points on the circumference)?

8.4 The Binomial Theorem

A less obvious but powerful use of combinatorial counting is for proving identities. If one counts something correctly in two different ways, obtaining two different expressions, then the expressions must be equal in value. We will study this technique more thoroughly in Chapter 11, but the following identity is useful now.

Theorem 8.10. *For integers n and k, $1 \leq k \leq n$,*

$$C(n, k) = C(n - 1, k) + C(n - 1, k - 1).$$

Proof 1. By definition, there are $C(n, k)$ different k-element subsets of the set $[n] = \{1, 2, \ldots, n\}$. This collection of subsets can be partitioned into two groups: those that contain the element 1 and those that do not. There are $C(n-1, k-1)$ subsets containing k elements in the former group and $C(n - 1, k)$ subsets containing $k - 1$ elements in the latter group. The sum of these two numbers must therefore be $C(n, k)$, as desired.

Proof 2. We can also prove this identity algebraically by using the fact that $C(n, k) = \frac{n!}{k!\,(n-k)!}$. See Problem 6. $\qquad\qquad\qquad\qquad\qquad\qquad\qquad\qquad\qquad\qquad\qquad\quad\square$

The recursive nature of Theorem 8.10 allows us to construct a table of values of $C(n, k)$, often called **Pascal's Triangle**; the first five rows appear below. Labeling the first row as row 0, we see that $C(n, k)$ is the kth entry of row n:

$$
\begin{array}{c}
1 \\
1 \quad 1 \\
1 \quad 2 \quad 1 \\
1 \quad 3 \quad 3 \quad 1 \\
1 \quad 4 \quad 6 \quad 4 \quad 1
\end{array}
$$

The numbers in the table may look familiar to the algebra student who has learned to expand binomial expressions of the form $(a + b)^n$. Indeed, the values of $C(n, k)$ are often called **binomial coefficients**, and we now explain why.

Consider the expansion of $(a_1 + b_1)(a_2 + b_2)(a_3 + b_3)$. The product of these three binomials is

$$a_1 a_2 a_3 + a_1 a_2 b_3 + a_1 b_2 a_3 + a_1 b_2 b_3 + b_1 a_2 a_3 + b_1 a_2 b_3 + b_1 b_2 a_3 + b_1 b_2 b_3.$$

Notice there are $2^3 = 8$ terms in the expansion. This is because each term of the expansion is the product of three factors, where the ith factor in each product is one of the two terms (a_i or b_i) in the ith binomial of the original expression, for $i = 1, 2, 3$. Observe that if we drop the subscripts, the original expression becomes $(a + b)^3$, and collecting like terms yields the expansion $a^3 + 3a^2 b + 3ab^2 + b^3$.

How could we have determined these coefficients by counting? Consider the term $3a^2 b$. The coefficient 3 denotes that there were 3 terms in the expansion that contained exactly two a's and one b, namely aab, aba, and baa. Why three terms, combinatorially? Out of the three original binomials, we needed to choose two of them that would contribute an a to the product, with the other factor contributing a b. There are $C(3, 2) = 3$ ways this could be done. A similar argument holds for each of the coefficients, and indeed we see that the four coefficients of 1, 3, 3, and 1 equal $C(3, 0)$, $C(3, 1)$, $C(3, 2)$, and $C(3, 3)$, respectively.

The above example generalizes to the so-called Binomial Theorem. Notice our use of the $\binom{n}{k}$ notation. There is not a hard-fast rule about this, but we tend to use the $C(n, k)$ form when emphasizing a counting interpretation and $\binom{n}{k}$ when referring to binomial coefficients.

Theorem 8.11 (Binomial Theorem). *For any positive integer n,*

$$(a + b)^n = \sum_{k=0}^{n} \binom{n}{k} a^{n-k} b^k.$$

Proof. Each term in the expansion of $(a + b)^n$ will be the product of a's and b's, k of the former and $n - k$ of the latter, for some k between 0 and n. Moreover, for fixed k, $0 \le k \le n$, there will be $\binom{n}{k}$ terms with this specified composition, because an a must be the chosen term from exactly k of the n binomials $(a + b)$. When these terms are collected, the resulting sum is the one given. □

Example 106. Expand the expression $(2x - 3y)^5$.

Solution. Let $a = 2x$ and $b = -3y$ and apply the Binomial Theorem to obtain

$$(2x - 3y)^5 = \sum_{k=0}^{5} \binom{n}{k}(2x)^{5-k}(-3y)^k$$

$$= (2x)^5 + \binom{5}{1}(2x)^4(-3y) + \binom{5}{2}(2x)^3(-3y)^2$$

$$+ \binom{5}{3}(2x)^2(-3y)^3 + \binom{5}{4}(2x)(-3y)^4 + (-3y)^5$$

$$= 32x^5 - 240x^4y + 720x^3y^2 - 1{,}080x^2y^3 + 810xy^4 - 243y^5. \quad \square$$

The Binomial Theorem can be generalized to more than two terms, and the resulting expression utilizes multinomial coefficients. This topic will be explored in Chapter 11. For now, we could expand an expression such as $(a+b+c)^3$ by letting $a+b = x$ and using the Binomial Theorem on $(x + c)^3$. (See Exercise 4.)

While the Binomial Theorem may be utilized by the algebra student who needs to expand binomials, a more useful purpose in discrete mathematics is to prove identities. The following identity has an elegant counting proof, which we will give in Section 11.4, but here we see that it follows immediately from the Binomial Theorem by letting $a = b = 1$.

Theorem 8.12. *For each positive integer n, $\sum_{k=0}^{n} \binom{n}{k} = 2^n$.*

This identity provides yet another proof that an n-element set A has 2^n subsets. The sum simply adds up the number of subsets of cardinality k for $k = 0$ (the empty set) to $k = n$ (A itself). The identity also demonstrates one of many noteworthy patterns in Pascal's Triangle: the sum of the numbers in each row is a power of 2. Specifically, the sum of the binomial coefficients in row n of Pascal's Triangle is 2^n. For $n = 3$, $1 + 3 + 3 + 1 = 2^3$.

Exercises and Problems.
(1) Verify Theorem 8.12 for the case $n = 5$.
(2) What is the coefficient of x^4y^6 in the expansion of $(5x + y)^{10}$?
(3) What is the coefficient of x^5y^{14} in the expansion of $(2x - 3y)^{19}$?
(4) Expand $(a + b + c)^3$ using the Binomial Theorem only (and properly placed parentheses).
(5) How many paths are there from the top entry of 1 in row 0 of Pascal's Triangle to the entry $C(6, 2) = 15$, where each step in the path must be down to the left or down to the right? Generalize.

(6) Give an algebraic proof (by showing the equivalence of the factorial representations of the two sides of the equation) of Theorem 8.10: For integers n and k, $1 \leq k \leq n$,

$$\binom{n}{k} = \binom{n-1}{k} + \binom{n-1}{k-1}.$$

(7) (a) Prove that for $n \in \mathbb{N}$, $\sum_{k=0}^{n} \binom{n}{k} 4^k = 5^n$.

(b) Generalize this result by making appropriate substitutions in the Binomial Theorem.

(8) Use induction to prove the Binomial Theorem:

$$(a+b)^n = \sum_{k=0}^{n} \binom{n}{k} a^k b^{n-k}.$$

(9) What is the term with the largest coefficient in the expansion of $(2x + 3y)^{100}$?

(10) What is the sum of all coefficients in the expansion of each expression?

(a) $(x+y)^{200}$

(b) $(x-y)^{200}$

(c) $(2x - 3y)^{200}$ (Express the answer in summation form.)

---

(11) (a) Substitute $a = 1$ and $b = -1$ in the Binomial Theorem to obtain a combinatorial identity.

(b) Interpret the identity in (a) in terms of Pascal's Triangle. Specifically, what does it prove about the entries in each row?

(c) Prove that for $n \geq 1$,

$$\binom{n}{0} + \binom{n}{2} + \binom{n}{4} + \cdots$$

and

$$\binom{n}{1} + \binom{n}{3} + \binom{n}{5} + \cdots$$

each equal 2^{n-1}.

(d) Show that there exist as many subsets of an n-element set containing an odd number of elements as there are subsets containing an even number of elements.

(12) Let $n \in \mathbb{N}$. Show that the binomial coefficients are always ordered in the following way:

$$\binom{2n}{0} < \binom{2n}{1} < \cdots < \binom{2n}{n}$$

and

$$\binom{2n}{n} > \cdots > \binom{2n}{2n-1} > \binom{2n}{2n}.$$

8.5 Additional Problems

This section contains a variety of problems that can be solved using techniques found throughout the chapter.

(1) Suppose set A contains seven points lying on a line and set B contains ten points lying on a parallel line. How many

   (a) segments have one endpoint in A and one endpoint in B?
   (b) triangles have their vertices among these seventeen points?
   (c) non-self-intersecting quadrilaterals have their vertices among these seventeen points?

What are the answers if we have $|A| = m$ and $|B| = n$ points on two parallel lines, where $m, n \in \mathbb{N}$?

(2) Three six-sided dice are rolled. The dice are identical, but nonstandard. Each die has the integers 1, 3, 9, 81, 243, and 729 on its six faces. How many different sums are possible?

(3) A math professor has four different discrete math books, three different geometry books, and five different calculus books to offer as prizes to the top three students in a math competition. The winner gets his or her choice of the twelve books. The second place student then gets to choose from among the remaining eleven books, etc. How many different orderings of three books can be awarded as prizes in each of the following cases?

   (a) All three award winners choose a discrete math book.
   (b) One book of each type is chosen.
   (c) No discrete math books are chosen.
   (d) At least one discrete math book is chosen.

(4) There are two main highways between cities A and B linked by ten secondary roads, as shown (crudely) in Figure 8.7. How many routes are there from A to B that do not pass through any junction twice?

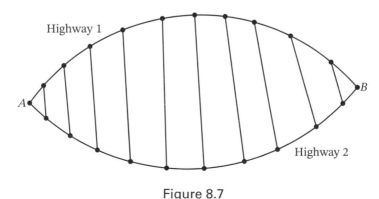

Figure 8.7

(5) (a) A man has $n = 10$ distinct types of candy. He wants to give five pieces to his daughter and five pieces to his son. In how many ways can he do this?
   (b) A man has $n = 10$ identical pieces of candy and wants to put them in two identical bags such that neither bag is empty. In how many ways can he do it?

(c) A woman has $n = 10$ distinct types of candy and wants to put them into two identical bags such that neither bag is empty. In how many ways can she do this?

(d) Answer the above questions for general n.

(6) Given sets A and B, $|A| = s \geq 1$, $|B| = t \geq 1$. Note any necessary constraints on s and t.

(a) How many relations from A to B are there?

(b) How many relations from A to B have their domain consisting of exactly one element of A ?

(c) How many functions from A to B are there?

(d) How many one-to-one functions from A to B are there?

(e) How many functions from A to B are bijections?

(f) How many functions from A to B have their range consisting of exactly one element of B (constant functions)?

(g) How many relations from A to B have their range consisting of exactly one element of B ?

(h) For how many relations ϕ from A to B is the following true:

$$\forall a \in A\, \exists b \in B,\ a\phi b?$$

(i) For how many relations ϕ from A to B is the following true:

$$\exists a \in A\, \exists b \in B,\ a\phi b?$$

(j) For how many relations ϕ from A to B is the following true:

$$\forall a \in A\, \forall b \in B,\ a\phi b?$$

Note: We did not ask the reader to count the number of onto functions from A to B. This is a difficult question, and we will address it in Chapter 11.

(7) In how many ways can fourteen distinct books be placed on five distinct shelves if the order of books on a shelf matters?

(8) What is the sum of all 4-digit numbers that can be formed out of digits from $\{1, 3, 5, 8, 9\}$ if no digit repeats?

(9) What is the greatest number of regions into which n lines can divide the plane?

(10) What is the greatest number of regions into which n planes can divide 3-space?

(11) What is the greatest number of regions into which n circles can divide the plane?

(12) How many terms are there in the expansion of $(w + x + y + z)^{20}$, after combining like terms?

(13) An international committee consists of nine (or n) members. Committee materials are stored in a safe. How many "locks" should the safe have, how many "keys" should be made for these locks, and how must these keys be distributed among the committee members, if the safe can be opened if and only if at least six (or k) members of the committee are present? Can you answer for general n and k?

A public aquarium has ten species of aquatic life that it wishes to feature in large tanks. However, because some of the species prey on others, they cannot all live in the same tank. Below we label the species from *A* to *J* and show each one's "enemy" species that cannot be placed in the same tank with this species. What is the minimum number of tanks the aquarium needs to accommodate all ten species?

Species	Enemies	Species	Enemies
A	B, E, H, J	F	E, I
B	A, C, G, I	G	B, D, J
C	B, D	H	A, E
D	C, E, G, J	I	B, F
E	A, D, F, H, J	J	A, D, E, G

9

Graph Theory

Thus you see, most noble Sir, how this type of solution to the Königsberg bridge problem bears little relationship to mathematics, and I do not understand why you expect a mathematician to produce it, rather than anyone else, for the solution is based on reason alone, and its discovery does not depend on any mathematical principle....

—Leonhard Euler (1707–1783)

In "continuous" mathematics such as algebra, trigonometry, and calculus, a *graph* is often defined to be the set of ordered pairs that satisfy an equation. Less formally, a graph can refer to the visual representation obtained by plotting the ordered pairs in a Cartesian plane. Graphs have proven to be robust, yet straightforward, mechanisms for representing and organizing complex relationships. For example, graphs may display two-way transit between locations, such as cities, intersections, airports, or network nodes, via bridges, flights, or network connections. In other applications, graphs may represent relationships—similarities or differences—between people, products, or characteristics.

Graphs are simply examples of relations as defined in Chapter 5. The distinctive nature of "discrete" graphs can be illustrated by the digraphs (short for directed graphs) defined in Section 5.1. For a relation R on a set A, each element of A is indicated by a labeled point (or vertex); the ordered pair $(a, b) \in R$, for $a, b \in A$, is represented by an edge drawn from a to b. Also recall that if R is a symmetric relation, whenever there is an edge from a to b, there must also be an edge from b to a. For graphs of these symmetric relations, we replace each "double edge" with a single undirected arc or line segment. In discrete mathematics, a *graph* will always refer to the visual representation of the symmetric relation just described.

9.1 The Language of Graphs

Graph theory provides tools to sort and analyze the connections between items and collections of items. In preparation for presenting these tools, some common language needs to be developed, which will facilitate concise communication.

9.1.1 Introductory Terminology. A **graph** G is a collection V of objects, called **vertices**, along with a collection E of unordered pairs of distinct vertices from V, called **edges**. V is called the **vertex set** of G, and E is called the **edge set** of G; they are sometimes denoted $V(G)$ and $E(G)$, respectively. We often write $G = (V, E)$ to emphasize that a graph is completely defined by its vertices and edges.

Example 107. Let $G = (V, E)$ where $V = \{A, B, C, D, E\}$ and $E = \{\{A, B\}, \{B, D\}, \{B, E\}, \{D, E\}\}$. The graph contains five vertices and four edges and may be drawn as in Figure 9.1.

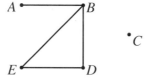

Figure 9.1. Graph G for Example 107

By convention, for edge $\{v, w\} \in E$, we require that vertices v and w be distinct. Therefore, no loops from a vertex to itself will appear in a graph, or from the viewpoint of Section 5.2, the associated relation must be irreflexive. In some circumstances, there is a restriction that at most one edge can link any pair of vertices of G; in this case, G is called a **simple graph**. This is the case in Example 107. If a situation calls for multiple edges joining a pair of vertices of G, then G is called a **multigraph**.

For vertices $v, w \in V(G)$, if v and w are linked by an edge—equivalently, $\{v, w\} \in E(G)$—we say that v and w are **adjacent**. If $\{v, w\} \notin E(G)$, we say that v and w are **nonadjacent**. In Example 107, A and B are adjacent, but A and D are nonadjacent. We often use the phrase "A is adjacent to B", with the understanding that this implies that B is adjacent to A. Vertex C in Example 107 is not adjacent to any other vertices; such a vertex is said to be **isolated**.

The **degree** (or **valence**) of a vertex is the number of edges containing that vertex. For a simple graph, this is the number of vertices adjacent to that vertex. The degree of vertex v in a graph G is denoted $\deg_G(v)$, or simply $\deg(v)$ when the context is clear. In Example 107, B is adjacent to A, D, and E; hence $\deg(B) = 3$. Isolated vertices, such as C, have degree 0. The list of degrees of vertices is an important feature of a graph, useful for both theory and applications of graphs. Beware that if one were to extend the notion of a graph to include loops at a vertex, the definition of degree would need to be modified.

Most concepts in elementary graph theory are easily described using the basic definitions given above, along with a heavy reliance on set theory language. For example, if v and w are adjacent vertices, they are sometimes called **neighbors**. The **neighborhood** of a vertex v in a graph G is the set $N(v)$ of all vertices in $V(G)$ with which v is

adjacent:
$$N(v) = \{w \in V(G) : \{v, w\} \in E(G)\}.$$
In Example 107, $N(B) = \{A, D, E\}$, while $N(C) = \emptyset$. Notice that in general $|N(v)| = \deg(v)$. The concept of neighborhood has obvious meaning in graphs arising from geographic applications, but it is also meaningful for distinguishing the closest relatives of an item in other contexts. Graphs for which there is regularity in the size of the neighborhoods have special interest. A graph G is said to be **regular** if $\deg(v)$ is the same number for each $v \in V(G)$. Clearly, the graph in Figure 9.1 is not regular. In Figure 9.2, each vertex has degree 4, making the graph regular.

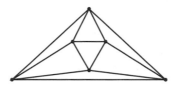

Figure 9.2

The connection to set theory also provides inspiration for producing new graphs from a given graph. For instance, consider what might be meant by a "subset" of a graph. A **subgraph** of a graph $G = (V, E)$ is any graph $G_0 = (V_0, E_0)$ where $V_0 \subseteq V$ and $E_0 \subseteq E$. There is complete freedom in selecting vertices and edges from G for the subgraph, provided that every edge selected for E_0 has its "endpoints" likewise selected for V_0. Figure 9.3 exhibits G and two of its many possible subgraphs.

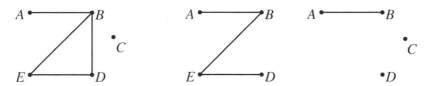

Figure 9.3. Graph G and two of its subgraphs

Another idea inspired by set theory is to interchange adjacent and nonadjacent pairs of vertices. For graph $G = (V, E)$, let the **complement** of G, written $\overline{G}$, denote the graph with vertex set V and edge set $\overline{E}$, which is the complement (as in Section 3.4) with respect to the universe of all unordered pairs of distinct vertices in V. Figure 9.4 reproduces the graph from Example 107 and compares it with its complement.

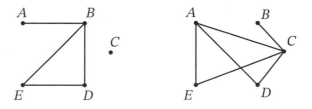

Figure 9.4. Graph G and its complement $\overline{G}$

9.1.2 Families of Graphs. There are four families of graphs—complete graphs, null graphs, paths, and circuits—that will be used to demonstrate concepts and provide ready examples throughout the chapter. They include graphs with the least and greatest possible numbers of edges along with graphs having common, basic shapes.

Let K_n denote the graph with $n \geq 1$ vertices and every pair of vertices adjacent; K_n is called the **complete graph** on n vertices. The name indicates that no more edges could be drawn in a simple graph. Such a graph is also called a **clique**, though this name is usually used to denote subgraphs that are complete. For example, the triangle subgraph on three vertices all adjacent to each other is called a 3-*clique*. It should be evident that every simple graph on n vertices will be a subgraph of K_n, including the three remaining families of graphs to be discussed.

What is the maximum number of edges in a simple graph with n vertices? Based on the examples in Figure 9.5, and others that we hope you will choose to draw, make a conjecture about the number of edges in K_n (a formula involving n).

Figure 9.5. K_4 and K_5

Notice that K_n is a regular graph, as each vertex has degree $n - 1$. In the accompanying exercises, we explore a variety of ways to show that K_n has $\frac{n(n-1)}{2}$ edges. The graph K_n appears in one solution to the so-called Handshaking Problem: *How many handshakes occur in a group of n people, where each person shakes hands one time with each other person* (see Exercise 6). In this scenario, each vertex of K_n represents a person and each edge represents a handshake.

The complement of a complete graph has no edges. Therefore, $\overline{K}_n$ is a graph consisting of n isolated vertices. Graphs in this family are sometimes called **null** graphs; examples appear in Figure 9.6.

Figure 9.6. $\overline{K}_3$ and $\overline{K}_4$

For $n \geq 1$, let P_n denote a **path** graph, in which $n - 1$ edges link n vertices in the manner pictured in Figure 9.7.

Figure 9.7. P_3 and P_4

More precisely, P_n is the graph where the vertices can be labeled $v_1, v_2, \dots, v_n$ so that

$$E(P_n) = \{\{v_k, v_{k+1}\} : 1 \leq k \leq n - 1\}.$$

In an arbitrary graph, a subgraph that is a path with no repeated vertices is called a **simple path** . This will be the main focus of the next section.

For $n \geq 3$, let C_n denote a member of the family of **circuit** graphs; each contains n vertices and n edges, as pictured in Figure 9.8. The vertices can be labeled in such a way that $E(C_n) = E(P_n) \cup \{\{v_1, v_n\}\}$.

Figure 9.8. C_3 and C_4

9.1.3 Connected Graphs.
The complete graphs, circuit graphs, and path graphs have the noticeable trait that each graph is "in one piece". Specifically, there is a path between any two vertices of the graph. Such a graph G is said to be **connected**. A graph that has two vertices not connected by a path is said to be **disconnected**. The number of connected pieces of a disconnected graph is an important property of the graph. Formally, the **components** of a graph G are the **maximally connected** subgraphs of G; i.e., no connected subgraph of G contains a component as a proper subgraph. As a trivial example, $\overline{K}_n$ is a disconnected graph with n components. For "untangled" graphs, connectedness usually can be determined by a quick glance rather than examining paths between vertices. For example, the graph in Figure 9.9 is disconnected with three components.

Figure 9.9. A graph with three components

A **cut edge** of a graph is an edge whose removal from the graph's edge set increases the number of components of the graph. The graph in Figure 9.9 has three cut edges, namely each edge in the middle component. Cycle graphs do not have cut edges, because the cycle is a single component and the removal of an edge does not disconnect the graph.

9.1.4 Graph Isomorphisms.
For some examples in this section, the vertices are named with letters or numbered subscripts, while other examples leave the vertices unlabeled. A simple graph in which each vertex is assigned a distinct label is called a **labeled graph**, and a graph with no labels assigned to vertices is called **unlabeled**. Labeled graphs fulfill the obvious need for identifying vertices, and therefore edges, as evidenced by the presentation of certain concepts in this section. Throughout the chapter, one must stay alert and detect whether theorems and exercises refer to labeled graphs or unlabeled graphs, particulary when the result involves counting (see Problem 22, for instance). One immediate combinatorial observation is that the n vertices of a graph can be assigned n distinct labels in $n!$ ways.

Labels also provide a means to formally define the notion of two graphs being structurally similar. For instance, it is clear that graph E in Figure 9.10 has the same structure as C_5. For a thorough justification, recall that $V(C_5) = \{v_1, v_2, v_3, v_4, v_5\}$ and $E(C_5) = \{\{v_1, v_2\}, \{v_2, v_3\}, \{v_3, v_4\}, \{v_4, v_5\}, \{v_5, v_1\}\}$, and consider the correspondence: $v_1 \rightarrow I$, $v_2 \rightarrow L$, $v_3 \rightarrow J$, $v_4 \rightarrow H$, $v_5 \rightarrow K$. The key issue is to verify that any two vertices that are adjacent in C_5 will correspond to vertices that are adjacent in graph E, and conversely. Notice that the graphs in Figure 9.10 all look different at first glance. However, there are actually three pairs of graphs for which the graphs in each pair are essentially "the same", other than the placement or labeling of the vertices. Can you find them? We will identify the three pairs after providing the definition of what we mean by "the same".

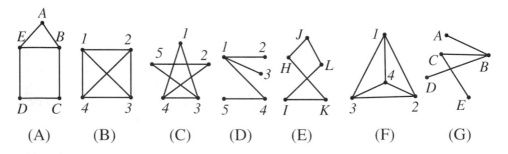

(A) (B) (C) (D) (E) (F) (G)

Figure 9.10. Seven graphs to consider

We define simple graphs G_1 and G_2 to be **isomorphic**—meaning "same structure" —if there is a one-to-one correspondence between $V(G_1)$ and $V(G_2)$ such that a pair of vertices in $V(G_1)$ are adjacent in G_1 if and only if the corresponding pair of vertices in $V(G_2)$ are adjacent in G_2. When G_1 and G_2 are isomorphic, we write $G_1 \cong G_2$. The discussion above shows that $C_5 \cong E$ in Figure 9.10.

Example 108. Graphs B and F in Figure 9.10 have labels assigned to the vertices in such a way that $\{v_i, v_j\} \in E(B)$ if and only if $\{v_i, v_j\} \in E(F)$. Therefore, $B \cong F$.

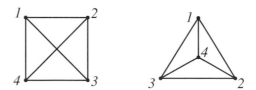

Graphs B and F in Figure 9.10 are isomorphic

Notice that B is the usual rendering of the complete graph K_4. Any graph isomorphic to a complete graph must also be a complete graph.   □

When two graphs are isomorphic, the number of vertices, number of edges, number of 3-cycles, length of longest path, number of components, degree sequence, etc., must be identical. If two graphs fail to share any structural property, then they are not isomorphic. However, to formally prove that two graphs are isomorphic, some type of adjacency-preserving correspondence between vertices must be demonstrated.

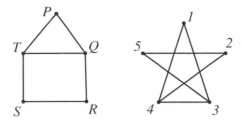

Graphs A and C are isomorphic

Graphs A and C from Figure 9.10 are isomorphic, as are graphs D and G. One might be able to "see" this by looking past the superficial differences in the graphs, but a rigorous justification would demonstrate an edge-preserving bijection between vertices of the two graphs. One such function for the former pair would be $\phi(P) = 1$, $\phi(Q) = 3$, $\phi(R) = 5$, $\phi(S) = 2$, and $\phi(T) = 4$, though there are others. Notice that adjacency is preserved in this bijection; for example, $\{P, Q\}$ is an edge in graph C, and $\{\phi(P), \phi(Q)\} = \{1, 3\}$ is an edge in graph C. You will be asked to prove that graphs D and G are isomorphic in Exercise 15.

Example 109. In Figure 9.11, G_1 (left graph) has an isolated vertex, while the degree of each vertex in G_2 (right graph) is at least one. For any one-to-one correspondence between the vertices of G_1 and G_2, the isolated vertex in G_1 will correspond to a vertex in G_2 that has a neighbor. Since adjacency cannot be preserved as required by the definition, G_1 and G_2 are not isomorphic. In general, a disconnected graph cannot be isomorphic to a connected graph.   □

Figure 9.11. Nonisomorphic graphs

9.1.5 Adjacency Matrices. In Chapter 5, the matrix representation of a relation was shown to have certain advantages for computer storage and for computation of various features of the relation. The same is true for graphs; a matrix is the primary mechanism for storing and processing the information in a graph. The **adjacency matrix**, or simply the **matrix**, of a simple graph G having n vertices, $v_1, v_2, \ldots, v_n$, is the $n \times n$ matrix with (i, j)-entry equal to 1 if v_i and v_j are adjacent, and 0 if v_i and v_j are nonadjacent. It is denoted M_G. As G represents a symmetric relation, M_G must be a symmetric matrix (the (i, j)-entry must equal the (j, i)-entry). The convention prohibiting loops at a vertex implies that the diagonal entries of the matrix must all be 0; i.e., the (i, i)-entry is 0 for $i = 1, 2, \ldots, n$.

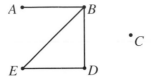

Figure 9.12. Graph G

For the graph in Figure 9.12, the matrix is

$$M_G = \begin{bmatrix} 0 & 1 & 0 & 0 & 0 \\ 1 & 0 & 0 & 1 & 1 \\ 0 & 0 & 0 & 0 & 0 \\ 0 & 1 & 0 & 0 & 1 \\ 0 & 1 & 0 & 1 & 0 \end{bmatrix},$$

noting that the rows and columns of the matrix are indexed alphabetically (A through E). Notice that the number of 1's in M_G is twice the number of edges in G. With a few moments' thought, it will be clear that $M_{\overline{K}_n}$ is an $n \times n$ matrix consisting of all 0's, while the entries of M_{K_n} are all 1's except for the 0's on the diagonal. The matrix of P_4, with vertices labeled in a natural order, is

$$M_{P_4} = \begin{bmatrix} 0 & 1 & 0 & 0 \\ 1 & 0 & 1 & 0 \\ 0 & 1 & 0 & 1 \\ 0 & 0 & 1 & 0 \end{bmatrix}.$$

The degree of a vertex of G is easily computed from M_G; simply add the entries in the corresponding row (or column, by symmetry). Therefore, the sum of degrees of all vertices in G will equal the total number of 1's in M_G. But each edge of G is represented twice in M_G due to symmetry; if vertex i is adjacent to vertex j, then vertex j is adjacent to vertex i. Here is a formal statement and proof of this result.

Theorem 9.1. *Suppose G is a graph with n vertices, $v_1, v_2, \ldots, v_n$, and e edges. Then*

$$\sum_{k=1}^{n} \deg(v_k) = 2e;$$

i.e., the sum of the degrees of all vertices is twice the number of edges.

Proof. For each edge $\{v, w\} \in E(G)$, the edge is counted once in $\deg(v)$ and once in $\deg(w)$. Therefore, the sum of degrees double-counts each edge of G. □

The following useful corollary is obtained by recalling the elementary ideas in Theorem 2.4. Its proof is left to Problem 21.

Corollary 9.2. *In any graph, there must be an even number of vertices with odd degree.*

Exercises and Problems.

(1) List the degree of each vertex in P_4 and P_6.

(2) The graph in Example 107 is not a complete graph. How many additional edges would need to be added to make it a complete graph?

(3) Explain why any graph with a vertex of degree 1 has a cut edge.

(4) For what value(s) of n will K_n have a cut edge?

(5) Which of the following families consist of regular graphs: $\overline{K}_n, K_n, C_n, \overline{C}_n, P_n, \overline{P}_n$?

(6) Each part of this exercise explores a different way to determine the number of edges in K_n.

    (a) Make direct use of the fact that the degree of each vertex in K_n is $n-1$ to show that K_n has $\frac{n(n-1)}{2}$ edges.

    (b) Show that $|E(K_n)| = C(n, 2)$. See Section 8.3.

    (c) Argue that the number of edges in K_n is $\frac{n(n-1)}{2}$ by counting the number of 1's in the adjacency matrix of the graph K_n by means of a summation.

    (d) Explain why the number of edges in K_n provides an answer to the Handshaking Problem: How many handshakes occur in a group of n people, where each person shakes hands one time with each other person?

(7) Suppose G is a simple graph with n vertices and m edges. How many vertices and edges will its complement $\overline{G}$ have?

(8) For a simple graph G on n vertices, describe the matrix $M_G + M_{\overline{G}} + I_n$.

(9) Consider a graph G with matrix

$$M_G = \begin{bmatrix} 0 & 1 & 1 & 1 & 1 \\ 1 & 0 & 0 & 1 & 0 \\ 1 & 0 & 0 & 1 & 0 \\ 1 & 1 & 1 & 0 & 0 \\ 1 & 0 & 0 & 0 & 0 \end{bmatrix}$$

with vertices labeled $v_1, v_2, \ldots, v_5$.

    (a) Find $\deg(v_1)$.

    (b) Determine the total number of edges in G.

    (c) Draw the labeled graph.

(10) Assuming a natural ordering of the vertices in C_5, such as clockwise, write down the matrix M_{C_5}.

(11) A **complete bipartite** graph, denoted $K_{m,n}$, consists of $m + n$ vertices, $A_1, A_2, \ldots,$ $A_m, B_1, B_2, \ldots, B_n$, with adjacency defined by the following neighborhoods: $N(A_i) = \{B_1, B_2, \ldots, B_n\}$ for $1 \leq i \leq m$ and $N(B_j) = \{A_1, A_2, \ldots, A_m\}$ for $1 \leq j \leq n$. Examples appear in Figure 9.13, where the A_i and B_j comprise the top and bottom rows of vertices, respectively.

    (a) How many edges does $K_{m,n}$ have?

    (b) What are the degrees of the vertices in $K_{m,n}$?

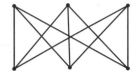 

Figure 9.13. $K_{3,3}$ and $K_{4,2}$

(12) Draw $\overline{K}_{3,3}$, the complement of $K_{3,3}$.

(13) Write a one-sentence proof that the degree of every vertex in K_n is $n-1$. Focus on the definitions of degree and complete graph.

(14) Show that C_4 and $K_{2,2}$ are isomorphic.

(15) Show that graphs D and G in Figure 9.10 are isomorphic.

(16) Find two nonisomorphic graphs on eight vertices for which each vertex has degree 2. This shows that two graphs with the same degree sequence are not necessarily isomorphic.

---

(17) How many labeled subgraphs does P_4 have?

(18) Construct a simple graph on six vertices and eleven edges that has a cut edge. Can you accomplish the task if twelve edges are present?

(19) Show that C_5 and $\overline{C}_5$ are isomorphic.

(20) Suppose $n \geq 3$. For what value(s) of n is C_n isomorphic to $\overline{C}_n$? Explain.

(21) Prove Corollary 9.2 using Theorem 9.1 and Theorem 2.4.

(22) Consider the problem of determining the total number of distinct, labeled graphs that can be drawn on n vertices. For instance, if $n = 2$, only two graphs are possible (K_2 and $\overline{K}_2$). For $n = 3$, there are 8 different graphs that can be drawn with vertices A_1, A_2, and A_3; we suggest that you draw them. Several different methods for determining this number are suggested below.

(a) For any graph G on n vertices, G is a subgraph of K_n. Hence, $E(G) \subseteq E(K_n)$. Reason that the number of graphs on n vertices equals the number of subsets of $E(K_n)$, which is $|\mathcal{P}(E(K_n))|$. Compute this number.

(b) As in part (a), any graph G on n vertices consists of some of the edges in K_n. Hence, for *each* edge found in K_n, one need only ask whether or not (two choices) that edge occurs in G. Conclude that the number of graphs on n vertices can be determined using the Multiplication Rule of Chapter 8, and determine this number.

(c) In a slight modification of the counting in part (b), consider the $n \times n$ adjacency matrix of 0's and 1's for any graph G on n vertices. Following the direction of Problem 6, as the matrix must be symmetric, this leaves fewer than half of the entries of the matrix for which there are two choices each (0 or 1). Determine the number of entries in question and apply the Multiplication Rule to obtain the number of graphs. (See Exercise 6(c).)

(23) Determine the number of labeled subgraphs of K_n.

(24) Prove that if graphs G_1 and G_2 are isomorphic, then their complements $\overline{G}_1$ and $\overline{G}_2$ are isomorphic.

(25) How many nonisomorphic graphs are there with ten vertices and 45 edges? How many with ten vertices and 43 edges?

9.2 Traversing Edges and Visiting Vertices

From the origins of graph theory to the present day, many applications of graphs have been geographic in nature. Vertices represent locations or regions to visit, while edges represent the means for people, planes, packets of digital information, etc., to travel between vertices. In this section, we find that applications and techniques differ significantly depending on whether the emphasis is on arriving at the destinations or on traversing the thoroughfares that link the locations.

9.2.1 Terminology. Recall that a **path** in a simple graph G is a sequence of vertices $v_1, v_2, \dots, v_{k+1}$ such that $\{v_i, v_{i+1}\} \in E(G)$ for $1 \le i \le k$. This path is said to have **length** k, indicating the number of edges in the path. No edge may be repeated in a path, and consecutive vertices must be distinct, but no other restriction is made on repeated vertices in a path.[1] If $v_1, v_2, \dots, v_{k+1}$ are distinct vertices, it is called a **simple path**. A closed path, meaning $v_1 = v_{k+1}$, is called a **circuit**. A circuit with no repeated vertices, except $v_1 = v_{k+1}$, is often called a **cycle**, and if the cycle has length k, it is called a k-**cycle**.

For the graph in Figure 9.14, vertices represent cities and edges represent round-trip flights between cities. An example of a path of length two from Dallas to NY is Dallas, Chicago, NY. A circuit of length four starting and ending at LA is LA, Chicago, NY, Miami, LA. Each of these paths is simple.

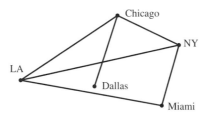

Figure 9.14

When describing paths in multigraphs, a sequence of vertices is not sufficient, as there is ambiguity regarding which edge joining the pair of vertices is to be traversed. In this case, vertices and edges of the multigraph G must be labeled, and a path has the form $v_1, e_1, v_2, e_2, \dots, v_k, e_k, v_{k+1}$ where $e_i = \{v_i, v_{i+1}\} \in E(G)$ for $1 \le i \le k$. Historically, the first use of graph theory was a multigraph in which a special path was desired.

---

[1] In some literature, a sequence of adjacent vertices with no repeated edges and no loops is called a **trail**. Other sources refer to any finite sequence of vertices linked by edges as a **walk** or a **chain**. Still others reserve **path** and **cycle** for directed graphs; see Chapter 5. We suggest that you honor your instructor's preferences for this terminology.

9.2.2 Euler Paths and Circuits. The story is told in mathematical lore of a town in 18th-century Prussia named Königsberg. Parts of the town were situated on both sides of the Pregel River, as well as on two islands in the river. The islands and river banks were linked by seven bridges in the manner shown in Figure 9.15. The story goes that residents of Königsberg enjoyed daily strolls about town with a particular interest in crossing the ornate bridges. Having competitive spirits, they undertook the challenge of crossing each of the seven bridges without recrossing any bridge (and staying out of the river!). Furthermore, the residents of Königsberg wondered, if they must traverse each bridge exactly once, could it be done in such a way that they ended on the same landmass on which they began their walk? If so, did their success depend on where the walk began? If the *Königsberg Bridge Problem* is new to the reader, experiment with tracing a few paths across land and bridges to see if this can be accomplished. For instance, the following path crosses all but one of the bridges: $A, a, B, b, A, c, C, d,$ A, e, D, f, B.

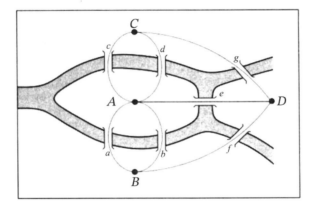

Figure 9.15. The Königsberg Bridges

The hero of this story is the Swiss mathematician Leonhard Euler (pronounced "oiler"), arguably the greatest mathematician of his time. With some imagination, one can picture him riding into Königsberg on a shiny steed wielding a novel, mathematical weapon and slaying this problem, to the relief and admiration of the citizens. No less exciting, Euler abstracted the essential characteristics of the land and bridges, thus creating what we now call a graph. Each landmass is represented by a vertex, as if the island or riverbank shrunk to a single point. As each bridge connects two landmasses, edges in the graph represent bridges. The multigraph formed by the Königsberg bridges is drawn in Figure 9.16 with the edges (bridges) labeled $a, b, \ldots, g$ and vertices (landmasses) labeled A, B, C, and D. Euler determined elementary criteria for whether a graph can be drawn without retracing an edge; such paths and circuits are named in his honor.

An **Euler path** in a graph G is a path that uses every edge of G. It necessarily traverses every edge exactly once, and it may or may not revisit vertices. An **Euler circuit** is a circuit that uses every edge of G; i.e., it traverses each edge exactly once and ends at the starting vertex. A graph G is said to be **Eulerian** if it contains an Euler circuit that includes all vertices of G. Thus, Eulerian graphs are necessarily connected, though this is not true of all graphs containing an Euler circuit.

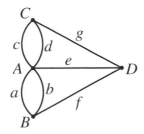

Figure 9.16. The Königsberg Bridge Graph

In Figure 9.17(a), A, B, C, D, B, E, A is an Euler path and an Euler circuit. In Figure 9.17(b), E, A, B, C, D, B, E, D is an Euler path, but not an Euler circuit. Figure 9.17(c) contains no Euler paths. In considering the families of graphs defined in Section 9.1, it is clear that the path graph P_n for $n \geq 2$ contains an Euler path and the circuit graph C_n for $n \geq 3$ contains an Euler circuit. The reader should verify that K_5 is Eulerian, while K_4 does not contain an Euler path, let alone an Euler circuit.

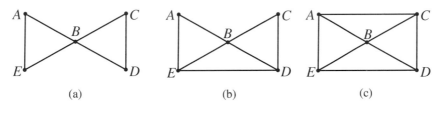

Figure 9.17

It is easy to envision real-world scenarios in which Euler paths would appear useful. If edges indicate roads or streets, with vertices representing intersections, an Euler path provides an efficient means for snow plows, street sweepers, or postal carriers to cover every road. An Euler circuit has the added benefit of returning to the point of origin without retracing any part of the route. In childhood puzzles and certain artistic ventures, the goal is to trace a figure without lifting your pencil or retracing any part of the figure. If the goal can be achieved, as in Figure 9.2, the figure is called a *unicursive drawing*.

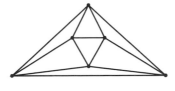

Figure 9.18

In Figure 9.18, the connection to graph theory and Euler paths is obvious. In other figures with curved or jagged lines, simply place a vertex at each intersection and define the connecting lines to be edges regardless of their shapes. In each of these applications, there is an emphasis on traveling or tracing the pathways with little concern for arriving at intermediate destinations.

Euler's criterion for determining the existence of an Euler path or circuit in a graph is easy to state and implement, since it depends only on whether degrees of vertices are even or odd integers. We will illustrate the process for finding an Euler circuit shortly, but the formal proof is somewhat technical, and we delay it until Section 9.5.

Theorem 9.3 (Euler's Theorem). *Let G be a connected graph on at least two vertices.*

(1) *If G contains no vertices with odd degree, then G contains an Euler circuit and G is Eulerian.*

(2) *If G contains exactly two vertices with odd degree, then G contains an Euler path that begins and ends at these two vertices.*

(3) *If G contains more than two vertices with odd degree, then G does not contain an Euler path (or an Euler circuit).*

From Corollary 9.2, no graph can have only one vertex with odd degree; thus all cases for a connected graph have been accounted for in the theorem. Furthermore, an Euler circuit in a connected graph utilizes all vertices, so such a graph is necessarily Eulerian. Thus, we can summarize Theorem 9.3 as follows:

Theorem 9.3 (restated). A connected graph on at least two vertices is Eulerian if and only if all vertices have even degree, and it contains an Euler path (but not an Euler circuit) if and only if exactly two vertices have odd degree.

In Figure 9.17, the degree of each vertex in (a) is even; hence an Euler circuit is present. The degree of each vertex in Figure 9.17(b) is even except for vertex D and vertex E, which are the endpoints of the Euler path. In Figure 9.17(c), there are four vertices with odd degree, thus eliminating the possibility of an Euler path. The denizens of Königsberg also have a definitive answer to their bridge-traversing question. The degree of each of the four vertices in Figure 9.16 is odd; there is no Euler path. There are an odd number of bridges touching each landmass; hence they cannot cross each bridge exactly once. The citizens of Königsberg never stood a chance.

Let us now develop a procedure for finding an Euler circuit when every vertex has even degree. Our explanation will also lead us to the formal proof of Theorem 9.3, which is given in Section 9.5. The approach is to start at any vertex, say v_1, and build a cycle α_0 by consecutively choosing a neighbor, via an unused edge, of the current vertex in the sequence being built. Since each vertex has even degree, once a vertex is entered via an edge, there must be an unused edge available for leaving the vertex, to continue the cycle, until eventually arriving back at v_1. In the leftmost graph of Figure 9.19, this first cycle, α_0, is indicated by the vertex sequence $(v_1, v_2, v_3, v_4, v_5, v_1)$. If α_0 is itself an Euler circuit, we are done. If not, proceed to the first vertex in α_0 that has degree greater than 2 (because this vertex will have some incident edges that were not part of α_0) and repeat the above process, again choosing unused edges, to obtain another circuit, α_1. In Figure 9.19, this vertex is v_2, and $\alpha_1 = (v_2, B, C, v_3, D, v_2)$; now replace v_2 in α_0 with the vertices in α_1 to create a longer circuit of distinct edges.

We then proceed around the vertices of α_0, continuing to take "detours" as necessary, until arriving back at v_1. Each such detour will provide a cycle α_i to replace a vertex of α_0. Notice that the detour from vertex v_4 in Figure 9.19 might follow the cycle v_4, E, F, H, v_4, leaving vertices I and J unused, or the detour could follow the longer cycle $\alpha_2 = v_4, E, F, H, E, I, J, H, v_4$. Either cycle is fine, as one should notice the recursive nature of this process will lead to all edges appearing in an eventual Euler circuit. (We will address this recursive part more carefully by using induction in the formal proof in Section 9.5.) In Figure 9.19, the Euler circuit found by this process is

$$v_1, v_2, B, D, v_3, D, v_2, v_3, v_4, E, F, H, E, I, J, H, v_4, v_5, v_1$$

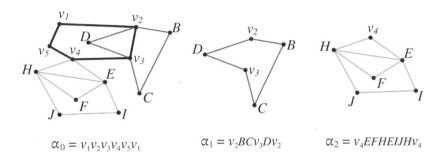

$$\alpha_0 = v_1 v_2 v_3 v_4 v_5 v_1 \qquad \alpha_1 = v_2 BC v_3 D v_2 \qquad \alpha_2 = v_4 EFHEIJH v_4$$

Figure 9.19

9.2.3 Hamilton Paths and Circuits.
In contrast to Eulerian graphs, where the emphasis is on edges and importance is given to every pair of adjacent vertices, we now turn the focus to visiting all of the vertices in a graph. The setting is still geographic, and edges are certainly the means for traveling between vertices; however, reaching each destination is stressed. This time the historical figure is Sir William Rowan Hamilton (1805–1865). In 1857, he invented a popular puzzle named the "Icosian Game" in which the players' goal was essentially to visit each vertex (representing cities of the world) of the dodecahedral graph exactly once and return to the starting city. Such paths now bear his name.

A **Hamilton path** in a graph G is a path that visits every vertex of G exactly once. Alternatively, a Hamilton path can be described as a simple path containing every vertex of G. A **Hamilton cycle** of G is a cycle containing every vertex of G. Since a Hamilton cycle is a circuit that visits every vertex exactly once, accepting that the final vertex is the starting vertex, it is sometimes referred to as a **Hamilton circuit**.

Notice that the path graph P_n contains a Hamilton path for $n \geq 2$, while C_n and K_n contain Hamilton cycles when $n \geq 3$. In Figure 9.20(a), the graph contains a Hamilton path (A, E, B, C, D) but does not contain a Hamilton cycle due to the need to pass through B multiple times. The graph in Figure 9.20(b) contains the Hamilton cycle A, B, C, D, E, A. Notice that it does not matter if some edges are untraversed. The graph in Figure 9.20(c) contains several Hamilton circuits (starting and ending at A) and reinforces the notion that a greater number of edges facilitates greater mobility in visiting vertices.

Applications of Hamilton paths and cycles are widespread. When vertices are cities and edges are transportation routes, a Hamilton path is a means for a music group to tour each city, a traveling salesperson to meet each customer, or an inspector to visit each site. In a street network, an Eulerian example might involve snow plows needing to traverse every street, while a Hamiltonian example might be a taxi driver using just

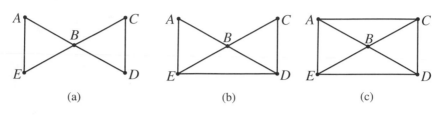

Figure 9.20

sufficiently many streets to deliver a tourist to all of the points of interest. For each use of Hamilton paths, the emphasis is on the locations, not on traveling all of the routes.

Any graph admitting a Hamilton path or cycle must be connected. Furthermore, in a graph with n vertices, any Hamilton path consists of n vertices connected by $n - 1$ edges, and any Hamilton cycle is a sequence of $n + 1$ vertices (with the first and last being the same vertex) connected by n edges. In general, determining whether a graph contains a Hamilton path is significantly more difficult than determining whether it contains an Euler path. The following standard theorems provide sufficient conditions for a simple graph to contain a Hamilton cycle. Each result guarantees the existence of a Hamilton cycle based on an abundance of edges, thus allowing the freedom of movement needed to visit every vertex. With such results we anticipate all converses to be false, and this should be readily established with a few doodles of graphs on scrap paper. Finally, it is interesting to note that the hypotheses depend on the degrees of vertices, as was the case for Euler circuits. As before, the proof is somewhat technical, and we delay it until Section 9.5.

Theorem 9.4 (Ore's Theorem). *Suppose G is a simple graph with n vertices, n $\geq$ 3. If for every pair of nonadjacent vertices v_1 and v_2 of G, $\deg(v_1) + \deg(v_2) \geq n$, then G contains a Hamilton cycle.*

Notice that the sum of the degrees of any two nonadjacent vertices in C_4, a graph on $n = 4$ vertices, is four, so by Theorem 9.4, C_4 contains a Hamilton cycle. (Of course this fact is obvious without using the theorem.) However, for any cycle on $n \geq 5$ vertices, the degree sum of any two nonadjacent vertices is still four, which is less than n, even though C_n has (is) a Hamilton cycle. Thus, C_n provides a counterexample to the converse of Ore's Theorem when $n \geq 5$.

The following theorem is much easier to apply, but its hypothesis is more restrictive.

Corollary 9.5. *For a simple graph G with n vertices, n $\geq$ 3, if the degree of each vertex is at least n/2, then G contains a Hamilton cycle.*

This is an immediate corollary of Theorem 9.4 since graphs that satisfy the hypothesis of Corollary 9.5 must satisfy the hypothesis of Theorem 9.4. For the four-cycle, each vertex has degree $n/2 = 2$, which confirms that C_4 contains a Hamilton cycle. Neither graph on five vertices in Figure 9.21 satisfies the hypothesis of Corollary 9.5; hence the corollary is silent on whether they admit Hamilton cycles. Here again, C_5 serves as a counterexample to the converse of Corollary 9.5.

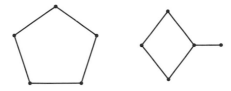

Figure 9.21

A Hamilton cycle can also be ensured if the graph has enough edges, without specifically referring to the degrees of any vertices.

Corollary 9.6. *For fixed $n \geq 3$, let $m = \frac{1}{2}\left(n^2 - 3n + 6\right)$. If G is a simple graph with n vertices and at least m edges, then G contains a Hamilton cycle.*

We omit the proof of this result, though in Problem 13 you are asked to show that this result is the best possible by constructing a graph on n vertices and $m - 1$ edges that does not have a Hamilton cycle.

When there is a choice of edges in forming a Hamilton cycle, there is flexibility to select a route that is optimal in some desired sense. For instance, if one knew the costs for traveling on individual edges, then the least expensive Hamilton cycle would be sought. A **weighted graph** has a number assigned to each edge, called the **weight** of that edge. The weight could describe the cost or distance for traversing that edge. For instance, in the weighted graph of Figure 9.22, vertices could represent cities with weights indicating the cost of a one-way bus ticket between pairs of cities. Weighted graphs are well suited for representing symmetric relations where the cost of the adjacency is highlighted. They will be employed in diverse ways in this chapter.

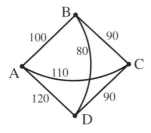

Figure 9.22. A weighted K_4 graph

A **minimum Hamilton cycle** of a weighted graph with n vertices is a Hamilton cycle for which the sum of the n weights of the edges forming the cycle is the smallest possible. A minimum Hamilton cycle is not uniquely determined; this can be due to the ordering of vertices based on the choice of starting point or by reversing the direction. We typically fix the starting (and ending) vertex and consider the forward and reverse orientation of the same edges to be equivalent, but Hamilton cycles using different edges can also have the same minimum total cost.

Applications involving minimum Hamilton cycles are often likened to the setting of a salesperson needing to visit every city in a territory exactly once for the least total cost; hence they are referred to as *Traveling Salesperson Problems*. In Figure 9.22, where vertices are cities and weights are the cost of a bus ticket, suppose we require the salesperson to start at city A, visit the other three cities, and return home for least cost. The reader should verify that the Hamilton cycle A, B, D, C, A has total cost 380, while all other Hamilton cycles except the reverse order have total cost 400. Therefore, A, B, D, C, A is a minimum Hamilton cycle.

In Chapter 13, we examine algorithms for achieving a minimum Hamilton cycle in a complete graph. For the small example in Figure 9.22, we essentially employed the most labor-intensive method, the Brute-Force Algorithm, which considers all possible Hamilton cycles, computes the total weight of each, and selects a Hamilton cycle with smallest total weight.

Exercises and Problems.

(1) Which would a tour guide find more useful: an Euler circuit or a Hamilton circuit?

(2) List all Hamilton cycles starting at A in the weighted graph in Figure 9.23. There are three different ones, as you do not need to consider the cycle formed by reversing the order of vertices in a previous cycle. Which of the three is a minimum Hamilton cycle?

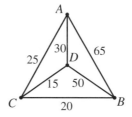

Figure 9.23

(3) Explain why there is no need to develop methods for finding minimum cost Euler circuits.

(4) Which of the complete graphs, K_n for $n \geq 3$, are Eulerian? Write a careful justification of why this is true.

(5) Consider the Petersen graph (see Figure 9.24). Can you find a Hamilton path? A Hamilton circuit? An Euler circuit? Give the vertex sequence for each one you found.

(6) Does the graph in Figure 9.25 have a Hamilton circuit? Does it have an Euler circuit? Explain in each case.

(7) Which of the drawings in Figure 9.26 are unicursive? Does it matter where you start drawing?

(8) A gated community contains 16 city blocks in a 4×4 square formed by five North-South streets and five East-West streets.

   (a) Draw a graph of this street network, where vertices are intersections and edges are one-block portions of streets that join intersections.

   (b) Suppose a security guard must patrol every street in the community. Is there a route to accomplish this without patrolling any street more than once? If so, can the patrol start and end at the same location?

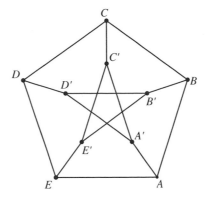

Figure 9.24

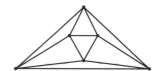

Figure 9.25

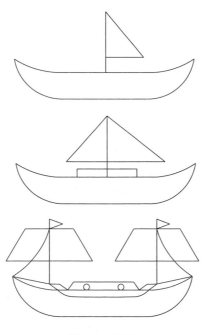

Figure 9.26

(c) Suppose a security guard must separately patrol both sides of every street in the community. Is there a route to accomplish this without patrolling any side of any street more than once? If so, can the patrol start and end at the same location?

(d) Suppose the four streets forming the perimeter of the community only need to be patrolled on the interior side of the street, while all other streets must be patrolled on both sides. Is there a route to patrol only the necessary sides of all streets without patrolling any side of any street more than once? If so, can the patrol start and end at the same location?

(e) Suppose there is a traffic light at each of the 25 intersections, and suppose that a worker must service each light. Describe a route for visiting each traffic light (intersection) exactly once. Can such a route start and end at the same intersection?

---

(9) What is the minimum number of edges necessary to guarantee a Hamilton path by Corollary 9.6 if $n = 10$.

(10) Prove the following: if G is acyclic (contains no cycles) and has no isolated vertices, then at least one vertex of G has degree 1.

(11) Recall the complete bipartite graphs $K_{m,n}$.

(a) Under what conditions on m and n will $K_{m,n}$ have an Euler circuit?

(b) Under what conditions on m and n will $K_{m,n}$ have an Euler path?

(c) Under what conditions on m and n will $K_{m,n}$ have a Hamilton circuit?

(d) Under what conditions on m and n will $K_{m,n}$ have a Hamilton path?

(12) What is the greatest number of edges a graph G on $n = 6$ vertices can have without having a Hamilton circuit if

(a) no restrictions on the graph are given?

(b) no vertices of degree 1 are permitted?

(13) Prove that the bound in Corollary 9.6 is the best possible for assuring a Hamilton cycle, by showing that for fixed n, one can find a graph on n vertices and $m = \frac{1}{2}\left(n^2 - 3n + 6\right) - 1$ edges that has no Hamilton cycle.

(14) Without using Ore's Theorem itself, prove that the hypothesis of the theorem—that the degree sum of any two nonadjacent vertices is at least n—is sufficient for showing that a graph is connected.

9.3 Vertex Colorings

We have seen that graphs are useful for depicting relationships between objects. The graph will contain all of the information essential to the original relationship, and various questions of interest may be pursued on the graph itself, even though the context has been stripped away. The problems of finding Hamilton and Eulerian circuits demonstrated this phenomenon in a travel context, where location or distance was important. In our next application, we use graphs to help solve *conflict problems*. The relationship between the original problem and the corresponding graph is less spatial, because the vertices and edges of a conflict graph (also called an *interference graph*) do not relate to position or roads traveled.

9.3.1 Conflict problems. In a typical conflict problem, a series of meetings must be scheduled for a set of people, but inevitably conflicts arise due to some meetings requiring some of the same people in attendance. Another scenario requires the storage of various substances, where some substances must be placed in different compartments due to the danger of them being mixed together. In each of these scenarios, the objects being scheduled/stored may be represented by vertices of a graph. An edge will be placed between two vertices if the objects (committees/substances) corresponding to those vertices are in conflict with each other (share a person/are dangerous together). The graph theory problem to solve, then, is to partition the vertices into classes so that no two vertices in a class share an edge.

Example 110. Suppose that Anderson, Baker, Charles, Darwin, and Ebert need to attend four committee meetings, scheduled on the hour, starting at noon. How many meeting times must be arranged if the committees are composed of the following individuals, denoted by the initial of each:

$$\text{Committee 1: } \{A, D, E\}, \quad \text{Committee 3: } \{A, B\},$$
$$\text{Committee 2: } \{C, E\}, \quad \text{Committee 4: } \{C, D\}.$$

Solution. The conflict graph associated with these committees is shown in Figure 9.27; the vertices represent committees, and an edge is placed between two vertices if the associated committees cannot meet at the same time. The graph-theoretic task is to assign a color (denoting a meeting time) to each vertex in such a way that adjacent vertices are assigned different colors. Apparently, three meeting times are needed, and here is one way to schedule them:

$$\begin{array}{ll}
\text{12:00 (white vertex):} & \text{Committee 1,} \\
\text{1:00 (black vertices):} & \text{Committees 2 and 3,} \\
\text{2:00 (gray vertex):} & \text{Committee 4.} \qquad \square
\end{array}$$

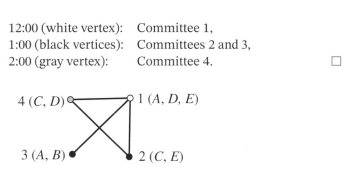

Figure 9.27. Conflict graph for Example 110

Notice that, in the above example, each color corresponds to a class (subset) of vertices. In any such problem, the assignment of different colors to vertices that are "in conflict" is known as *coloring the vertices*. Beyond their application to scheduling problems, vertex colorings have become a prominent area of graph theory in their own right. In this section we introduce the topic and give a few applications; we will give a more theoretical treatment in Chapter 13, where we examine chromatic polynomials.

A **proper vertex coloring** of a simple graph G using k colors is an assignment of an element from a set of size k to each vertex, subject to the requirement that adjacent vertices must be assigned different elements (from the set of k "colors"). If a graph G can be properly colored with k colors, we say G is k-**colorable**. The **chromatic number** of G, denoted $\chi(G)$, is the smallest value of k for which G is k-colorable.

The conflict graph G in Figure 9.27 is clearly 3-colorable, since we exhibited a proper 3-coloring. Moreover, one needs at least three colors due to the 3-clique of vertices 1, 2, and 3, where each is adjacent to the other two. Thus, the fewest number of colors needed for a proper coloring of G is 3, so $\chi(G) = 3$.

Example 111. In computer science, one technique for optimizing the execution time of compiler code is *register allocation*. The values of variables are stored in registers where they may be accessed quickly. Not all variables are used at the same time, so several of them may be assigned to the same register. However, variables in the same register may not be accessed at the same time without corrupting their values. For optimum efficiency, as few registers should be used as possible. Suppose that variables $A, B, \ldots, F$, and G in a program are each needed in the following steps of a 6-step program:

$$A: 1, 2, 5 \quad B: 2, 4, 6 \quad C: 1, 3, 4, 5 \quad D: 4, 6 \quad E: 3, 5 \quad F: 1, 4, 6 \quad G: 2, 3, 5$$

(1) Create an interference graph G, where the vertices are variables and an edge is placed between two vertices if the corresponding variables are ever accessed during the same step.
(2) Give the graph a proper vertex coloring, using as few colors as possible.
(3) What does each color represent in this particular problem?
(4) What is the value of $\chi(G)$ and how does it apply to this problem?

Solution. Notice that variable A cannot be assigned to the same register as variables B, C, E, F, and G, so there must be an edge between A and each of those vertices. Proceeding similarly with each vertex, we obtain the graph in Figure 9.28.

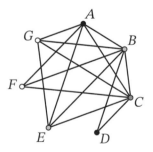

Figure 9.28. Interference graph for Example 111

One proper coloring, as shown, is to use four colors, with color classes $\{A, D\}$, $\{B, E\}, \{C\}$, and $\{F, G\}$. Since A, B, C, and G are mutually adjacent (creating a 4-clique), one must always use at least four colors, and thus $\chi(G) = 4$. In this problem, each color represents a different storage register, so four registers are needed to execute the steps of the program. □

From our discussion above, we see that n colors are necessary and sufficient for a proper coloring of a complete graph on n vertices. The following theorem gives the chromatic numbers for several families of graphs. The proofs are not difficult, and since they give good practice with using the definitions, we leave their proofs as an exercise.

Theorem 9.7. (1) *If G has at least one vertex, $\chi(G) \geq 1$.*

(2) *If G has at least one edge, $\chi(G) \geq 2$.*

(3) *If G is a **bipartite graph** (a subgraph of a complete bipartite graph) on at least 2 vertices, $\chi(G) \leq 2$. (These graphs include paths and trees, which are discussed in Section 9.4.)*

(4) *$\chi(K_n) = n$.*

(5) *$\chi(C_n) = 2$ (n even) or 3 (n odd).*

We use the theorem above in our final example of an application of the chromatic number.

Example 112. Six regional All-Star teams have been invited to a tournament where a custom jersey style is provided for each team. To avoid confusion, if two teams compete with each other in the tournament, their jerseys must be different styles; otherwise, their jerseys can be the same.

(a) If the tournament has a Round-Robin format (each team plays every other team), how many jersey styles are needed?

(b) If Teams 1, 3, and 5 have a separate Round-Robin tournament, and Teams 2, 4, and 6 do as well, how many jersey styles are needed?

(c) Will the answer change if the winner of the "odd" bracket must play the winner of the "even" bracket?

(d) How many jersey styles are needed if the tournament is organized so that each even-numbered team only plays each of the odd-numbered teams, and vice versa?

Solution. The problem may be modeled with a graph by assigning a vertex to each team and placing an edge between two teams that play each other. The number of jersey styles needed will be the chromatic number of the graph G describing the tournament. In (a), $G = K_6$, and of course six jersey styles are needed. In (b), G contains two disjoint 3-cliques as subgraphs, and since each of them has chromatic number 3, three jersey styles are sufficient. In (c), we add one more edge connecting a vertex from each one of the subgraphs in (b). This may increase the chromatic number to 4 if the winners of each bracket had already been assigned the same jersey style. In (d), $G = K_{3,3}$, and since $\chi(G) = 2$, two jerseys are sufficient. □

9.3.2 Map Colorings and Planar Graphs. Historically, a primary application of vertex colorings was the coloring of maps. Certainly one desires for adjacent countries or states of a map to be colored differently if they share a common border, in order that the border can easily be distinguished. Each map has a corresponding graph, constructed by assigning a vertex to each country/state and placing an edge between vertices corresponding to two countries if those countries share a common border. In Figure 9.29, we see a map of seven countries on the left, with the corresponding conflict graph on the right, where edges represent adjacent countries in the map.

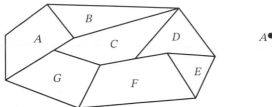 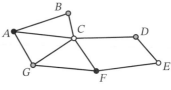

Figure 9.29

The chromatic number of the graph on the right is 3. Thus the countries in the map on the left can be colored in three colors in such a way that bordering countries will be colored differently, as in Figure 9.30.

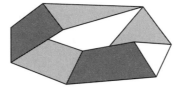

Figure 9.30

Our suggestion that the map on the left was equivalent in some sense to the graph on the right in Figure 9.29 bears more scrutiny. Does such an equivalence mean there is a one-to-one correspondence between maps and graphs? If so, we should also be able to start with a graph (defined by a set of vertices and edges) and construct a corresponding map. We could try to do this by letting each vertex denote a "capital" of a country and view each edge as a road between two capitals; then, we would need to create a map in which a border is drawn between two countries if their capitals are adjacent in the graph. Can such a map always be created? A moment's thought should reveal that there will be a difficulty if two edges of the graph cross. If each edge is viewed as an actual road crossing the border from one country to another, then there cannot be an intersection with another road that is connecting two other countries. Otherwise the first two countries would not share a common boundary. Notice that the graphs created by connecting two capitals with an edge (road) have the property that none of their edges cross. This is a critical feature of a map. However, a rendering of a graph in which edges do not cross is not always possible! For example, consider K_5, the complete graph on five vertices. Each vertex is adjacent to the four others, and with so many edges, the task of making the edges not cross is futile. The reader is invited to try this. Our best attempt is shown in Figure 9.31, but we came up one edge short, as edge $\{A, C\}$ is not present.

The class of graphs for which there is a drawing in which edges do not cross is so important that they have a name: *planar graphs*. Thus, the graph in Figure 9.29 is planar, while it appears that K_5 is not planar (though we have not proven it yet).

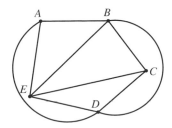

Figure 9.31. K_5 is not planar

Using our construction above, there is a one-to-one correspondence between maps and planar graphs, so all questions about coloring countries on a map so that adjacent countries are not colored the same can be rephrased in terms of proper vertex colorings of planar graphs. This was a long introduction to bring us to one of the most celebrated theorems in all of graph theory, the Four Color Theorem.

Theorem 9.8 (Four Color Theorem). *If G is a planar graph, $\chi(G) \leq 4$.*

The conjecture that any map with connected states/countries could be colored properly using at most four colors was made in the mid-19th century. Over the years, various proofs were attempted, to no avail, but no counterexamples were found. In the late 19th century, Percy Heawood gave an elementary proof that at most five colors were needed [**11**]. Finally, in 1977, University of Illinois mathematicians Kenneth Appel and Wolfgang Haken announced with much excitement that they had proven the statement. They classified the infinite set of possible maps into 1,936 different configurations (isomorphism classes) and used over 1,000 hours of computing time to check the configurations. While the computer-assisted proof was met with some skepticism initially, the results were verified independently (also by computer) in later years, and while the proof is still not considered ideal by mathematicians, there is now little doubt that the result is correct.

One problem related to planar graphs, often appearing in puzzle books, is the *Gas-Water-Electricity puzzle.* In this problem, there are three houses and three utility buildings, and the task is to create routes from each house to each of the three utilities in such a way that the routes do not cross. As we will see in Chapter 13, this is not possible in the plane, since the graph $K_{3,3}$ is not planar. However, in the meantime, the reader is invited to attempt the task on a nonplanar surface, such as a torus!

Exercises and Problems.

(1) Provide a proper vertex coloring for each graph in Figure 9.32, using as few colors as possible for each. In each case, give the chromatic number of the graph.

(2) According to Theorem 9.7, one needs eight colors to properly color the vertices of K_8. Why does this not contradict the Four Color Theorem?

(3) A k-partite graph is a graph where the vertices can be partitioned into k subsets, where there are no edges between two vertices of any one subset. What can be said about the chromatic number of a k-partite graph G?

(4) For each map shown in Figure 9.33, create the corresponding planar graph and determine the chromatic number of the graph.

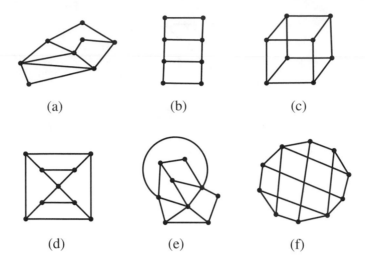

Figure 9.32. Graphs for Exercise 1

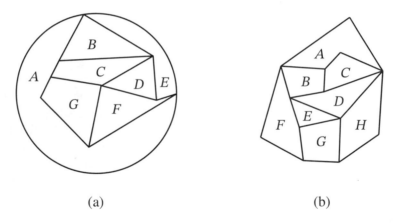

Figure 9.33. Graphs for Exercise 4

(5) For $n \geq 4$, the wheel graph, W_n, is defined as the the graph obtained by adding an additional vertex to C_{n-1}, the cycle graph on $n-1$ vertices, and placing an edge between this new vertex and each vertex of the cycle. Determine the chromatic number of W_n.

(6) Draw an example of a planar graph G for which $\chi(G) = 4$, where G does not have K_4 as a subgraph.

(7) Prove Theorem 9.7.

For each problem from 8 to 11, create an appropriate conflict graph to answer the question, and explain the meanings of the vertices, edges, and vertex colors. Our hope is to get to a point where students seeing questions like these will think, "Oh, that is really just a graph coloring problem!"

(8) The authors of this book want to have some discreet meetings with their friends Paul, Herb, Joel, Richard, László, Belá, Susanna, Ken, and Donald. For reasons we can't go into now, we don't want to have any two of those friends at the same meeting if they share a common letter in their first names. How many meetings must we set up if each of them must attend a meeting? (All three authors will be at each meeting, but the authors' names are irrelevant.)

(9) Answer the opening hook problem for the chapter, which asks you to find the minimum number of tanks needed.

(10) Suppose that two radio stations need to be assigned different frequencies if their broadcasting locations are within 120 miles of each other. How many different frequencies must be used for the seven stations whose distances from each other are shown in Table 9.1?

Table 9.1. Distances for Problem 10

	A	B	C	D	E	F	G	H
A	0	75	160	100	175	90	135	230
B	75	0	100	50	130	110	120	200
C	160	100	0	70	50	130	80	140
D	100	50	70	0	80	80	65	150
E	175	130	50	80	0	115	50	85
F	90	110	130	80	115	0	60	135
G	135	120	80	65	50	60	0	90
H	230	200	140	150	85	135	90	0

(11) An integer is said to be *squarefree* if it is not divisible by any square larger than 1. In other words, an integer is squarefree if it is prime or the product of distinct primes. The following integers are squarefree:

$$
\begin{aligned}
x_1 &= 105 & (&= 3 \cdot 5 \cdot 7), \\
x_2 &= 91 & (&= 7 \cdot 13), \\
x_3 &= 1{,}265 & (&= 5 \cdot 11 \cdot 23), \\
x_4 &= 3{,}434 & (&= 2 \cdot 17 \cdot 101), \\
x_5 &= 2{,}737 & (&= 7 \cdot 17 \cdot 23), \\
x_6 &= 286 & (&= 2 \cdot 11 \cdot 13), \\
x_7 &= 3{,}333 & (&= 3 \cdot 11 \cdot 101), \\
x_8 &= 1{,}482 & (&= 2 \cdot 3 \cdot 13 \cdot 19), \\
x_9 &= 5{,}757 & (&= 3 \cdot 19 \cdot 101).
\end{aligned}
$$

Let N be the product of all of the given x_i. What is the smallest number k for which $N = y_1 \cdot y_2 \cdots y_k$, where each y_k is squarefree and is the product of some (at least one) x_i. Create an appropriate conflict graph to answer the question, and explain the meanings of the vertices, edges, and vertex colors.

(12) Create a graph G for which $\chi(G) = 5$ but G has no 5-cliques as subgraphs.

(13) The popular Sudoku puzzle can be represented as a graph coloring problem. The standard version of the game is a 9×9 grid, and the solver must place digits from 1 through 9 in each of the 81 squares in such a way that each row, column, and 3×3 subgrid (nine nonoverlapping ones) contains exactly each digit exactly once.

Let the 81 vertices of the graph represent the squares of the puzzle, and label the vertex associated with row i and column j as $v_{i,j}$.

(a) What rule is used to assign edges to the graph?

(b) How many colors should be used to color the vertices, and what does each "color" represent?

(c) In a typical puzzle, enough of the squares are initially filled in with digits in such a way that there is a unique way to complete the puzzle. What is the analog for the graph coloring in this problem?

---

(14) Provide a construction for a family of graphs G_n for which $\chi(G_n) = n$, $n \geq 3$, but G has no n-cliques as subgraphs.

9.4 Trees

Trees figure prominently in discrete mathematics and its applications, and the topic arises from two seemingly different approaches. To a graph-theorist, a tree is simply a special type of graph in the sense that it is "minimally connected", meaning that no superfluous edges are present beyond those needed to preserve connectedness. In a connected graph, there is at least one path between any two vertices. Observe that if a second path exists between any two vertices, then a circuit would be formed by merging the two paths, and such a graph would have more than the minimum number of edges needed for connectedness. Therefore, a tree has no circuits; recall that a graph with no circuits is called **acyclic**.

Alternatively, from the perspective of relations, a tree is a special type of directed graph for which one vertex is "central" in the sense that there is a unique path from the central vertex, or root, to any other vertex. A family tree is a familiar example, where the root is the earliest common ancestor. The differences between the two approaches are not substantial. In the first approach, the relation is symmetric, and there is a unique path (of undirected edges) between any two vertices. In the second approach, the relation is asymmetric, and there is a unique path (of directed edges) from the root to any other vertex. We will explore the vocabulary, properties, and applications of trees from these approaches in separate sections in order for them to be studied independently, if desired.

9.4.1 Undirected Trees. A **tree** is a graph that is connected and acyclic. Alternatively, a tree can be defined to be a graph with a unique path between any two vertices. Family trees are aptly named; vertices are family members with an edge indicating that one member is the offspring of the other. A decision tree charts a series of yes-or-no questions to arrive at a unique diagnosis of a medical condition or a car that won't start. A counting tree provides a visual means of demonstrating the Multiplication Rule of Chapter 8 by enumerating all of the options in an organized manner.

Figure 9.34 illustrates two examples of trees. Each is connected and admits no circuits, even if a region appears to be enclosed by parts of edges.

Figure 9.35 illustrates graphs that are not trees. The first graph is acyclic, but it is not connected. Such a graph is called a forest; a **forest** is simply a graph without cycles. The second is connected but contains a circuit.

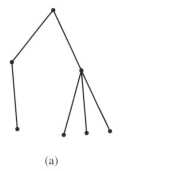
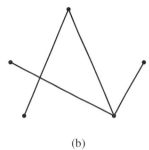

(a)   (b)

Figure 9.34

Figure 9.35

The path graph P_n is a connected graph on n vertices with no circuits, so it is a tree. Furthermore, these graphs have $n - 1$ edges, and the addition of another edge would complete a circuit in the graph. It turns out that these statements hold for all trees, as the number of edges in a tree is uniquely suited to fulfill both parts of its definition.

Theorem 9.9. *Let G be a simple graph with n vertices.*

(1) *If G is connected, then G has at least $n - 1$ edges.*

(2) *If G is acyclic, then G has at most $n - 1$ edges.*

Proof. (1) We prove the result by strong induction on m, the number of edges. For fixed $n \geq 1$, the statement is trivially true when $m = 0$. Assume that a simple, connected graph on n vertices and k edges, for $k \geq 0$, satisfies $n - 1 \leq k$. Suppose G is a simple, connected graph with n vertices and $k + 1$ edges, for $k \geq 0$. Let H be the graph formed by removing any edge from G. Note that $|V(H)| = n$ and $|E(H)| = k$. If H is connected, then by the inductive hypothesis, $n - 1 \leq k$. Conclude that $|V(G)| - 1 \leq k < |E(G)|$. If H is not connected, then it has two components, with n_1 and n_2 vertices and m_1 and m_2 edges, respectively. Then, by the inductive hypothesis, $n_i \leq m_i + 1$, for $i = 1, 2$. Therefore,

$$n = n_1 + n_2$$
$$\leq (m_1 + 1) + (m_2 + 1) = (m_1 + m_2) + 2 = k + 2.$$

Conclude that $|V(G)| - 1 = n - 1 \leq k + 1 = |E(G)|$. (An alternate proof using strong induction on the number of vertices is described in Exercise 22.)

(2) In order to use induction on n, let $P(n)$ be the statement that every acyclic graph on n vertices has at most $n - 1$ edges. Since the only simple graph with one vertex

is acyclic and has zero edges, $P(1)$ is true. Assume $P(k)$ is true for some arbitrary $k \geq 1$. Suppose G is an acyclic graph on $k + 1$ vertices; we need to show it has at most k edges. Let v be a vertex of degree 1 in G, as guaranteed by Exercise 10 in Section 9.2. Then $G - v$ (the graph obtained by removing v and the edge to which it is incident) is an acyclic graph on k vertices; by the inductive hypothesis, $G - v$ has at most $k - 1$ edges. Therefore, $|E(G)| = |E(G - v)| + 1 \leq (k - 1) + 1 = k$. $\square$

Corollary 9.10. *Any tree on n vertices has n − 1 edges.*

We conclude that if an additional edge is inserted into a tree, a circuit will result. If an edge is removed from a tree, the resulting graph will be disconnected.

Corollary 9.11. *Suppose G is a simple graph with n vertices. Any two of the following will imply the third and yield that G is a tree:*

(1) $|E(G)| = n - 1$.

(2) *G is connected.*

(3) *G is acyclic.*

9.4.2 Directed and Rooted Trees. In the introduction to Section 9.4, the key property of trees was described in the more general setting in which the relations were not necessarily symmetric and were represented by digraphs. In this case, each vertex is linked by a unique, directed path from the central vertex, or "root", and consequently, the root is uniquely determined.

A **directed tree** is a digraph that contains a vertex v_0, called the **root** of the tree, for which there exists a unique path from v_0 to every other vertex. The terminology used in describing the properties of trees is a mix of metaphors arising from botanical trees and family trees where genealogy is indicated by a directed edge from parent to child. In Figure 9.36, the tree is drawn to indicate that v_0 is the root vertex. We say that v_0 is the **parent** of v_1, v_2, and v_3 and that v_1, v_2, and v_3 are the **offspring** (or **children**) of v_0. Similarly, v_1 is the parent vertex of v_4 and v_5. If a path exists from one vertex to another vertex, such as v_3 to v_8, the latter is called a **descendant** of the former. Vertices having the same parent are called **siblings**. In Figure 9.36, vertices v_2, v_4, v_5, v_7, and v_8 have no offspring; such vertices are called **leaves**.

The **level** of a vertex in a tree indicates the length of the unique path from the root to that vertex. The root is the only vertex with level 0. In Figure 9.36, v_2 has level 1, v_6 has level 2, and v_7 has level 3. The highest level of a tree—the length of the longest path from the root—is called the **height** of the tree. The height of the tree in Figure 9.36 is 3.

Upon recalling the notions of in-degree and out-degree from Chapter 5, some general statements can be made about directed trees. First, the root has in-degree 0 and is the only vertex with in-degree 0. All other vertices have in-degree 1 due to the uniqueness of the path from the root. The leaves of a directed tree are precisely the vertices with out-degree 0. These facts lead to an easy calculation of the number of edges in a directed tree.

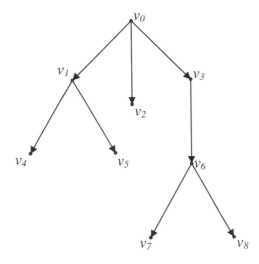

Figure 9.36

Theorem 9.12. *A directed tree on n vertices has n − 1 edges.*

Proof. In any digraph, the sum of the in-degrees must equal the total number of edges. Of the n vertices in the tree, one vertex has in-degree 0 while $n - 1$ vertices have in-degree 1. Thus, the total number of edges is $n - 1$. □

The distinction between directed and undirected trees is not profound despite the relations having a fundamental difference in regard to symmetry. An undirected tree is obtained from a directed tree by simply replacing each directed edge with an undirected edge. Given an undirected tree, a directed tree is obtained by selecting one vertex as the root and replacing all edges with edges directed away from the root. For undirected trees, any vertex may be distinguished as the root, and levels of vertices may be reckoned by the distance from this root. Such simple graphs are often called **rooted trees**. Familial terminology (parent, offspring, etc.) and arboreal terminology (leaves, forest, etc.) are used for rooted trees, as well as for directed trees.

A **binary tree** is a rooted (or directed) tree for which every vertex has at most two offspring. A **complete binary tree** is one for which each vertex that is not a leaf has exactly two offspring. These definitions may be generalized to k-**ary trees** where each vertex has at most k offspring.

Binary decision trees are important tools utilized in many fields—computer science, medicine, and game theory, to name a few. In each case, the result of a given question has two possible outcomes, which leads to more such questions or to a decision. Vertices of the tree that are not leaves represent questions, edges represent answers to the parent question (such as Yes or No), and leaves represent decisions or final answers. A simple example is the *Guess Who?* game by Milton-Bradley in which a player asks a series of yes-or-no questions to determine which of 24 people has been selected as the mystery person. Questions are concerned with physical attributes or names, such as: "Does the person wear glasses?" or "Does the first name start with a vowel?" A record of questions that a player *would* ask in response to each yes-or-no answer is modeled by a complete binary tree, provided that every mystery person can be deduced in a unique manner given this set of questions. The leaves of the tree correspond to identifying questions of the form "The mystery person is Maria, right?" where there is certainty of an affirmative answer, because the player deduced the identity of the mystery person in the previous question.

An efficient way to play the game would be to ask a question that eliminates half of the remaining people, provided this is possible. Using such an approach, if the game included sixteen candidates, then answers to five questions would determine a unique path on five vertices, ending in a leaf, and the mystery person could be deduced and identified. In order to guarantee identification of the mystery person from the group of 24 people used in the actual game, six questions are required. The decision tree with this efficient strategy will have height five, and all 24 leaves will be found on levels 4 and 5. Less efficient strategies require more questions and decision trees with greater height. For those willing to explore further, it soon becomes apparent that the height of a binary tree is computed using the base-2 logarithm.

Theorem 9.13. *The minimum height of a binary tree with n vertices is* $\lceil \log_2(n+1) \rceil - 1$.

Proof. Consider a binary tree with n vertices and height h. Notice that the number of vertices at level k is at most 2^k, for $k = 0, 1, \ldots, h$. Therefore, $n \leq 1 + 2 + \cdots + 2^h = 2^{h+1} - 1$. As $n + 1 \leq 2^{h+1}$, we apply the logarithm to obtain $\log_2(n+1) \leq h+1$. Now, $\log_2(n+1)$ need not be an integer, but since $h+1 \in \mathbb{N}$, we conclude $\lceil \log_2(n+1) \rceil \leq h+1$. The result follows. $\square$

In a complete binary tree, if leaves appear only in levels $h-1$ and h, then all previous levels consist of parents with two offspring; such trees achieve the minimum height possible. This models the efficient strategy for *Guess Who?* described above. With 24 leaves on the tree, the tree will have 47 vertices (apply Problem 19). Hence, the minimum height of the tree is $\lceil \log_2(47+1) \rceil - 1 = \lceil 5.58 \rceil - 1 = 6 - 1 = 5$, as previously found.

9.4.3 Spanning Trees. We have seen that trees are special graphs in which each vertex is reachable from any other vertex in a unique manner. The prohibition of circuits minimizes the number of edges, thereby eliminating redundancy or ridiculous relationships, depending on the application. If a group of cities must be interlinked by pipelines (or utilities or roadways), a tree structure provides an efficient means to do so. When planning such a minimal structure, one should first create a graph showing all possible connections, then select the edges needed to form a tree based on a desired criterion.[2]

A **spanning tree** of a connected graph G is a tree whose vertex set is $V(G)$ and whose edge set is a subset of $E(G)$. The word "spanning" is indicative of the fact that every vertex is used in the tree. We see from Corollary 9.10 that if G has n vertices, then any spanning tree of G must have $n-1$ edges. For the labeled graph G in Figure 9.37, three choices for spanning trees are shown. It should be easy to verify that G has four more distinct spanning trees, giving a total of seven.

Given a connected graph that is not a tree, there will be at least one circuit present. If any edge of this circuit is removed, the remaining subgraph will be connected. This

[2]Redundancy can be beneficial, of course. The failure of any one link in a tree will destroy the connectivity of the network. The Internet is an example of a high level of redundancy; worldwide connections can be preserved despite numerous cable failures.

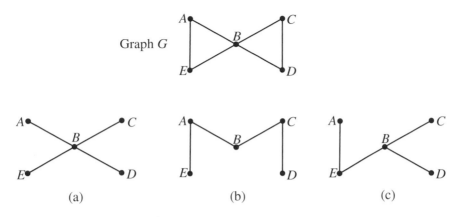

Figure 9.37

process may be continued until no circuits remain and the subgraph is a spanning tree of the original graph. Alternatively, one may select edges from $E(G)$ to construct a spanning tree. If a connected graph G has n vertices, then $n - 1$ edges can be chosen sequentially so that the next edge to be chosen does not form a circuit with previously chosen edges. By Corollary 9.11, the result will be a tree; it is a spanning tree of G.

We employ the first method for finding a spanning tree to the graph of five American cities connected by round-trip flights in Figure 9.14 of Section 9.2. Perhaps an airline's cost-cutting measure requires offering a minimal number of flights, while still enabling passengers to travel from one city to another, even if multiple connecting flights are needed. First, break the 4-cycle by eliminating an edge arbitrarily, say {LA, Miami}. Next, break the remaining 3-cycle by removing any edge, for instance {LA, NY}. The resulting graph is a spanning tree, shown in Figure 9.38 with four (bolded) edges.

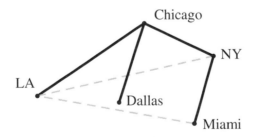

Figure 9.38

Spanning trees for directed graphs can be similarly obtained, and the result is a directed tree. It is necessary that the starting digraph have the property that there exists a vertex v_0 for which there is at least one (directed) path from that vertex to every other vertex. To obtain the spanning tree, remove edges while preserving this path property from v_0 (the root).

Realistically, the criteria used for selecting edges of a graph G for inclusion in a spanning tree of G is based on some cost of incorporating those edges. If G is geographic in nature, edges with shorter distances would be preferred. For a connected, weighted graph G, a **minimum spanning tree** (or MST) of G is a spanning tree for which the sum of the weights of the edges in the tree is the smallest possible. Two algorithms for finding MSTs will be studied in Chapter 13. Here we discuss a common-sense method for finding an MST; it is well suited for smart shoppers.

Example 113. Consider the weighted graph shown on the left in Figure 9.39 and the goal of finding a minimum spanning tree for this graph.

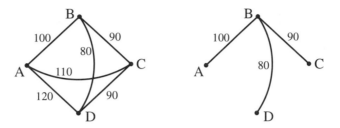

Figure 9.39. A weighted K_4 graph and MST

It stands to reason that the edge of least weight, $\{B, D\}$, with weight 80, should be included in the MST; select this edge. The next cheapest edge (weight 90) would also be a good choice, but here there is a tie between edges $\{B, C\}$ and $\{C, D\}$. We cannot choose both, since this would form a circuit with the selected edges, so we pick one randomly; let us choose $\{B, C\}$. The next cheapest, allowable edge is $\{A, B\}$ with weight 100; select it. We have created an MST for G, and it appears as the tree on the right in Figure 9.39. Since G has four vertices and we have chosen three edges, the selection of any other edge will form a circuit of selected edges. The total weight of the MST is 270, and this is the smallest possible total weight for an MST of G. It is notable that G admits a second MST with total weight 270. This MST would have been found if we had chosen edge $\{C, D\}$ instead of edge $\{B, C\}$ in the tree above.

The method employed in Example 113, picking the cheapest edges while avoiding circuits, is known as Kruskal's Algorithm for finding minimum spanning trees. It will be described in careful detail in Chapter 13.

Exercises and Problems.

(1) Which of the graphs in Figure 9.40 are trees? Which are forests, but not trees?

(2) (a) Consider graphs with three vertices labeled a, b, and c. How many different trees can be drawn with these vertices? Sketch them.

  (b) How many different trees can be drawn with four labeled vertices?

  (c) There are 125 different trees with five labeled vertices. Form a conjecture for the number of distinct, labeled trees with n vertices.

(3) The following statements about trees are false. Provide brief corrections to make meaningful, true statements about trees.

  (a) A graph which contains a path from every vertex to every other vertex is a tree.

  (b) A tree will have the same number of edges as vertices.

  (c) Every graph with n vertices and $n - 1$ edges is a tree.

  (d) A tree with n vertices can have up to n leaves.

  (e) Every graph has a spanning tree.

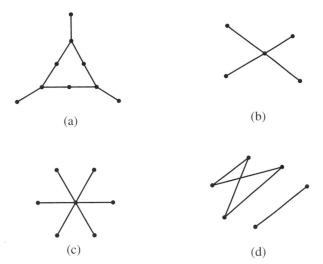

(a)   (b)

(c)   (d)

Figure 9.40. Graphs for Exercise 1

(4) Provide brief justification for each of the following statements.

   (a) Each component of a forest is a tree.
   (b) Given a forest with k components, $k - 1$ edges can be inserted into the graph in order to produce a single tree.
   (c) If every edge of a graph is a cut edge, then the graph is a tree.
   (d) The number of leaves in a directed tree with $n \geq 2$ vertices is between 1 and $n - 1$, inclusive.
   (e) If a connected graph is not a tree, then it admits at least two spanning trees.

(5) Suppose a connected graph G has 13 vertices and 20 edges. How many edges need to be removed to obtain a spanning tree of G?

(6) (a) Given a complete graph on five vertices, how many edges must be removed to obtain a spanning tree?
   (b) How many edges must be removed from a complete graph on n vertices to obtain a spanning tree?

(7) How many edges are there in a forest on n vertices with k components?

(8) Sketch a spanning tree for each graph in Figure 9.41.

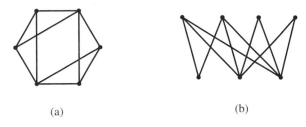

(a)   (b)

Figure 9.41

(9) Let T be a tree with n vertices.

   (a) Determine the sum of the degrees of all vertices.

   (b) What is the largest possible degree for a vertex of T? How many vertices of T can have this degree? (Use part (a).)

   (c) Suppose n is even. What is the largest possible degree for which T can have two vertices with this degree?

(10) Suppose T is a tree with the largest degree of any vertex being Δ. What can be said about s, the number of leaves in T?

(11) Write a formal proof that a tree with $n \geq 2$ vertices must have at least two vertices of degree 1.

---

(12) For the graph in Figure 9.42, determine how many spanning trees are possible. Explain.

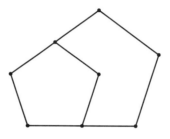

Figure 9.42

(13) Any directed tree T is the digraph representation of a relation R. Which of the following properties of relations are satisfied by any relation R corresponding to a directed tree?

   (a) Reflexive

   (b) Irreflexive

   (c) Symmetric

   (d) Asymmetric

   (e) Transitive

   (f) If aRb and bRa, then a is not related to c.

(14) (a) Let R be an irreflexive, asymmetric relation that satisfies the following property: if aRb and bRc, then a is not related to c. Provide an example to show that the digraph of R need not be a directed tree.

   (b) If a relation satisfies all three conditions in part (a) and its digraph has exactly one vertex of in-degree 0, must its digraph be a directed tree?

(15) Arthur Cayley (1821–1895) showed that the bond graph of every saturated hydrocarbon is a tree. In the bond graph of a molecule, vertices indicate atoms, and edges indicate chemical bonds between atoms.

   (a) Search for and draw the bond graph for each of the following: ethane, propane, and butane. Verify that each is a tree.

(b) Search for and name a molecule whose bond graph is not a tree. Explore the question of how chemical properties of a compound are affected by whether or not the bond graph has circuits.

(c) Saturated hydrocarbon molecules consist of hydrogen and carbon atoms for which each hydrogen atom bonds with exactly one other atom (carbon), each carbon atom bonds with four other atoms, and the maximum number of hydrogen atoms is present for the given number of carbon atoms.

To prove that the bond graph must be a tree, assume by way of contradiction that it is not a tree. Since we know which type of atoms are contained in any circuit (why?), delete an edge from this circuit and follow the rules for bonds in a saturated hydrocarbon. The contradiction comes from the number of hydrogen atoms exceeding the maximum.

(16) The Mathematics Genealogy Project (genealogy.math.ndsu.nodak.edu) provides the data needed to construct a digraph whereby an edge from one mathematician to another indicates that the former has served as the doctoral advisor of the latter. For instance, there is an edge from Karl Weierstrass to Sofya Kovalevskaya.

Construct a directed tree whose root is Carl F. Gauss with at least two leaves corresponding to living discrete mathematicians (such as your Discrete Math instructor or the authors). Choices may need to be made to avoid circuits (why?). Begin by searching for the discrete mathematician's name, and then follow the link for his or her advisor.

(17) In a certain city, the rules for driving in a High Occupancy Vehicle (HOV) lane are as follows: on a nonholiday weekday, between 6:00 a.m. and 9:00 a.m. and between 3:00 p.m. and 6:00 p.m., a car must have at least two occupants.

(a) Create a decision tree for a set of yes-or-no questions—about holidays, weekdays, times of day, and numbers of occupants—to make a determination of whether it is permissible to drive in the HOV lane.

(b) Does this tree have the least possible height for accomplishing this task? Explain.

(18) Consider a two-player number-guessing game in which Player 1 chooses a number between 1 and 10, and Player 2 must try to guess it. In response to an incorrect guess, Player 1 answers "Higher" or "Lower", and Player 2 guesses again. The game continues until Player 2 guesses the correct number.

(a) Devise a plan with specific numbers for how Player 2 could make an initial guess and respond to all of Player 1's higher-or-lower answers. Create a decision tree that records all of these guesses, from the initial guess, to the responses to each incorrect guess, to the last guess, which Player 2 knows is the chosen number. The ten vertices will correspond to guesses about particular numbers, and edges indicate whether the answer is "lower" or "higher". The leaves of the tree are numbers for which Player 2 is certain that they are the chosen numbers based on the elimination process.

(b) Consider other guessing strategies that are more or less efficient and hence may have decision trees of lesser or greater height, respectively.

(19) Suppose T is a complete binary tree on n vertices.

    (a) Prove that the number of vertices n is an odd number and the number of edges is an even number.

    (b) Prove that the number of leaves in T is $\dfrac{n+1}{2}$.

(20) Suppose T is a complete k-ary tree on n vertices for $k \geq 2$.

    (a) Determine the statements analogous to Problem 19(a) about the number of vertices and edges in T.

    (b) Determine the number of leaves in T in a manner analgous to Problem 19(b).

(21) (a) Suppose a complete binary tree has i vertices that are not leaves (often called **internal vertices**). How many vertices are in the tree?

    (b) If a complete k-ary tree has i vertices that are not leaves, how many vertices are in the tree?

(22) Use strong induction on n to prove Theorem 9.9(1): if G is a simple, connected graph with n vertices, then G has at least $n-1$ edges.

---

(23) Let $d_1, d_2, \dots, d_n$ be a sequence of positive integers. Prove that $d_1 + d_2 + \cdots + d_n = 2(n-1)$ if and only if (d_i) is the degree sequence of some tree.

(24) In this problem we outline an inductive proof of the following result. If $(d_1, d_2, \dots, d_n)$ is the degree sequence of a tree, then the number of labeled trees on n vertices having this degree sequence is

$$\binom{n-2}{d_1-1, d_2-1, \dots, d_n-1} = \frac{(n-2)!}{\prod(d_i-1)!}.$$

    (a) Prove the base case of the given statement (using the right-hand side representation for the number of trees) when $n = 2$, in which case the sequence necessarily is $(1, 1)$.

    (b) Let $(d_1, d_2, \dots, d_n)$ be the degree sequence of a tree. Since the number of labeled trees with any given degree sequence does not depend on the order in which the degrees are given, without loss of generality assume the degrees are ordered so that $1 = d_1 \leq d_2 \leq \cdots \leq d_n$. Let T denote the set of all labeled trees with this degree sequence. For $2 \leq i \leq n$, let $A_i \subseteq T$ be the set of trees containing the edge $\{v_1, v_i\}$. Prove that these A_i partition T.

    (c) Use the inductive hypothesis to show that

$$|A_i| = (d_i - 1)\binom{n-3}{d_1-1, d_2-1, \dots, d_n-1}.$$

    (d) Calculate and simplify $\sum_{i \geq 2} |A_i|$ to finish the proof.

(25) In this problem we derive Cayley's formula for T_n, the number of distinct, labeled trees on n vertices for $n \geq 2$. A similar proof is attributed to Jim Pitman in [**1**]. In this proof, we determine a_n, the number of ways to add $n-1$ directed edges in sequence to the graph $\overline{K_n}$ in order to form a rooted tree, in two different ways.

   (a) Argue that $a_n = T_n \cdot n!$.

   (b) Show that $a_n = \prod_{k=1}^{n-1} n(n-k)$ by counting the number of choices available each time a directed edge is added.

   (c) Prove that $T_n = n^{n-2}$.

(26) Use Problem 24 to give another proof of Cayley's formula for the number of labeled trees on n vertices.

9.5 Proofs of Euler's and Ore's Theorems

9.5.1 Proof of Euler's criteria for an Euler circuit.
Prior to reading this subsection, the reader is invited to reread the discussion following Theorem 9.3 in Section 9.2, since the process we used to construct the Euler circuit for the graph in Figure 9.19 indicates how we will proceed here. The result of the following lemma may not seem surprising after reviewing that process; it is central to the proof of Theorem 9.3.

Lemma 9.14. *If G is a connected graph and the edge set of G can be partitioned into the edge sets of a disjoint union of cycles of G, then G has an Euler circuit.*

Proof. We prove the statement by strong induction on the number of edges of G. In the base case, G is a 3-cycle, and the result clearly holds. Assume the statement is true for graphs with fewer than m edges for some arbitrary $m \geq 4$, and let G be a connected graph on m edges, where the edge set of G can be partitioned into the edge sets of disjoint unions of cycles of G.

Let $\alpha_0 = (v_1, v_2, \dots, v_k, v_1)$ be one of the cycles in the disjoint union of cycles of G. Let G' be the graph with vertex set $V(G)$ and edge set $E(G) \setminus E(\alpha_0)$ and let $G_1, G_2, \dots, G_t$ be the nontrivial connected components of G'. Each G_i is connected and has fewer than m edges; furthermore, because the edges of G could be partitioned into disjoint cycles, the same is true for each G_i. Therefore, by the inductive hypothesis, each G_i has an Euler circuit. Since G was connected, each G_i shares at least one common vertex with α_0; for each i, we choose one of them to denote as v_{s_i}, $1 \leq i \leq t$. Assume these vertices are ordered as they appear in α_0; i.e., $s_1 \leq s_2 \leq \cdots \leq s_t$. (For example, in the leftmost graph of Figure 9.19, two nontrivial components remain after the edges of the bold cycle α_0 are removed, so $t = 2$. We choose vertex v_2 (v_3 would also be an option) as the shared vertex of α_0 with G_1 and v_4 is the only vertex shared with G_2. Thus, $s_1 = 2$, and $s_2 = 4$.)

Let α_i, for $1 \leq i \leq t$, denote the sequence of vertices in an Euler circuit of G_i starting and ending with v_{s_i}. Then, to build an Euler circuit for G from the cycle α_0, we simply travel around α_0, inserting the vertices from each α_i when we reach v_{s_i}. The sequence of vertices

$$(v_1, v_2, \dots, v_{s_1-1}, \alpha_1, v_{s_1+1}, \dots, v_{s_i-1}, \alpha_i, v_{s_i+1}, \dots, v_{s_t-1}, \alpha_t, v_{s_t+1}, \dots, v_k, v_1)$$

form an Euler circuit of G, which completes the proof. $\qquad\square$

We now prove the following version of Euler's Theorem.

Theorem (Theorem 9.3). *A connected graph has an Euler circuit if and only if all vertices have even degree, and it contains an Euler path (but not an Euler circuit) if and only if exactly two vertices have odd degree.*

Proof. The necessity of having all even-degree vertices in order for G to have an Euler circuit follows from these two facts: (i) in such a (directed) circuit, the in-degree and out-degree of each vertex are equal, and (ii) the degree of each vertex in G will be the sum of its in-degree and out-degree.

The sufficiency of having all even-degree vertices follows immediately from Lemma 9.14, once we prove the following claim.

Claim. *If all vertices of a graph G_1 have even degree, then its edges can be partitioned into the edge sets of a disjoint union of cycles of G_1.*

Proof. We prove this claim in a constructive fashion. Choose a vertex v_1 with $\deg(v_1) \neq 0$, and start to form a sequence of distinct vertices, $v_1, v_2, v_3, \ldots$, where $\{v_i, v_{i+1}\} \in E(G_1)$. As was argued in the proof of Lemma 9.14, since all vertices have even degree, one will always be able to find an unused vertex to continue the sequence until eventually reaching a vertex v_k that is incident to v_1. The sequence $v_1, v_2, \ldots, v_k, v_1$ forms a cycle C_1, and if the edges of this cycle are removed from G_1, the resulting graph will be a graph G_2 whose vertices are all even. The process can be repeated on G_2 to find a second cycle, C_2, whose edges are distinct from those of C_1. The process can be repeated, and since edges are removed from G_i at the ith iteration and since $E(G_1)$ is assumed to be finite, eventually we reach an n for which G_n is the null graph. The edges of the cycles $C_1, C_2, \ldots, C_n$ partition the edges of G_1, as desired, which completes the proof of the claim.   ◇

We omit the details of the proof that G contains an Euler path, but not an Euler circuit, if and only if G has exactly two vertices of odd degree. The idea is very similar to the proof above. The main difference is that the first and last vertices of an Euler path are the only vertices in the path with odd degree. Thus, one can follow the same process by starting at an odd-degree vertex and ending at the other odd-degree vertex (with "detour" circuits along the way, if necessary, as before).   □

9.5.2 Proof of Ore's Theorem.

Theorem (Theorem 9.4). *Suppose G is a simple graph with n vertices, $n \geq 3$. If for every pair of nonadjacent vertices v_1 and v_2 of G, $\deg(v_1) + \deg(v_2) \geq n$, then G contains a Hamilton cycle.*

Proof. Our proof has two parts. We first show that, under the given hypotheses, a graph must have a Hamilton path. We then show that the existence of that Hamilton path guarantees a Hamilton circuit.

Suppose G is a graph of order n, with $\deg(u) + \deg(v) \geq n$ for all nonadjacent vertices u and v, and let $P = (v_1, v_2, \ldots, v_m)$ be a path in G. Assume $m < n$. We claim we can find a longer path than P in G; proving this claim will show that G must have a Hamilton path.

If v_1 or v_m has a neighbor not on P, then we can append that vertex at the beginning or end of P to create a longer path, as desired. So, assume all of their neighbors lie on P and let u be a vertex not on P.

Claim. *There is a vertex v_i such that $v_{i+1} \in N(v_1)$ and $v_i \in N(u)$.*

Proof. Let $N_P(u)$ denote the neighbors of u that lie on P and $N_{\overline{P}}(u)$ denote the neighbors of u not on P. Then,

$$n \leq \deg(u) + \deg(v_1) = |N_P(u)| + |N_{\overline{P}}(u)| + \deg(v_1)$$
$$\leq |N_P(u)| + (n - m) + \deg(v_1).$$

Hence, $|N_P(u)| \geq m - \deg(v_1)$.

Let B be the set of vertices that are immediate predecessors (along P) of vertices from $N(v_1)$. Note that since $B = \{v_i : v_{i+1} \in N(v_1)\}$, $|B| = \deg(v_1)$. Therefore,

$$|B \setminus \{v_1\}| + |N_P(u)| \geq (\deg(v_1) - 1) + (m - \deg(v_1)) = m - 1.$$

Since $B \setminus \{v_1\}$ and $N_P(u)$ are both subsets of the $(m - 2)$-element set $\{v_2, v_3, \dots, v_{m-1}\}$, by the Pigeonhole Principle,[3] $B \setminus \{v_1\}$ and $N_P(u)$ share a common vertex v_i in that set. It follows that $v_{i+1} \in N(v_1)$, which proves the claim. ◇

Having proven the claim, we have found a path in G longer than P, namely

$$(u, v_i, v_{i-1}, \dots, v_1, v_{i+1}, \dots, v_m).$$

We may therefore assume G has a Hamilton path $P = (v_1, v_2, \dots, v_n)$.

If v_1 and v_n are neighbors, then the Hamilton path extends to a cycle, and we are done. Otherwise, by letting $v_n = u$ in the claim above and $m = n$ in its proof, there is a vertex v_i such that $v_{i+1} \in N(v_1)$ and $v_i \in N(v_n)$. It follows that $(v_1, v_2, \dots, v_i, v_n, v_{n-1}, v_{n-2}, \dots, v_{i+1}, v_1)$ is a Hamilton cycle in G, which completes the proof of Ore's Theorem. □

---

[3]The Pigeonhole Principle will be covered in Section 11.2, but the interpretation here is simply that if A and B are subsets of C and if $|A| + |B| > |C|$, then $A \cap B \neq \emptyset$.

A group of people sits in a circle, each holding an even number of pieces of candy. Each person simultaneously gives half of his or her own candy to the person on the right. Any person who ends up with an odd number of pieces of candy selects one piece from a large bowl of candy in the center of the circle, so that once again all persons have an even number of pieces. Show that eventually all persons will have the same number of pieces of candy. Is this statement also true if at each iteration each person gives half of his or her candy to each person to the left and right?

10

Invariants and Monovariants

If you change the way you look at things, the things you look at change.

—Wayne Dyer (1940-2015)

Many problems in discrete math involve a set of configurations, or states, depicting some event. The permissible moves for changing the configuration in question from one state to another are provided. Examples are abundant in games and puzzles (checkers, chess, Rubik's cube, etc.). A typical question, then, asks whether or not a specific state is attainable: "Can Player 1 beat Player 2 in six moves?" or "Can the cube be solved?" In this short chapter, we study the notions of invariants and monovariants, which can be used to solve such problems. We start with several examples, postponing the general discussion until later.

10.1 Invariants

Example 114. The integers $1, 2, \ldots, 9, 10$ are written on a blackboard. Felix picks any two of the numbers, erases them, and writes the absolute value of their difference on the board. He repeats this procedure with the resulting nine numbers, and so on. After he does this nine times only one number remains on the board. Can the number be 4?

First Thoughts. For example, consider the following sequences. Each row represents the numbers left after a move is made.

```
1 2 3 4 5 6 7 8 9 10
2 3 4 5 6 7 8 9 9     (10 − 1 = 9, so replace 10 and 1 with 9)
3 4 5 7 8 9 9 4       (6 − 2 = 4)
4 5 7 8 9 4 6         (9 − 3 = 6)
5 7 8 4 6 5           (9 − 4 = 5)
5 4 6 5 1             (8 − 7 = 1)
4 6 1 0               (5 − 5 = 0)
2 1 0                 (6 − 4 = 2)
2 1                   (1 − 0 = 1)
1
```

Several more attempts will show that not only is the final number not a 4, but it is never even. This is what we will prove. ◇

Solution. Consider the procedure Felix performs as a "move". Originally there are five odd numbers on the board. How does this quantity change with each move? If two even numbers are chosen, their difference is even, and the number of odd numbers on the board does not change after the move. If an even number and an odd number are deleted, then their difference is odd, and the number of odd numbers on the board does not change after the move. If two odd numbers are deleted, then the number of odd numbers on the board decreases by two after the move. Therefore, each time a move is made, the number of odd numbers on the board either does not change or decreases by two. Since its initial value is 5, an odd number, it will be odd after every move. If 4 were the last number on the board, then the number of odd numbers on the board at that moment was zero. But zero is an even number. Therefore it is not possible to have 4 as the last number. □

Further Thoughts. We have proven that the final number cannot be an even integer. For your consideration: Is it possible to obtain each of the odd integers from 1 to 9 as the last number? ◇

Example 115. Grace tore up a sheet of paper into eight pieces, then tore some of these into eight pieces, and so on, indefinitely. If we refer to the tearing up of a piece of paper into eight new ones as a "move", must Grace obtain exactly 2,017 pieces after some number of moves?

Solution. Yes, she must. Each time a move is performed, the total number of pieces is increased by seven ($-1 + 8 = 7$). Therefore, after each move, if we divide the total number of pieces by 7, we should always get the same remainder. Since we started with one piece, the number of pieces after each move must always be congruent to 1 modulo 7, and all such positive integers, including $2,017 = 288 \times 7 + 1$, must be obtained. □

Example 116. Is there a social network containing a group of fifteen people in which each person is a friend of exactly three other people in the network?

Solution. No, there is not. Clearly, every network can be built by starting with fifteen isolated individuals (no connections at all) and then adding one friendship connection at a time between two individuals (a "move"). Let us number the individuals 1, 2, ... , 15.

For $1 \leq i \leq 15$, let d_i denote the number of friends of the ith individual, and let $d = d_1 + d_2 + \cdots + d_{15}$. Note that d and each d_i will change as we make moves. In the initial network, all d_i are 0, so $d = 0$. After a move is performed, exactly two of the d_i values increase by 1, and therefore d increases by 2 with every move. Thus the value of d is even after any number of moves. In the network we are describing, each d_i is 3; hence, $d = 3 \times 15 = 45$, which is an odd number. Therefore such a social network does not exist.   □

Further Thoughts. This is not the only—or the most natural—solution to the question. Find an easy way to justify the answer using graph theory.   ◇

Before proceeding further, let us analyze the solutions of these three examples. In each case, we interpreted the problem as a sequence of allowable "moves" between elements of a certain set S, the set of "states" or "configurations" associated with the problem. Each time, the moves were defined within the solution. What about the various sets S?

In Example 114, the set S can be defined as a set of all collections of nonnegative integers (repetitions of elements in a collection are allowed).

In Example 115, S is the set of all positive integers.

Example 116 differed from the previous two in that no procedure is mentioned in the problem, and it was our task, in modeling the problem, to introduce moves and states. We defined the set of states S to be the set of all possible networks with fifteen phones.

The only requirement for S is that it has to contain all states that can be produced from the original one by allowable moves. In Examples 114 and 115 above, our stated set S contains some states that cannot be reached by allowable moves, which is permissible. But the modeling of the problems as a sequence of moves between states was not the key feature of the solutions. The most important component was the introduction of a numerical function on S with the property that, as one moved from state to state via allowed moves, the function's value *did not change*.

We call such a function an **invariant** and denote it by *Inv*. Thus $Inv(s) = Inv(s')$ for any states $s \in S$ and $s' \in S$ for which s' is reachable from s by a move (or sequence of moves).

In Example 114, if A_s is the set of odd numbers on the board at a given state s, then *Inv* can be defined as $Inv(s) = |A_s| \bmod 2$. Thus, $Inv(s) = 0$ if s contains an even number of odd numbers, and $Inv(s) = 1$ otherwise. This "mod 2" function is also often called the **parity** function; in this example, $Inv(s)$ is the parity of the number of odd numbers in s. Another invariant one could choose for this example is the parity (even or odd) of the sum of the numbers on the board.

In Example 115, *Inv* can be defined as the number of pieces of paper, modulo 7.

In Example 116, *Inv* can be defined to be the parity of d, where $d = d_1 + d_2 + \cdots + d_{15}$.

In the examples above for which a negative answer was obtained, the value of *Inv* on the initial state was different from the value on the state described in the question. In such cases, this state could not be reached via allowable moves. Let us continue with more examples.

Example 117. On a table, there are seven lettered tiles (A through G) lined up in a neat row, in reverse order, as shown below.

$$\text{G \quad F \quad E \quad D \quad C \quad B \quad A}$$

At each step, one may switch the positions of any two adjacent letters. Can this list of seven letters be put in alphabetical order in exactly 50 steps?

First Thoughts. What invariant can we find here? After a little experimentation one finds that although it is not difficult to place the pieces in alphabetical order, the number of moves needed to do so appears to always be odd. If we can prove this, then the answer to the question will be "no".   ◇

Solution. We model the problem as we did the previous ones, letting the set S of states be the set of permutations of these seven letters (of which there are 7!) and calling an interchange of any two adjacent letters in a permutation an "allowable move". What invariant can we find here?

We call a pair of letters from a list an **inversion** if the two letters appear in the wrong order (meaning that an alphabetically later letter comes *before* an alphabetically earlier one). To form an inversion, the letters do not have to be adjacent in the list. For example the list BDEACGF has 6 inversions: B and A; D and A; D and C; E and A; E and C; and G and F.

For any permutation $s \in S$, let $i(s)$ denote the number of inversions in s. (Recall that i counts *all* pairs that are in the wrong order, not just adjacent pairs.) First, notice that the list GFEDCBA has $C(7, 2) = 21$ inversions, because every pair of letters occurs in the wrong order; therefore, $i(\text{GFEDCBA}) = 21$.

Now, neither i nor the parity of i is invariant. But the careful observer will notice that at each step, i will change by *exactly* 1, either increasing or decreasing. This is the hook that we need! Since i is required to change by 1 at each step, i will change parity at each step; i.e., as we perform allowable moves by transposing pairs of adjacent letters, i goes from odd to even to odd to even, etc. When the number of moves completed is even, i is odd, and vice versa. Thus, after n moves, the sum $i + n$ must be odd. If it were possible to achieve the alphabetized list in exactly $n = 50$ moves, we would have $i(\text{ABCDEFG}) + n = 0 + 50 = 50$, which is even. This means that it is not possible to produce the reversed list in exactly 50 steps.   □

Further Thoughts. We saw here that if two adjacent items in a permutation are switched, then the number of inversions will change by exactly one, thereby changing the parity of the permutation. It turns out that the transposition of *any two items* in a permutation, adjacent or not, changes its parity. This is because we can "simulate" a switch of nonadjacent elements by several adjacent switches. For suppose two nonadjacent elements x and y have exactly m elements between them in the permutation

$$(\ldots, x, a_1, a_2, \ldots, a_m, y, \ldots).$$

Let us switch x with a_1, then with a_2, then with a_3, and so on until we switch it with a_m, and then with y. This gives a total of $m + 1$ adjacent switches so far. Then we move y to the left via adjacent switches with $a_m, a_{m-1}, \ldots, a_1$ until it is where x started out. This yields an additional m transpositions and results in a switch of the originally nonadjacent elements x and y in $2m + 1$ adjacent switches. But as we showed above, each of these adjacent switches changed the parity; since $2m + 1$ is odd, the parity was changed an odd number of times, resulting in a net parity change for the whole operation.   ◇

Example 118. The famous "Fifteen Puzzle" by Sam Loyd[1] starts with a 4×4 grid containing fifteen 1×1 square blocks, numbered 1 through 15, and one empty block, like the grids shown below. If one starts with the configuration shown on the left, is it possible to slide the blocks around to achieve the "solved" position shown on the right?

1	2	3	4
5	6	7	8
9	10	11	12
13	15	14	

$\longrightarrow$

1	2	3	4
5	6	7	8
9	10	11	12
13	14	15	

Solution. We can view the empty space as a 16th virtual block having number 16 in it, and every allowable move can be thought of as a switch with this "block 16". Then, instead of working with the 4×4 board, we can imagine that all sixteen blocks are lined up in a 1×16 row formed by placing the 1×4 rows end-to-end. Thus every board position corresponds to one of the 16! permutations of these sixteen blocks, and every move on the grid corresponds to a switch of a certain element of the permutation with our block 16.

We will look at this puzzle in both of the forms mentioned above, as a 1×16 permutation and as a 4×4 block puzzle. We will show that if one starts with a permutation $a = (1, 2, \ldots, 13, 15, 14, 16)$, it will be impossible to obtain the "standard" permutation $b = (1, 2, \ldots, 13, 14, 15, 16)$ by using allowable moves only.

Suppose the desired position *can* be achieved, and consider the number of moves that the alleged solution uses. Looking at this as a 4×4 block puzzle, we can conclude that the number of moves must be even. For, since 16 starts and ends its journey around the board in the same position on the board (the lower right corner), it moves left as many times as it moves right, and it moves up as many times as down. Since every move really does move 16 in one of these directions, the total number of moves is even.

Now let us consider the permutation. Each of the moves in the board corresponds to a switch in the permutation, so our alleged solution must correspond to an even number of switches in the permutation. According to the discussion preceding this example, each switch changes the number of inversions of the permutation. Hence an even number of switches leads to a permutation of the same parity. However, the number of the inversions in a is one, and the number of inversions in b is zero, so they are of different parity. The obtained contradiction implies that the required transformation is not possible.   □

When the method of invariants is first introduced as a problem-solving tool, the question that inevitably follows is, "How do we find an appropriate invariant for a given problem?" There is no simple answer. Certainly creativity is required, and in that sense the technique is somewhat artistic. With practice, one can improve in using this method—and in problem-solving in general—by following examples and by trying, failing, and sometimes succeeding. Although not widely utilized, the method of invariants can produce solutions in unexpected settings, so it is a good strategy to keep in mind.

---

[1]More on this puzzle can be found in *Matters Mathematical* [**12**] by Herstein and Kaplansky.

Many more examples and useful discussions can be found in the article by Ionin and Kurlyandchik [**14**] and in the book by Fomin, Genkin, and Itenberg [**9**].

Exercises and Problems.

Ever tried. Ever failed. No matter. Try again. Fail again. Fail better.

—Samuel Beckett (1906–1989)

(1) There are eleven glasses on a table, all standing upside down. A "move" consists of turning any four of them over. Using only this type of move, is it possible to get all the glasses right-side up?

(2) The numbers $1, 2, \ldots, 2{,}017$ are written on the board. Two numbers are erased and replaced by $s \bmod 13$, where s is the sum of the two numbers. This operation is repeated until one number is left. What is the number?

(3) You have a 4×4 square grid with -1's in the top row, -1's in the leftmost column, and $+1$'s in the other nine positions. At each move, you may change the sign of all the entries of any row or the entries of any column. Show that you can never reach a position where all entries are positive. Is it possible for all entries to be negative?

-1	-1	-1	-1
-1	$+1$	$+1$	$+1$
-1	$+1$	$+1$	$+1$
-1	$+1$	$+1$	$+1$

(4) Begin with an 8×8 checkerboard and prune it by cutting out two opposite corners. Is it possible to tile the remaining 62-square board with 1×2 dominos? (Each domino covers exactly two adjacent squares, no gaps or overlaps allowed.)

(5) You are given 25 copies of a 1×4 tile, with which to cover all the squares of a 10×10 square grid, without gaps or overlaps. Can this be done?

(6) Start with a null graph on n vertices. Now start building a graph by adding edges, with the restriction that you do not create any cycles. Let k denote the number of connected components at any stage, and let e denote the number of edges which have so far been placed at any stage. Show that the quantity $c = k + e$ is an invariant. Use this fact to prove that a tree (a connected graph on n vertices which has no cycles) must have exactly $n - 1$ edges.

(7) You are given 3 rows of coins, with cent values laid out as follows:

1	5	10	25
1	10	5	25
25	10	5	1

Starting at the first row, you switch the positions of any adjacent pair of coins in that row. Then you go to the second row, and again you switch the positions of any adjacent pair that you wish. Then you go to the third row, switch, then back to the first row, switch, then the second row, switch, and so on for as long as you wish. Your goal is to have each row end up in increasing order, the way the first row is initially. Can this be done?

10.2 Monovariants

The problems in this section and our method for solving them are similar in spirit to those in the previous section. As before, we start with several examples, postponing the general discussion.

Example 119. Consider a rectangular array with m rows and n columns whose entries are real numbers. It is permissible to reverse the signs of all the numbers in any row or column. Prove that after a (finite) number of these operations we can make the sum of the numbers in each line (row or column) nonnegative.

Solution. Consider the following algorithm for solving this problem. If there is no line with a negative sum, we are done. If there is a line with a negative sum, change the signs of all numbers in this line.

Changing the signs of all numbers in one row may affect the sum of the numbers in some columns, and then changing the numbers in those columns may affect the sums in many rows, and so on. While it is conceivable that this process will run in circles, leading to no better situation than that with which we started, we will show that this algorithm must actually terminate.

Let M be the sum of all mn numbers from the table. Suppose the sum of numbers in some line is $x < 0$. Then reversing signs of all numbers in this line, we get another array with the sum of all its entries being $M - 2x$, which is greater than M since x is negative. Therefore each step of our algorithm leads to a new table with a strictly larger sum.

And now we can see why our algorithm cannot run forever. Our operation can produce only finitely many arrays, because each of the mn entries can take on only two values (differing in their sign), and so all in all there could be only *finitely* many different tables obtained this way. Therefore there are only finitely many different values of M. If the algorithm does not stop, it will produce infinitely many ever-increasing, and therefore distinct, values of M, which is a contradiction. □

Example 120. Given $2n$ points on the plane such that no three points lie on one line, define a *configuration* of the points to be any set of n segments in which each segment connects two points from the set and each point is used on exactly one segment. Prove there is a configuration for which no two segments intersect.

Solution. Consider the following algorithm for solving this problem. Start with a random intersection that uses all $2n$ points. If no two segments of the configuration intersect in the plane, then we are done. Otherwise, pick a pair of intersecting segments, say $\overline{AB}$ and $\overline{CD}$, and apply the following "move": erase them, and draw two new segments $\overline{AC}$ and $\overline{BD}$. (See the left figure in Figure 10.1.) This process can be repeated as long as there is a pair of intersecting segments. Clearly, at each move we eliminate one point of intersection, but other intersections may be formed.

We will show that the algorithm must eventually terminate. For each configuration we can compute M, the sum of the lengths of all n segments. Since the number of configurations is finite, it is no more than $\binom{\binom{2n}{2}}{n}$, M can take only finitely many values. We will show that the value of M strictly decreases with every move.

Since other segments are not affected by the move, it suffices to show that $AC + BD < AB + CD$ (i.e., the sum of the diagonal lengths of a quadrilateral exceeds the sum

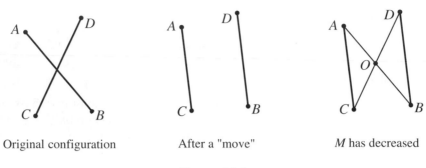

Original configuration   After a "move" M has decreased

Figure 10.1

of the lengths of two opposite sides). Let O be the intersection of $\overline{AB}$ and $\overline{CD}$. (See the right figure in Figure 10.1.) Then

$$AB + CD = (AO + OB) + (CO + OD)$$
$$= (AO + CO) + (BO + DO) > AC + BD.$$

This last part follows from the triangle inequality. We have proven that the algorithm must terminate, for otherwise it would produce infinitely many different values of M.

$\square$

Example 121. Consider a set of n people, of whom some are friends and some are enemies. Prove that they can always be divided into two groups such that each person has at least as many friends in the other group as in his or her own group. (We assume that A is a friend of B if and only if B is a friend of A; that is, friendship is a symmetric relation.)

Solution. This problem can be easily rephrased in terms of coloring of vertices of a graph:

> *Consider a graph on n vertices. Prove that all of its vertices can be colored with two colors, say red and blue, such that for every vertex at most 1/2 of its neighbors have the same color as the vertex itself.*

We let the vertices of the graph correspond to the n people. Two vertices are joined by an edge if and only if the corresponding people are friends. Vertices of the same color correspond to people belonging to the same group.

There is a surprisingly simple algorithm for constructing such a coloring. First, assign colors to the vertices at random. If the condition is satisfied, then we are done. If not, pick a vertex for which more than half of its neighbors are the same color as itself and change the color of this vertex. Repeat until no such vertices exist. We then have the desired coloring!

The remaining question is, "Does this algorithm ever stop?" To see that it does, first we observe that the number of different colorings of the vertices (using two colors) is finite. For each coloring, count M, the number of edges in the graph whose endpoints are colored in different colors, and note that $M \leq 2^{\binom{n}{2}}$. We now show that after applying our move, the value of M strictly increases.

Suppose a chosen vertex, V, has k neighbors, of which $s > k/2$ are colored the same as V. Changing the color of V adds s to M and subtracts $k - s$ from M. Therefore, for the new coloring, there are $M + s - (k - s) = M + 2s - k > M$ edges whose endpoints are colored in different colors. This proves that eventually we will reach a point where we cannot find a vertex for which $s > k/2$; if not, the algorithm would produce infinitely many different values of M, which contradicts that M is bounded above by the number of edges in the graph. $\qquad\square$

Before proceeding further, let us analyze the solutions of Examples 119–121. As we did in the previous section on invariants, we interpreted each problem as a sequence of allowable moves between elements of a *finite* set, S, of states or configurations associated with the problem. As before, each state in S must be producible from the original state by allowable moves.

In Example 119, S can be defined as the set of all $m \times n$ arrays of real numbers obtained from a given $m \times n$ array by arbitrary changes of signs of its entries.

In Example 120, S is the set of all figures obtained by joining n pairs of given $2n$ points by segments.

In Example 121, S is the set of all different separations on the set of n people in two groups, or the set of all colorings of the vertices of the graph in colors red and blue.

Sometimes, as in Example 119, both S and the allowable moves were given in the problem. However, in Examples 120 and 121 it was up to the solver to create them.

Here is where we deviate slightly from the previous section on invariants. There, we introduced a numerical function *Inv* on each set of states, with the important property that *Inv* was constant over the given set S. Here, in each case we defined a function M which had the useful property of being strictly monotone with respect to the moves. More precisely, for every pair of states $(s, s') \in S$ for which s' is obtained from s by an allowable move, $M(s) > M(s')$ (monotone increasing), or for every pair of states $(s, s') \in S$ for which s' is obtained from s by an allowable move, $M(s) < M(s')$ (monotone decreasing). We call such a function M **monovariant**.

In Example 119, $M(s)$ was the sum of all entries of the array s, and it was an increasing function.

In Example 120, $M(s)$ was the sum of lengths of all segments in figure s, and it was a decreasing function.

In Example 121, $M(s)$ was the number of edges in a graph s whose endpoints were colored in different colors, and it was an increasing function.

In each example, the termination of the algorithm followed from the same basic argument: *it is impossible to make an infinite number of allowable moves on a finite set of states if the underlying function is monovariant.*

Example 122. A nonconvex polygon is subjected to the following operation: if the part of the polygon connecting (clockwise) some vertex A to another vertex B lies completely in the exterior of the polygon, then this part is rotated 180° about the midpoint, M, of $\overline{AB}$. See Figure 10.2. Prove that after a finite number of such operations the resulting polygon will be convex.

Solution. The strictly increasing monovariant here is the area of the polygon, which clearly increases after each move. It is a bit trickier to prove that the set of possible states is finite. This is accomplished by observing that each side of the polygon is a segment of given length and given slope and noting that any permissible move simply

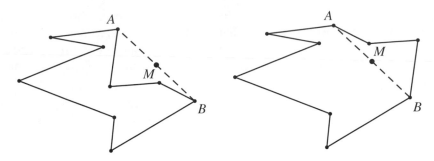

Figure 10.2

rearranges this set of segments, preserving their lengths and slopes. Since there are a finite number of ways in which these segments can be arranged to form a polygon, there are only a finite number of possible states. □

Example 123. Finitely many squares of an infinite black and white "checkerboard" grid are initially colored black. After one second, and each second thereafter, each square of the grid takes the color of the majority of the following three squares: the square itself, its top neighbor, and its right-hand neighbor. Prove that some time later there will be no black squares remaining in the grid.

Solution. Draw a rectangle bounded by grid lines which contains all the black squares. Introduce rectangular coordinate axes so that the origin is at the lower left corner of the rectangle. See Figure 10.3.

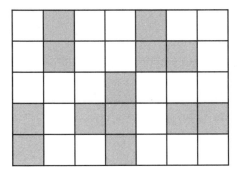

Figure 10.3

A little thought will reveal that at no moment will there ever be a black square outside of the rectangle. Thus the set of states can be taken to be the set of all colorings of the squares within the rectangle using black or white, which is finite. To define a monovariant M, compute the sum of the x- and y-coordinates for the upper-right corner of each black square, and let M be the greatest of these sums. This sum must decrease after each move, and therefore M will eventually be zero. At this moment, none of the squares will be black. □

Another method, called "Going to Extremes", is closely related to the method of monovariants. Briefly described, this method reduces to the following: consider the value of a function (essentially, our monovariant) on a finite set of all possible states, and then consider some state s that minimizes (or maximizes) this function. Show that if this state did not have the desired property, then a state with a smaller (or larger) function value could be constructed, contradicting our choice of s.

For example, in Example 123, among all achievable states, consider the one which maximizes the sum of all entries in the array. This maximum exists since the number of states is finite. If the corresponding array had a line with negative sum, then switching the signs of the elements along that line would yield an array with a larger sum (of all entries), a contradiction.

More examples and useful discussions can be found in [7].

Exercises and Problems.

(1) A group of persons sits in a circle, each holding an even number of pieces of candy. Each person simultaneously gives half of his or her own candy to the person on the right. Any person who ends up with an odd number of pieces of candy selects one piece from a large bowl of candy in the center of the circle, so that once again all persons have an even number of pieces. The process is then repeated.

    (a) Show that eventually all persons will have the same number of pieces of candy.

    (b) Will all persons eventually have the same number of pieces of candy if, at each iteration, each person gives half of his or her candy to each person to the left and right?

(2) Each member of a parliament has no more than three enemies among other members. Prove that the parliament can be split into two houses such that each member has no more than one enemy in the same house. (Assume that being an enemy is a symmetric relation.)

---

(3) Given n red and n blue points in the plane, no three of them collinear, prove that one can draw n nonintersecting segments joining red points to blue points.

(4) Consider n points, no three of which are collinear, and n lines, no two of which are parallel, in the plane. Prove that a perpendicular segment can be dropped from each point to one of the lines, one perpendicular per line, such that no two segments intersect.

(5) A dinner party of size $n \geq 4$ is seated around a circular table. If the ages, a, b, c, and d, of four consecutive dinner guests satisfy the inequality $(a - d)(b - c) > 0$, then the guests with ages b and c must change seats. Prove that the guests can be rearranged in this manner only a finite number of times.

(6) For fixed positive integers e and v, let $\mathcal{G} = \mathcal{G}(v, e)$ represent the set of all graphs with v vertices and e edges. For a graph $G \in \mathcal{G}$, call G **2-path extremal** if no other graph in $\mathcal{G}$ contains more paths of length two than G does. Prove that if a graph G has nonadjacent vertices x and y with $\deg(x) > \deg(y)$ and y has a neighbor z that is not a neighbor of x, then G is not 2-path extremal. What does this mean about the structure of 2-path extremal graphs?

---

(7) King Arthur summoned $2n$ knights to his court. Each knight has no more than $n - 1$ enemies among the knights present. Prove that Merlin can seat the knights at the Round Table in such a way that no two enemies will sit next to each other.

(8) Prove the following corollary to Ore's Theorem using techniques from this chapter: *Let G be a simple graph on $n \geq 3$ vertices, where each vertex has degree at least $n/2$. Prove that G contains a Hamilton cycle.*

(9) Let $k \geq 2$ be an integer. A graph is called *k-partite* if its vertex set V can be partitioned in such a way that no two vertices in a partition class are adjacent to each other. Prove each of the following statements.

    (a) Every graph with e edges contains a bipartite (2-partite) subgraph with at least $e/2$ edges.

    (b) Every graph with e edges contains a 3-partite subgraph with at least $\frac{2e}{3}$ edges.

    (c) For every $k \geq 2$, every graph with e edges contains a k-partite subgraph with at least $\frac{(k-1)e}{k}$ edges.

Remark. Clearly, part (c) of this problem is a generalization of parts (a) and (b). The following statement is a stronger version of part (c), but we do not know of a proof that uses monovariance. *Every graph on kn vertices and e edges, $2 \leq k \leq n$, contains a k-partite subgraph with at least $\frac{(k-1)e}{k}$ edges, with each partition class having n vertices.* ◇

(10) Consider a simple graph G with the vertex set $\{v_1, \dots, v_n\}$ such that $d_G(v_i) = d_i$ for all $i = 1, \dots, n$ and $d_1 \geq d_2 \geq \cdots \geq d_n$. Prove that either vertex v_1 is joined to vertices $v_2, v_3, \dots, v_{d_1+1}$ or one can rearrange some edges of G in such a way that in the obtained graph G', $d_{G'}(v_i) = d_G(v_i) = d_i$ for all $i = 1, \dots, n$ and vertex v_1 is joined in G' to vertices $v_2, v_3, \dots, v_{d_1+1}$.

Remark. This statement is the key idea in a proof of a theorem by Havel and Hakimi, which gives a necessary and sufficient condition for a sequence of n nonnegative integers to be **graphic**, i.e., to be a degree sequence of a simple graph: for $n > 1$, a sequence $d_1 \geq d_2 \geq \cdots \geq d_n$ is graphic if and only if the sequence

$$d_2 - 1, d_3 - 1, \dots, d_{d_1+1} - 1, d_{d_1+2}, d_{d_1+3}, \dots, d_n$$

is graphic.

This formulation is not as difficult as it may appear at first glance. The second degree sequence will have one less term than the given sequence. The first d_1 terms of the second sequence will each be one less, respectively, than the first d_1 terms appearing after the first term of the first sequence. After that point, the two sequences coincide. For example, the theorem asserts that the sequence $(5, 4, 4, 3, 3, 2, 2, 1)$ will be graphic if and only if the sequence $(3, 3, 2, 2, 1, 2, 1)$ is graphic if and only if the sequence $(2, 1, 1, 2, 1, 1)$ is graphic if and only if $(1, 0, 1, 1, 1)$ is graphic if and only if $(0, 1, 1, 0)$ is graphic, which it is. (Note that the terms of each sequence must be placed in decreasing order to form the next sequence.) ◇

A father has twenty identical one-dollar bills that he wants to place into four unmarked envelopes, one for each of his four children. In how many ways can he distribute the bills if the envelopes each must hold a different number of bills?

11

Topics in Counting

Never underestimate a theorem that counts something.
—from *A First Course in Abstract Algebra*
John B. Fraleigh (1930–)

11.1 Inclusion-Exclusion

We begin this section with a problem one might encounter in a puzzle book or middle school math contest. Try to solve it yourself before reading our solution.

Example 124. Customers at an ice cream shop are surveyed as to their taste preferences. Here are the responses.

- 48 of those surveyed like chocolate.
- 39 of those surveyed like strawberry.
- 49 of those surveyed like cookies & cream.
- 15 of those surveyed like strawberry and chocolate.
- 21 of those surveyed like chocolate and cookies & cream.
- 26 of those surveyed like strawberry and cookies & cream.
- 8 of those surveyed like all three flavors.
- 5 of those surveyed like none.

How many people completed the survey?

Solution. One approach to solving a problem like this is to use a Venn diagram. Letting C, S, and K denote the sets of persons who like chocolate, strawberry, and cookies & cream, respectively, we create the diagram shown in Figure 11.1(a). We can fill each region of the diagram with numbers corresponding to the number of elements in each set by moving from the inside out and filling in the regions representing the sets or intersection of sets with numbers giving the number of elements in that set.

We begin by putting an 8 in the region corresponding to $C \cup K \cup S$, since 8 persons liked all three flavors. Then, since $|C \cup S| = 15$, this leaves $15 - 8 = 7$ elements for

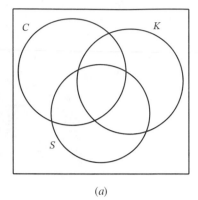 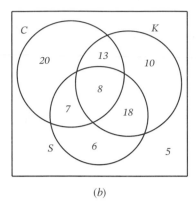

(a)   (b)

Figure 11.1

the part of $C \cup S$ that does not include K. Similarly, we place a 13 (= 21 − 8) and an 18 (= 26 − 8) in the other parts of $C \cup K$ and $S \cup K$, respectively. At this point, we have accounted for $7 + 8 + 13 = 28$ elements of C, which leaves $48 − 28 = 20$ elements for the remaining part of C, denoting those who like only chocolate. Similarly, $49 − 10 = 39$ persons like only cookies & cream and $39 − 33 = 6$ like only strawberry. The completed diagram is shown in Figure 11.1(b). Adding up all of the numbers in the diagram, we see that a total of

$$20 + 13 + 10 + 7 + 8 + 18 + 6 + 5 = 87$$

persons completed the survey.   □

The above method worked fine for this example with just three sets. However, with more sets and for more abstract examples, formalizing the arithmetic behind the above calculations will be helpful. Notice that in terms of the sizes of sets, it was sufficient to find $|C \cup K \cup S|$ (and then add 5).

Here is another problem, seemingly unrelated. Suppose your teacher returns your exams to you in random fashion, one exam per student. In how many ways can she do this if no student receives his or her own exam? Think about this problem if you wish, and we will return to it later in the section.

What the two problems above have in common is that in each case we need to count the number of items having specified properties, where the items holding these properties may overlap. Translating into mathematical language, we want to count the number of elements in the union of overlapping sets. (In the second problem, it may not be clear yet what properties or sets should be formed.) The general method for finding the number of elements in the union of sets utilizes the Principle of Inclusion-Exclusion; the applications of this principle are wide and powerful, as we will see.

Before stating the general case, we will derive the Inclusion-Exclusion formula for the number of elements in the union of three sets, S, C, and K. We already saw in Chapter 3 that

$$|A \cup B| = |A| + |B| - |A \cap B|.$$

The reason for subtracting $|A \cap B|$ is to avoid double-counting the elements that lie in both A and B. Similarly, the value of $|C| + |K| + |S|$ will generally exceed $|C \cup K \cup S|$, because elements that appear in $S \cap K$, $C \cap S$, and $K \cap S$ will be double-counted. Therefore we should subtract $|C \cap K|$, $|C \cap S|$, and $|K \cap S|$ from our total. However, in doing so, we have taken away too much! Consider the elements in $C \cap K \cap S$. Each one was included three times in the sum $|C| + |K| + |S|$ and was subsequently removed three times when the sizes of the intersections were subtracted. To rectify this, $|C \cap K \cap S|$ must be added back in. Putting this all together, we obtain

$$|C \cup K \cup S| = |C| + |K| + |S| - |C \cap K| - |C \cap S| - |K \cap S| + |C \cap K \cap S|.$$

This formula is verified in the above example, where we had

$$|C \cup K \cup S| = 39 + 49 + 48 - 26 - 15 - 21 + 8 = 82.$$

Adding in the five students who did not like any of the flavors, we obtain 87, as before.

Example 125. How many integers in $[1,000]$ are relatively prime with 3 and 5? (This example is a restatement of Problem 8(d) in Section 8.2.)

Solution. All 1,000 integers in $[1,000]$ are either relatively prime with 3 and 5 or are multiples of 3 or 5. We can therefore solve the problem by finding the number of integers in the latter group. Let A_3 be the subset of $[1,000]$ containing all multiples of 3, and similarly for A_5. Clearly, $A_3 = \{3, 6, \dots, 999\}$ and $A_5 = \{5, 10, \dots, 1,000\}$, so $|A_3| = 333$ and $|A_5| = 200$. The intersection of these two sets will be the multiples of 15 in $[1,000]$, namely $\{15, 30, \dots, 990\}$, so $|A_3 \cap A_5| = 990/15 = 66$.

Then, by the Principle of Inclusion-Exclusion for two sets, the number of integers in $[1,000]$ that are multiples of 3 or 5 is

$$|A_3 \cup A_5| = |A_3| + |A_5| - |A_3 \cap A_5| = 333 + 200 - 66 = 467.$$

Hence, the number of integers in $[1,000]$ relatively prime with 3 and 5 is $1000 - 467 = 533$. $\qquad\square$

The general case of finding $|A_1 \cup A_2 \cup \dots \cup A_n|$ for arbitrary sets follows the same process as above: sum the number of elements in the individual sets, subtract the numbers of elements in the pairwise intersections of the sets, add the numbers of elements in the intersection of any three of the sets, subtract the numbers of elements in the intersection of any four of the sets, etc. This counting method is captured in the following important theorem.

Theorem 11.1 (Inclusion-Exclusion). *Let A_1, A_2, ..., A_n be a nonempty collection of subsets of a finite set Ω. Then*

$$\left| \bigcup_{i \in [n]} A_i \right| = \sum_{1 \le i \le n} |A_i| - \sum_{1 \le i < j \le n} |A_i \cap A_j| + \sum_{1 \le i < j < k \le n} |A_i \cap A_j \cap A_k|$$
$$- \cdots + (-1)^{n-1} \left| \bigcap_{i \in [n]} A_i \right|.$$

Theorem 11.1 formally restates the procedure described in the previous paragraph for finding the size of the union of sets. Notice the special indexing that was used to denote the various intersections in each term and the $(-1)^n$ factor to ensure that the final term has the proper parity. Though the idea may be clear from the examples, the general statement is quite formidable to state precisely. Even in the statement of the theorem, while we gave the first three terms (each of which is a sum in its own right) and the last term, we did not clearly state the general term in the sum. For that reason, a rigorous proof is somewhat difficult. We therefore present and prove an alternate statement of the Inclusion-Exclusion Principle that does not use an ellipsis in lieu of stating the general term. Our proof is a slight adaptation of the one given by Jukna in [15].

Theorem 11.2 (Alternate form of Inclusion-Exclusion). *Let $A_1, A_2, ..., A_n$ be a nonempty collection of subsets of a finite set Ω. Let $A_\varnothing = \Omega$, and for any nonempty set I, $I \subseteq [n]$, let $A_I = \bigcap_{i \in I} A_i$. Then*

$$\left| \bigcap_{i \in [n]} \overline{A_i} \right| = \sum_{I \subseteq [n]} (-1)^{|I|} |A_I|.$$

Take time to carefully study Theorem 11.2 and Example 126, which accompanies it. Notice that the alternate version of the Principle of Inclusion-Exclusion gives an expansion of $|\Omega| - | \bigcap_{i \in [n]} \overline{A_i} |$, which is sufficient because of De Morgan's Law, which states that

$$\left| \bigcup_{i \in [n]} A_i \right| = |\Omega| - \left| \overline{\bigcup_{i \in [n]} A_i} \right| = |\Omega| - \left| \bigcap_{i \in [n]} \overline{A_i} \right|.$$

Part of the theorem statement that may look peculiar is the term $(-1)^{|I|}|A_I|$. Just what is that I doing there?! In the summation, I takes on all possible subsets of $[n]$; in this way, we choose all possible subsets of a given size in order to take their intersections. For example, if $I = \{1, 2, 5\}$, then $A_I = A_1 \cap A_2 \cap A_5$. This will be just one of the $\binom{n}{3}$ ways of selecting three of the A_i to intersect. Recall also that the general description of the method requires alternating between adding and subtracting various intersections. This requirement is implemented in the theorem by way of the factor $(-1)^{|I|}$; a given term is added or subtracted depending on the parity of I, i.e., depending on whether an even or odd number of the A_i are being intersected.

Proof. We define S as

$$S = \sum_{I \subseteq [n]} (-1)^{|I|} |A_I|.$$

Then

$$S = \sum_{I \subseteq [n]} \sum_{x \in A_I} (-1)^{|I|} = \sum_{x \in \Omega} \sum_{\{I : x \in A_I\}} (-1)^{|I|}.$$

Let $S_x = \sum_{\{I : x \in A_I\}} (-1)^{|I|}$. Then $S = \sum_{x \in \Omega} S_x$. We show that $S_x = 1$ for $x \in \bigcap_{i \in I} \overline{A_i}$ and $S_x = 0$ otherwise.

To see this, let $J_x := \{i : x \in A_i\}$. Note that $J_x = \varnothing$ if and only if $x \in \bigcap_{i \in I} \overline{A_i}$, and in this case $S_x = \sum_{I = \varnothing} (-1)^{|\varnothing|} = (-1)^0 = 1$. When $J_x \neq \varnothing$ (which happens if and only if $x \notin \bigcap_{i \in I} \overline{A_i}$), we have

$$S_x = \sum_{\{I : x \in A_I\}} (-1)^{|I|} = \sum_{\{I : I \subseteq J_x\}} (-1)^{|I|}$$

$$= \sum_{k=0}^{|J_x|} \sum_{\{I : I \subseteq J_x, |I| = k\}} (-1)^{|I|}$$

$$= \sum_{k=0}^{|J_x|} \binom{|J_x|}{k} (-1)^k = (1 - 1)^{|J_x|} = 0.$$

Therefore, $S = \sum_{x \in \Omega} S_x = | \bigcap_{i \in I} \overline{A_i} |$, which concludes the proof. $\square$

Example 126. Eight books, including two each of math, history, English, and biology books, are placed from left to right on a shelf. In how many ways can this be done if two books from the same subject area are not placed beside each other?

First Thoughts. As a way to begin thinking about how to answer a question like this, consider the much simpler situation in which only the two math books may not be placed beside each other. In this case, the simplest solution results from counting the number of orderings of the books in which the two math books are adjacent and subtracting that number from the total number of orderings of the eight books on a shelf. To count the number of ways in which the two math books are adjacent, we think of the two math books as one unit and use the Multiplication Rule: $(7!) \cdot 2$. (The factor of 2 accounts for the number of ways to swap the positions of the adjacent two math books within a given ordering of the eight books.) The number of ways of arranging any eight distinct books is 8!. Therefore, the number of ways to order the eight books on a shelf if we do not permit the two math books to be adjacent is

$$8! - (7!) \cdot 2 = 30{,}240.$$

Now, suppose we take the next step and assume that neither the two math books nor the two history books can be placed next to each other. We could again count the number of orderings in which the two math books are adjacent and count the number of orderings in which the two history books are adjacent and subtract those quantities from 8!. If we do this, however, we will be subtracting too much; in some of the arrangements in which the two math books are adjacent, the two history books are also adjacent, so we will be "double subtracting". This can be resolved by counting the number of orderings in which both the math books and the history books are adjacent and adding that back in. It can now be seen that the solution to the question originally asked is most easily obtained by using the Principle of Inclusion-Exclusion.   ◇

Solution. Let M, H, E, and B be the sets of orderings of the eight books in which the two math books are adjacent, the two history books are adjacent, the two English books are adjacent, and the two biology books are adjacent, respectively. We seek to find all orderings that do not overlap with any of these sets; i.e., we want to find

$$|\overline{M} \cap \overline{H} \cap \overline{E} \cap \overline{B}|.$$

Applying Theorem 11.2, using the set $\{M, H, E, B\}$, each term of the summation S will correspond to I equaling a different subset of $\{M, H, E, B\}$. There will be $2^4 = 16$ of them, broken down as follows:

(a) One subset with all four elements M, H, E, and B.

(b) $\binom{4}{3} = 4$ subsets with three elements from $\{M, H, E, B\}$.

(c) $\binom{4}{2} = 6$ subsets with two elements from $\{M, H, E, B\}$.

(d) $\binom{4}{1} = 4$ subsets with one element from $\{M, H, E, B\}$.

(e) One empty subset.

We will handle the five cases above separately. In part (a), when $|I| = 4$, the term in the sum is $(-1)^4 |M \cup H \cup E \cup B|$. The set in question here has both books of all four subject types adjacent. There are $4! = 24$ ways to order the four subjects, and within each subject, two ways to order the two books. Hence,

$$(-1)^4 |M \cup H \cup E \cup B| = (4!)(2^4) = 384.$$

In part (b), there are 4 ways to have $|I| = 3$. In each of these cases, three of the subjects must have their two books adjacent, and the two books in the fourth subject can be placed anywhere (adjacent or not). Thus we have five "blocks" to place (three subjects and two books). There are $5!$ ways to place these blocks, and within each of the three subject blocks, two ways to order the two books. Hence, the number of orderings of the books contributed by this term is

$$(-1)^3 (4)(5!)(2^3) = -3,760.$$

In part (c), there are 6 ways to have $|I| = 2$. In each of these cases, two of the subjects have their two books adjacent, and the two books in the other two subjects can be placed anywhere. Thus we have six "blocks" to place (two subjects and four books). There are $6!$ ways to place these blocks, and within each subject block, two ways to order the two books. Hence, the number of orderings of the books contributed by this term is

$$(-1)^2 (6)(6!)(2^2) = 17,280.$$

In part (d), there are 4 ways to have $|I| = 1$. In each case, one of the subjects has its two books adjacent, and the two books in the other three subjects can be placed anywhere. Thus we have seven "blocks" to place (one subject and six books). There are $7!$ ways to place these blocks, and within the subject block, two ways to order its two books. Hence, the number of orderings of the books contributed by this term is

$$(-1)^1 (4)(7!)(2^1) = -40,320.$$

In part (e), we do not require any of the subjects to have their books adjacent. Thus the eight books can be placed anywhere, and there are $8!$ ways this can happen. The number of ordering contributed by this term is $8! = 40,320$.

Putting the five cases together, we see that the number of book placements where no two books from the same subject are beside each other is

$$|\overline{M} \cap \overline{H} \cap \overline{E} \cap \overline{B}| = 384 - 3,760 + 17,280 - 40,320 + 40,320 = 13,904. \qquad \square$$

We now return to a problem mentioned at the beginning of the section. In how many ways can a teacher randomly return n exams to her students, one exam per student, so that no student receives his or her own exam? At its core, this problem is one about derangements. A **derangement of** $[n]$ is a permutation of the integers $1, 2, 3, \ldots, n$ so that i is not in the ith position for $1 \le i \le n$. For example, there are two derangements of $[3]$, namely $(2, 3, 1)$ and $(3, 1, 2)$.

Example 127. How many derangements are there of the set $[n]$?

First Thoughts. With Inclusion-Exclusion problems, one must be careful to define the sets appropriately. In this case, our universal set will be the $n!$ permutations of $[n]$. (Recall that a k-permutation of $[n]$ is a sequence of length k whose terms are distinct elements from $[n]$; here, we are using "permutation" to indicate an n-permutation of $[n]$.) We need to count the number of permutations that are not derangements. The idea will be to construct sets of orderings that are not derangements and count the number of orderings in the union of those sets. The set of all derangements will be the sequences in the complement of that union.   ◇

Solution. Let Ω denote the set of all $n!$ permutations of $[n]$. For each i, $1 \le i \le n$, let D_i denote those sequences in Ω for which i is in the ith position. The union of all of the D_i will be precisely the nonderangements in Ω. Therefore, by De Morgan's Law the number of derangements will be

$$|\overline{D_1} \cap \overline{D_2} \cap \cdots \cap \overline{D_n}| = n! - |D_1 \cup D_2 \cup \cdots \cup D_n|.$$

Let us determine the number of elements in the intersection of various D_i. By symmetry, all that matters is the number of sets involved in a given intersection, rather than the actual choice of sets. First, $|D_1| = (n-1)!$, because once 1 is placed in the first position, the other $n-1$ integers can be arranged in $(n-1)!$ ways. Similarly, $|D_i| = (n-1)!$ for $1 \le i \le n$.

Second, $|D_1 \cap D_2| = (n-2)!$, because there are $(n-2)!$ ways to arrange the integers 3 through n once the integers 1 and 2 are fixed in the first two positions. Similarly, $|D_i \cap D_j| = (n-2)!$ for $i \ne j$, and there will be $\binom{n}{2}$ such pairwise intersections.

By the same reasoning, $|D_1 \cap D_2 \cap \cdots \cap D_k| = (n-k)!$ for $1 \le k \le n$; likewise, there will be $(n-k)!$ elements in the intersection of any k of the D_i. There will be $\binom{n}{k}$ ways to choose k of the sets for the intersection.

Applying the Principle of Inclusion-Exclusion, we obtain

$$|D_1 \cup D_2 \cup \cdots \cup D_n| = n(n-1)! - \binom{n}{2}(n-2)! + \binom{n}{3}(n-3)!$$

$$+ \cdots + (-1)^{k+1}\binom{n}{i}(n-i)! + \cdots + (-1)^{n+1}.$$

The above sum simplifies nicely to

$$n! - \frac{n!}{2!} + \frac{n!}{3!} - \frac{n!}{4!} + \cdots + (-1)^{n+1}.$$

Subtracting this value from $n!$, we see that the number of derangements of $[n]$ has the amazingly simple form of

$$\frac{n!}{2!} - \frac{n!}{3!} + \frac{n!}{4!} - \frac{n!}{5!} + \cdots + (-1)^n = n!\left(\frac{1}{2!} - \frac{1}{3!} + \frac{1}{4!} - \cdots + \frac{(-1)^n}{n!}\right). \qquad \square$$

Further Thoughts. Those students who have studied infinite series may recognize that the number of derangements tends towards $n!/e$ as $n \to \infty$. Since there are $n!$ permutations of $[n]$, this proves that as n goes to infinity, the probability of a permutation being a derangement approaches $1/e \approx 0.37$. If the professor of a large class of students randomly gives back exams to her students, one exam per student, the probability that no student receives his or her own test is about 0.37.   ◇

Exercises and Problems.

(1) For this problem, we define a "word" to be a sequence of letters from our alphabet of 21 consonants and five vowels. How many 5-letter words

  (a) begin or end with a normal (a, e, i, o, or u) vowel?

  (b) begin or end with a normal vowel, with all letters distinct?

  (c) contain at least two normal vowels, with all letters distinct?

  (d) do not have any consecutive normal vowels or consecutive consonants?

(2) In how many ways can a professor return exams to ten students such that no student receives his or her own exam?

(3) Eight people get together for a gift exchange. Each person is given the name of one other person in the group and will buy a gift for that person. In how many ways can the names be assigned?

(4) In how many ways can eight (differently colored) 6-sided dice be rolled so that each number appears at least once? What is the probability that each number appears at least once?

(5) Use Inclusion-Exclusion to count the number of surjective (onto) functions from [5] to [3].

(6) How many integers in the set of integers from 1 to 1,000 (inclusive) are divisible by a prime number less than 10?

---

(7) A bridge hand consists of 13 cards from a standard 52-card deck. How many different hands contain at least one card in each of the four suits? What is the probability of getting dealt such a hand?

(8) Use Inclusion-Exclusion to show that the number of surjective functions from $[m]$ to $[n]$ is $\sum_{k=0}^{n-1}(-1)^k C(n,k)(n-k)^m$.

11.2 The Pigeonhole Principle

Despite its odd name and the fact that the principle has little to do with pigeons, the *Pigeonhole Principle* is well known and commonly used. While the statement itself may seem absolutely trivial, we will see how to use the principle in some very clever ways to solve problems that may not have obvious solutions.

Principle 11.3 (The Pigeonhole Principle, familiar version). *If one places n pigeons into m pigeonholes, $n > m$, then at least one pigeonhole will contain more than one pigeon.*

Stated in mathematical language, we have the following version.

Theorem 11.4 (The Pigeonhole Principle). *Let A be a finite set, partitioned into finite subsets $S_1, S_2, \ldots, S_m$. If $|A| = n > m$, then at least one of these m subsets contains more than one element.*

The principle essentially states that when dividing up objects (not necessarily pigeons!) into nonempty groups, one cannot have more objects than groups without having two or more objects in some group. Our first example of the Pigeonhole Principle was the second problem in Section 1.1, where we argued that there must be two people in Los Angeles who have the same number of hairs on their heads. We estimated that there were about 100,000 hairs on a human head and about 4 million people in Los Angeles. To put this problem into the language of the Pigeonhole Principle, we let the "pigeons" (set A) be the people in Los Angeles, and we let pigeonhole k (set S_k) contain all people who have k hairs on their heads, $0 \le k \le 100{,}000$. Since $|A| = 4{,}000{,}000 > 100{,}000$, for some j it is the case that S_j contains at least two people.

Example 128. Explain how we are sure that at a sold-out Ohio State football game in Columbus, Ohio, there will be at least two people who have the same two initials (first and last name) and were born in the same year.

Solution. For each attendee, let (F, L, Y) be the sequence in which F, L, and Y denote the first initial, last initial, and birth year, respectively, of the given attendee. For example, $(D, L, 1964)$ would denote a person with the initials D. L. born in 1964. Let us assume that all attendees are under 100 years old. By the Multiplication Rule there will be $26 \cdot 26 \cdot 100 = 67{,}600$ distinct sequences of the given type. Since Ohio State's stadium holds more than 100,000 people, we can be certain that when it is sold out there will actually be thousands of persons who share both initials and birth year with another person. In this solution, the "pigeons" are the people at the game and the "pigeonholes" are the distinct triples of the form (F, L, Y). $\quad\square$

Example 129. If seven distinct integers are selected from the set $[12]$, two of the seven must have the unlucky sum of 13.

Solution. Notice that the two numbers in each of the six pairs $(1, 12)$, $(2, 11)$, $(3, 10)$, $(4, 9)$, $(5, 8)$, and $(6, 7)$ add up to 13; consider these pairs the "pigeonholes". When selecting seven distinct integers from $[12]$, two of these "pigeons" must end up in the same pigeonhole. $\quad\square$

Example 130. Of 90 people seated (equally spaced) at a round table, more than half are women. Prove that there are two women who are seated diametrically opposite to one another.

Solution. Consider the 45 (unordered) pairs of diametrically opposite seats to be the "pigeonholes". Each woman (a "pigeon") is in the pigeonhole that contains her seat. Since there are at least 46 women around the table, one of the 45 pigeonholes must contain two of them. $\quad\square$

Example 131. During the year 2017, Gary drank at least one cup of coffee per day and exactly 500 cups of coffee in total. Prove that there must be a sequence of consecutive days in which Gary drank exactly 225 cups of coffee.

Solution. Let a_k denote the total number of cups of coffee Gary drank during the first k days of the year, $1 \leq k \leq 365$. It follows that

$$1 \leq a_1 < a_2 < a_3 < \cdots < a_{364} < a_{365} = 500.$$

We desire to find integers i and j, $1 \leq i < j \leq 365$, for which $a_i + 225 = a_j$.

Let us add 225 to each term in the string of inequalities to obtain

$$226 \leq a_1 + 225 < a_2 + 225 < \cdots < a_{365} + 225 = 725.$$

These two strings of inequalities contain 730 numbers, namely

$$a_1, a_2, \ldots, a_{365} = 500 \quad \text{and} \quad a_1 + 225, a_2 + 225, \ldots, a_{365} + 225 = 725,$$

and each of these 730 numbers must lie between 1 and 725. Therefore, two of the numbers must be equal. But which ones can be equal? Certainly none of the first 365 numbers can be equal, because $(a_1, a_2, \ldots, a_{365})$ is a strictly increasing sequence.

Similarly, none of the second group of 365 numbers can be equal. Therefore, it must be the case that there are an i and j, $1 \le i \le j \le 365$, such that $a_j = a_i + 225$. In this case, Gary drank exactly 225 cups of coffee between days $i + 1$ and j, inclusive. □

Example 132. On a 12-player basketball team, the sum of the points scored by those players after five games is 332. Prove that a starting lineup (five players) can be chosen so that the sum of their points scored is at least 139.

Solution. There are $\binom{12}{5} = 792$ five-person subsets of the twelve players. We do not know the sums of the points scored for all such subsets, but we can compute the sum of all such sums! Each of the twelve players' point totals belongs to exactly $\binom{11}{4} = 330$ such subsets, because that is the number of ways that one player can choose four other players to also be in the starting lineup. When we add the points from all 792 subsets, the total must be $330 \cdot 332$. Therefore the average of these 792 sums is $330 \cdot 332/792 \approx 138.3$. Since each sum is an integer, this proves that there must be a starting lineup for which the total of the points scored by the players is at least 139. □

Further Thoughts. In the previous example, we made no explicit mention of the Pigeonhole Principle. However, the principle can be applied to prove the general claim that any set of numbers must contain a number that is at least as high (or at least as low) as the mean of the set. This fact is stated in Principle 11.6, which is a consequence of the following generalization of the Pigeonhole Principle. ◇

Principle 11.5 (Generalized Pigeonhole Principle). *If one places n pigeons into m pigeonholes with respective capacities of $c_1, c_2, \ldots, c_m$ and $n > c_1 + c_2 + \cdots + c_m$, then at least one of the pigeonholes will contain more pigeons than its capacity.*

In the Generalized Pigeonhole Principle, by setting each c_i equal to $\lfloor (n-1)/m \rfloor$ and noting that $n > m \cdot \lfloor (n-1)/m \rfloor$, we obtain the following.

Principle 11.6 (Extended Pigeonhole Principle). *If one places n pigeons into m pigeonholes, then one of the pigeonholes will contain at least $\lfloor (n-1)/m \rfloor + 1$ pigeons.*

Example 133. Prove that in any set of 99 natural numbers, there is a subset of fifteen of them with the property that the difference of any two numbers in the subset is divisible by 7.

Solution. Let $A \subset \mathbb{N}$, with $|A| = 99$. Then A can be partitioned into the seven subsets $B_0, B_1, \ldots, B_6$, where each element of B_i is congruent to i modulo 7.

From here, we can apply Principle 11.5 by letting each of seven pigeonholes (one for each B_i) have capacity 14. Since we have 99 pigeons ($|A| = 99$), one of the pigeonholes must contain more than $\frac{99}{7} \approx 14.3$ pigeons. Since the number of pigeons must be an integer, one of the B_i contains at least 15 elements. (Alternatively, we may apply Principle 11.6 directly and conclude that one of the seven B_i contains at least $\lfloor (99-1)/7 \rfloor + 1 = 15$ elements.) The fifteen elements in this subset will all be congruent to each other modulo 7. □

Exercises and Problems.

(1) Suppose a drawer contains 10 brown socks, 5 blue socks, and 7 black socks. How many socks must be pulled out to guarantee that two of the ones drawn will match in color?

(2) There are sixteen teams in the American Conference of the National Football League. Each team plays sixteen games (not all against teams in the American Conference). What can you conclude about the worst record in the conference if no two teams end up with the same number of wins?

(3) There are 37 students in a class. Each got an A, B, C, or D on a test. Show that there are at least ten students who received the same grade.

(4) Prove that in a party with $n \geq 2$ people, there will always be two people who shake the same number of hands.

(5) Prove that among any seven integers, there must be two of them whose difference is divisible by six.

(6) Prove that among any 21 integers, there must be three of them that have the same last digit.

---

(7) In any group of twenty people, must there be two who know the same number of other people in the group? Assume that if A knows B, then B knows A, and no one knows himself or herself.

(8) Prove that if five points are selected from inside or on a square of side length 1, then there are two points whose distance is at most $\sqrt{2}/2$.

(9) Prove that if 101 points are selected from inside or on a square of side length 1, then there are two points whose distance is at most $\sqrt{2}/10$.

(10) Cells of an $n \times n$ square grid have been painted red, white, and blue.

    (a) For $n = 15$, prove that there must be (at least) two rows of cells containing the same number of squares of (at least) one of the colors. (This problem appeared in the 1994–1995 St. Petersburg Regional Mathematical Olympiad.)

    (b) Suppose that the numbers of cells painted red in the n rows are all different, and similarly for the numbers of white and blue cells. What are the possible values of n?

(11) Twelve points lie on or inside a convex region P with area 10. (A convex region is one in which the segment joining any two points of the region will lie within the region.) Prove that some subset of three of the points form the vertices of a (perhaps degenerate) triangle of area at most one.

---

(12) Prove that in any sequence $(x_1, x_2, \ldots, x_n)$ of integers there must be a subsequence of consecutive terms whose sum is a multiple of n.

(13) Prove that if the entire plane is colored in two colors, then there are two points at distance one from each other that have the same color.

(14) Seventeen rooks are placed on an 8×8 chessboard. Prove that there are at least three rooks that do not threaten each other. (This problem appears on the fantastic website *cut-the-knot.org*.)

(15) Let $A \subset [99]$ with $|A| = 50$, such that no two members of A have the property that one is a multiple of the other. Prove that A contains all odd integers between 50 and 100.

11.3 Multinomial Coefficients

In Section 8.4 we saw that the coefficient of $a^k b^{n-k}$ in the expansion of $(a+b)^n$ equals $\binom{n}{k}$, the number of ways of dividing n distinguishable objects into two labeled groups of sizes k and $n-k$. In this section we make a similar connection between the coefficients of $(a_1 + a_2 + \cdots + a_m)^n$ and the number of ways of dividing n distinguishable objects into m distinct groups with specified sizes. We begin with an example.

Example 134. How many ways can a group of 30 students be assigned to travel in a minibus, a van, and a car if the minibus holds eighteen passengers, the van seven passengers, and the car five passengers?

Solution. We can answer this question using binomial coefficients and the Multiplication Rule. There are $\binom{30}{18}$ ways to assign the passengers to the minibus. Once the minibus is filled, there are $\binom{12}{7}$ ways to assign seven of the remaining twelve students to the van. Then, there is $\binom{5}{5} = 1$ way to "choose" the (remaining) five passengers who will ride in the car. By the Multiplication Rule, $\binom{30}{18}\binom{12}{7}\binom{5}{5}$ different assignments can be made. If we rewrite this value in terms of factorials, a very nice final form emerges:
$$\binom{30}{18}\binom{12}{7}\binom{5}{5} = \frac{30!}{18!\,12!}\frac{12!}{7!\,5!}\frac{5!}{5!\,0!} = \frac{30!}{18!\,7!\,5!}. \qquad \square$$

Notice that the sum of the integers in the denominator of the final expression above (18, 7, and 5) is 30. A similar phenomenon occurs with the binomial coefficient $\binom{n}{k} = \frac{n!}{(n-k)!\,k!}$, where the sum of $n-k$ and k is n.

This example gives rise to the definition of the multinomial coefficient $\binom{n}{k_1,k_2,\dots,k_m}$. For positive integer n and nonnegative integers $k_1, k_2, \dots, k_m$ whose sum is n, we define the **multinomial coefficient** as follows:
$$\binom{n}{k_1, k_2, \dots, k_m} = \frac{n!}{k_1!\,k_2! \cdots k_m!}.$$

This value gives the number of ways of placing n distinguishable objects into m distinguishable groups of sizes $k_1, k_2, \dots, k_m$. This fact follows from straightforward factorial cancelations like those in Example 134.

Example 135. In a class of 25 students, the teacher decides to (randomly!) assign seven students an A, eight students a B, six students a C, and four students a D. In how many ways can this be done?

Solution. The 25 students and four grade types are both distinguishable, so the assignment of grades can be made in $\binom{25}{7,8,6,4} = \frac{25!}{7!\,8!\,6!\,4!} = 138{,}567$ ways. $\qquad \square$

Example 136. Twenty players arrive at a gym to play basketball, and they want to divide into four teams of five players each. In how many ways can this be done?

First Thoughts. At first glance, this problem seems similar to the previous one. However, here the teams are not distinguishable from each other. If the teams had been named A, B, C, and D and it mattered which players were on which labeled team, then the answer would simply be $\binom{20}{5,5,5,5}$. ◇

Solution. There are $\binom{20}{5,5,5,5}$ ways to divide the players into four labeled teams. Since the teams are not to be distinguished here, we must divide this number by $4! = 24$ to obtain $\frac{1}{24}\binom{20}{5,5,5,5} = \frac{20!}{(24)(5!)^4} = 488,864,376$. □

The Multinomial Theorem, presented next, is a natural extension of the Binomial Theorem. The Binomial Theorem gave a representation of $(a + b)^n$ in summation form, using binomial coefficients $\binom{n}{0}, \binom{n}{1}, \binom{n}{2}, \ldots, \binom{n}{n}$. In the Multinomial Theorem, $(a + b)$ is replaced by the sum of m terms, and the coefficients in the expansion are the multinomial coefficients just encountered.

Theorem 11.7 (Multinomial Theorem). *Let $a_1, a_2, \ldots, a_m$ be variables and let n be a positive integer. Then*

$$(a_1 + a_2 + \cdots + a_m)^n = \sum_{k_1 + k_2 + \cdots + k_m = n} \binom{n}{k_1, k_2, \ldots, k_m} a_1^{k_1} a_2^{k_2} \cdots a_m^{k_m}.$$

Proof. The proof uses strong induction on m. It is straightforward, though a bit tedious due to the notation. Here is an outline of the proof.

The statement is true when $m = 1$, because each side of the equation equals $(a_1)^n$. When $m = 2$, the statement is equivalent to the Binomial Theorem.

For the inductive step, we write

$$(a_1 + a_2 + \cdots + a_m + a_{m+1})^n = ([a_1 + a_2 + \cdots + a_m] + [a_{m+1}])^n.$$

Treating $N = a_1 + a_2 + \cdots + a_m$ as one term, we can apply the Binomial Theorem to $(N + a_{m+1})^n$:

$$(N + a_{m+1})^n = \sum_{k=0}^{n} \binom{n}{k} N^{n-k} a_{m+1}^{k}.$$

Then, within that expansion, the inductive hypothesis will be applied to the expansion of N^{n-k}, noting that for $0 \leq k \leq n$,

$$\binom{n}{k}\binom{n-k}{k_1, k_2, \ldots, k_m} = \binom{n}{k_1, k_2, \ldots, k_m, k}.$$

We leave the details to the reader. □

Example 137. Find $(a + b + c)^3$.

Solution. According to the theorem, we have

$$(a + b + c)^3 = \binom{3}{3,0,0}a^3b^0c^0 + \binom{3}{2,1,0}a^2b^1c^0 + \binom{3}{2,0,1}a^2b^0c^1$$

$$+ \binom{3}{0,3,0}a^0b^3c^0 + \binom{3}{1,2,0}a^1b^2c^0 + \binom{3}{0,2,1}a^0b^2c^1$$

$$+ \binom{3}{0,0,3}a^0b^0c^3 + \binom{3}{1,0,2}a^1b^0c^2 + \binom{3}{0,1,2}a^0b^1c^2$$

$$+ \binom{3}{1,1,1}a^1b^1c^1$$

$$= a^3 + 3a^2b + 3a^2c + b^3 + 3ab^2 + 3b^2c + c^3 + 3ac^2 + 3bc^2 + 6abc.$$

□

We close this section by making a connection between multinomial coefficients and the number of rearrangements (reorderings of the letters) of a given word. Before reading on, the reader is invited to consider why the number of rearrangements of "MAMMAL" equals the coefficient of the M^3A^2L term in the expansion of $(M+A+L)^6$.

The value of the coefficient in the Multinomial Theorem is $\binom{6}{3,2,1}$, which equals $\frac{6!}{3!\,2!\,1!}$, the number of rearrangements found by using the method explained in Section 8.3. However, our goal here is to give a combinatorial explanation for the equality, rather than simply notice the equivalence of the two expressions. In expanding the product

$$(M + A + L)(M + A + L)(M + A + L)(M + A + L)(M + A + L)(M + A + L),$$

each of the six factors must contribute either an M, an A, or an L towards each term of the sum. One such example would be for the 1st, 3rd, and 4th factors to contribute an M, for the 2nd and 5th factors to contribute an A, and for the 6th factor to contribute an L. This choice of variables from each factor is shown below, in bold:

$$① \qquad ② \qquad ③ \qquad ④ \qquad ⑤ \qquad ⑥$$
$$(\mathbf{M} + A + L)(M + \mathbf{A} + L)(\mathbf{M} + A + L)(\mathbf{M} + A + L)(M + \mathbf{A} + L)(M + A + \mathbf{L})$$

By selecting the bold variables above, in order, we form the word "MAMMAL".

As another example, choosing the bold variables below from each of the six factors leads to the word "AMMLMA":

$$① \qquad ② \qquad ③ \qquad ④ \qquad ⑤ \qquad ⑥$$
$$(M + \mathbf{A} + L)(\mathbf{M} + A + L)(\mathbf{M} + A + L)(M + A + \mathbf{L})(\mathbf{M} + A + L)(M + \mathbf{A} + L)$$

These two examples illustrate a natural one-to-one correspondence between the allocation of variables towards the M^3A^2L term in the expansion $(M + A + L)^6$ and the rearrangements of "MAMMAL". Thus, the coefficient of the M^3A^2L term will be precisely the number of ways we can choose 3 M's, 2 A's, and 1 L. There are $\binom{6}{3,2,1}$ ways to do this, and each one of them corresponds to a unique rearrangement of "MAMMAL".

Exercises and Problems.

(1) Compute the values of these multinomial coefficients.

(a) $\binom{12}{4,8}$

(b) $\binom{12}{2,3,7}$

(c) $\binom{10}{1,6,3}$

(d) $\binom{6}{1,1,1,1,1,1}$

(e) $\binom{6}{2,2,2}$

(2) Expand the following expressions using the Multinomial Theorem.

(a) $(a + b + c + d)^2$

(b) $(x - 3y + 2z)^3$

(3) What is the coefficient of xy^2z^4 in the expansion of $(x + y + z)^7$?

(4) What is the coefficient of $a^5b^2c^7$ in the expansion of $(a + b + c)^{14}$?

(5) What is the coefficient of $x^5y^2z^7$ in the expansion of $(2x - 3y + z)^{14}$?

(6) What is the coefficient of $x^3y^2z^5$ in the expansion of $(4x - 2y + 7z)^{10}$?

(7) Use multinomial coefficients to determine the number of rearrangements of the word *EXPONENTIATION*.

(8) Use multinomial coefficients to determine the number of rearrangements of the word *ANTIDISESTABLISHMENTARIANISM*.

(9) In how many different ways can the 52 cards of a standard deck of cards be placed into four indistinguishable piles of thirteen cards each?

(10) How many different bridge deals are possible? (In a bridge deal, each player receives thirteen cards from a standard deck, and the players are labeled as North, East, South, and West.)

(11) A youngster receives 22 different types of candy for Halloween. He decides to eat three pieces per day for the next six days and then eat four pieces on the seventh day. In how many ways can he allocate the pieces of candy to accomplish this feat?

---

(12) What is the coefficient of the term in the expansion of $(x^2 + y + z)^{10}$ for which the exponent of y is 3 and the exponent of z is 5?

(13) What is the sum of all coefficients in the expansion of each expression below?

(a) $(w + x + y + z)^{100}$

(b) $(x + 2y - 3z^2)^{100}$

(14) For fixed n, what is the sum of all multinomial coefficients of the form $\binom{n}{k_1,k_2,\ldots,k_m}$?

(15) How many 13-card bridge hands have five cards of one suit, four cards of another suit, three cards of another suit, and one card in the final suit?

(16) Show that the probability that each player receives one ace in a bridge deal is $\binom{13^4}{/} 524$. This value can be obtained after simplifying the result from a standard counting argument. However, the simplicity of the expression begs for a direct argument. Can you find it?

(17) For nonnegative integers n and k, prove that $\binom{kn}{n,n...,n}$ is divisible by $k!$.

11.4 Combinatorial Identities

In this short section, we illustrate the use of a **combinatorial proof** to show two quantities are equal. In this method, a counting question (with one answer) is posed and two or more different arguments are used to answer the question. Since the answers must match, an identity is proven. Such a proof is often more elegant and desirable than an algebraic proof. We examine a variety of examples here and in the problem set. Many more can be found in the book *Proofs that Really Count: The Art of Combinatorial Proof* [4].

There are at least three approaches for determining the number of possible lottery tickets in Example 100, where a player must select six numbers from $\{1, 2, ..., 40\}$ and then specify one of those numbers as the "red" number. Our method there yielded a total of $6 \cdot C(40, 6)$ different tickets. Alternatively, one could first choose the red number (40 ways to do this). Then, there would be $C(39, 5)$ ways to choose the remaining five numbers to complete the set of six, resulting in $40 \cdot C(39, 5)$ ways to complete the lottery ticket. Yet another method would be to select the five nonred numbers first, and there are $C(40, 5)$ ways to do this. The red number would then be chosen from the remaining 35 numbers. By the Multiplication Principle, this yields $C(40, 5) \cdot 35$ ways to form the ticket. The skeptic may verify that $6 \cdot C(40, 6) = 40 \cdot C(39, 5) = 35 \cdot C(40, 5)$ with a calculation.

We can generalize this specific example to give the following combinatorial identity.

Theorem 11.8. *For positive integer n and $1 \le k \le n$,*

$$k \cdot C(n, k) = n \cdot C(n-1, k-1) = (n-k+1) \cdot C(n, k-1).$$

Proof. Left to reader in Problem 2.   □

We saw the following identity previously, where we proved it with the Binomial Theorem. A combinatorial proof is very nice here, too.

Theorem (Theorem 8.12). *For any positive integer n, $\sum_{k=0}^{n} \binom{n}{k} = 2^n$.*

Proof. We use a combinatorial proof by answering the question "*How many subsets can be formed from an n-element set?*" in two different ways. On one hand, we have already seen in Theorem 8.5 that the answer is 2^n. On the other hand, since there are $\binom{n}{k}$ ways to form a subset with k elements, $0 \le k \le n$, we can simply sum the values of $\binom{n}{k}$ for $0 \le k \le n$ to obtain the given identity.   □

Further Thoughts. The technique of considering different cases by summing over numerous values of one variable is called **conditioning on the variable**. In the summation above, we conditioned on the number of elements in the subset. This technique is used in almost any combinatorial proof involving a summation. ◇

In Problem 15 of Section 5.4, we determined that the number of ways of climbing $n \geq 1$ stairs by taking only one or two steps at a time equals f_{n+1}, where f is the Fibonacci sequence $(1, 1, 2, 3, 5, 8, \ldots)$. The Fibonacci numbers abound in combinatorial identities, as do combinatorial arguments for proving them.

Example 138. Prove that $f_1 + f_2 + \cdots + f_{n-1} = f_{n+1} - 1$, for $n \geq 2$.

Proof. Consider the question, "*How many ways are there to climb n stairs by taking one or two steps at a time with at least one step of size two?*" On one hand, by eliminating the case where all steps are size one, we see there are $f_{n+1} - 1$ ways. Alternatively, condition on the number k for which the last step of size two is taken, covering stairs $k - 1$ and k. In such a scenario, all stairs from $k + 1$ to n are climbed one step at a time, while there are f_{k-1} ways to climb the first $k - 2$ stairs. Since k ranges from 2 to n, the total number of ways to climb the n stairs is

$$\sum_{k=2}^{n} f_{k-1} = f_1 + f_2 + \cdots + f_{n-1},$$

which establishes the identity. ☐

Example 139. Give a combinatorial proof of the following identity:

$$\binom{n}{0} + 2\binom{n}{1} + 4\binom{n}{2} + 8\binom{n}{3} + \cdots + 2^{n-1}\binom{n}{n-1} + 2^n\binom{n}{n} = 3^n.$$

First Thoughts. A short proof exists by letting $a = 2$ and $b = 1$ in the binomial expansion of $(a + b)^n$. For a combinatorial proof, the presence of 3^n suggests that we think of a problem where n consecutive decisions get made, 3 choices per decision, and find another way to count the total number of ways to represent the decisions. Try to think of such a problem before reading the scenario we created. ◇

Solution. Suppose you have n coins, each minted in a different year. In how many ways can you lay a sequence of them (of any length up to n, including the empty sequence) in a row, chronologically by mint date, where changing a coin from heads to tails, or vice versa, is considered to change the sequence?

We answer this question in two ways. One way to answer the question is to break the sequences into $n + 1$ cases, conditioning on the number of coins laid out. There are 2^k ways to create a head-tail sequence with k specific preordered coins, and there are $\binom{n}{k}$ ways to select k coins from the given set of n coins. Hence, there are $\binom{n}{k}(2^k)$ ways to select and orient (as heads or tails) k distinct coins. Letting k take all values from 0 to n, we see that one answer to the question appears in the left-hand side of the equation we are trying to prove.

On the other hand, we can think of this process as sequentially making a 3-way choice for each of the n coins once they are ordered by mint date: heads, tails, or put it in your pocket. There are three ways to make this choice for each coin, which yields 3^n total ways. Since both solutions give an answer to the question, we have proven the given identity. ☐

Example 140. Prove that

$$\left(\binom{n}{2}{2}\right) = 3\binom{n}{4} + 3\binom{n}{3}.$$

First Thoughts. When you see a familiar quantity like $\binom{n}{2}$ in an identity, you should take time to think of various counting scenarios in which the quantity arises. We know that $\binom{n}{2}$ counts the number of two-element subsets of an n-element set. On the left-hand side, we are thus counting the number of pairs of two-element subsets in an n-element set. Since the graph K_n has $\binom{n}{2}$ edges (edges are defined to be two-element subsets of the vertices), this problem can be answered in a graph context. What does the expression on the left count in K_n, and how else can this quantity be counted? ◇

Proof. Consider the question, "*How many labeled subgraphs of K_n have two edges?*" On one hand, based on the number of edges in a complete graph, the answer is $\left(\binom{n}{2}{2}\right)$.

An alternative counting method involves two cases: either the two edges of the subgraph share a vertex or they do not. When the two edges have a common vertex, the subgraph contains three vertices. There are $\binom{n}{3}$ ways to choose the three vertices from K_n, and once those vertices are chosen for the labeled subgraph, there are three ways to place two edges on them (since exactly one vertex will have degree 2). By the Multiplication Rule, this case includes $3\binom{n}{3}$ subgraphs. When the two edges are disjoint, four vertices are needed. There are $\binom{n}{4}$ ways to select those four vertices. Fix one of them, and notice that there are three possibilities for its neighbor; then the other two vertices form the other edge. This gives $3\binom{n}{4}$ subgraphs for this case, yielding a total of $3\binom{n}{3} + 3\binom{n}{4}$ subgraphs with two edges, which establishes the identity. □

Problems.

(1) For positive numbers a and b, find an appropriate area in two different ways to give a geometric proof that $(a+b)^2 = a^2 + 2ab + b^2$.

(2) Prove the following recurrence relations between the binomial coefficients for integers n and k, $1 \le k < n$. Try to prove each in two ways: using algebraic manipulations and by a combinatorial argument.

(a) $\binom{n}{k} = \frac{n}{k}\binom{n-1}{k-1}$.

(b) $\binom{n}{k} = \frac{n-k+1}{k}\binom{n}{k-1}$.

(c) $\binom{n}{k} = \binom{n-1}{k-1} + \binom{n-1}{k}$.

(3) Let m, k, and n be integers with $0 \le k \le m \le n$. Answer the following question in two different ways, thereby establishing a combinatorial identity. *In how many ways can one select m integers from $[n]$ and then color k of those m integers red and the rest of them green?* Verify that your identity works for $n = 20$, $m = 12$, and $k = 4$.

(4) Use a combinatorial argument to prove the following identity:

$$\frac{\sum_{k=0}^{n} k \binom{n}{k}}{2^n} = \frac{n}{2}.$$

(5) Show that the following identity holds for $n = 5$ and then use a counting proof to establish it for $n \geq 2$:

$$\sum_{k=1}^{n} \binom{n}{k}(k)(k - 1) = n(n - 1)2^{n-2}.$$

---

(6) Use an argument similar to the one in Example 138 to prove that

$$f_1 + f_3 + f_5 + \cdots + f_{2n-1} = f_{2n}.$$

(7) Consider the following equation, where f is the Fibonacci sequence:

$$f_{n+1}^2 - f_{n+1} = 2 \sum_{k=0}^{n-2} f_{k+1} f_{n-k} f_{n-k-1}.$$

(a) Show that the equation holds for $n = 6$.

(b) Prove the equation is an identity by answering the following question in two different ways: *"In how many ways can Leonardo climb n stairs on Monday and n stairs on Tuesday, taking one or two steps at a time, if he must use a different sequence of steps on the two days?"*

(8) For integers $1 \leq k \leq n$, consider the equation

$$\binom{k}{k} + \binom{k+1}{k} + \binom{k+2}{k} + \cdots + \binom{n}{k} = \binom{n+1}{k+1}.$$

(a) Show that the equation is valid for $k = 4$ and $n = 7$.

(b) Describe where the terms in the equation appear in a convenient form in Pascal's Triangle.

(c) Prove that the equation is a combinatorial identity.

(9) For positive integers n and k, prove that

$$n^k - n = (n - 1)(n^{k-1} + n^{k-2} + \cdots + n^2 + n)$$

by considering the number of nonconstant sequences of length k with entries from $[n]$.

(10) Use a technique similar to the one employed in Example 140 to prove that

$$\left(\binom{n}{2}{3} \right) = \binom{n}{3} + 16 \binom{n}{4} + 3 \binom{n}{3} \binom{n-3}{2} + 15 \binom{n}{6}.$$

(11) Problem 17 in Section 11.3 asked you to prove that $\binom{nk}{n,n,\ldots,n}$ is divisible by $k!$ for positive integers n and k. Give a combinatorial proof of this fact.

(12) Use a counting argument to prove that

$$\sum_{k=0}^{n} \left[\binom{n}{k} \right]^2 = \binom{2n}{n}.$$

(13) Use a combinatorial proof to prove that for positive integers n,

$$\sum_{k=0}^{\lfloor n/2 \rfloor} \binom{n}{2k} = 2^{n-1}.$$

(14) Use a combinatorial proof to prove that

$$\sum_{k \geq 0} \binom{n}{2k} = \sum_{k \geq 0} \binom{n}{2k+1},$$

assuming that $\binom{n}{j} = 0$ for $j > n$. (The argument is easier when n is even.)

11.5 Occupancy Problems

A goal of this section is to introduce several new functions arising from the broad class of counting problems known as *occupancy problems*. Unlike with combinations and permutations, however, closed-form expressions for some of the solutions are difficult to obtain. In the first two subsections we investigate the partitions of a set and the partitions of an integer. In the third subsection we examine their connection to the aforementioned occupancy problems.

11.5.1 Partitions of a set.

Example 141. A rhyming scheme with k sounds is a pattern of endings in which each of the k sounds ends at least one line of the rhyme. A scheme is typically denoted by a sequence of letters, with lines that rhyme represented by the same letter. This 5-line limerick[1] is one such example, following the rhyme pattern (a, a, b, b, a):

> There was a young lady named Bright,
> Who traveled far faster than light.
> She set out one day,
> In a relative way,
> And returned home the previous night.

How many different 5-line rhyming schemes can be employed with two sounds?

Solution. Suppose the sounds are a and b and the sound at the end of the first line is a. We now look at cases. There is only one pattern in which a occurs once, namely (a, b, b, b, b). There will be 4 patterns where a appears twice—one such example is the pattern (a, a, b, b, b)—since there are four choices for the second position of a. Similarly, there will be $C(4, 2) = 6$ patterns where a occurs three times and $C(4, 3) = 4$ patterns where a appears four times. In no patterns can a appear at the end of all five lines, since sounds a and b must each appear at least once. Altogether, we see there are $1 + 4 + 6 + 4 = 15$ rhyming patterns in which the first line ends with an a.

 If the first line ends in sound b instead of a, then similar counting to what is above will also give us fifteen cases. However, the corresponding rhyming schemes will actually be the same, with a and b simply interchanged in the pattern. Therefore, there are fifteen 2-sound rhyming schemes. □

---

[1] Various versions of this limerick have appeared over the years, but A. H. Reginald Buller is considered to be the original author, in 1923.

Further Thoughts. Alternatively, each rhyming scheme may be defined by partitioning the line numbers into two subsets in which the members of a subset are the line numbers that end in a particular sound, a or b. In this case, the limerick pattern would be described as $\{\{1, 2, 5\}, \{3, 4\}\}$, which for convenience we abbreviate as $\{125, 34\}$. Here are all 15 schemes, using this abbreviated format:

$$\{1, 2345\}, \{12, 345\}, \{13, 245\}, \{14, 345\}, \{15, 234\},$$
$$\{123, 45\}, \{124, 35\}, \{125, 34\}, \{134, 25\}, \{135, 24\}, \{145, 23\},$$
$$\{1234, 5\}, \{1235, 4\}, \{1245, 3\}, \{1345, 2\}. \qquad \diamond$$

To generalize the example, we define the **Stirling number of the second kind**[2], denoted $S(n, k)$, to be the number of ways of distributing n distinguishable objects into k indistinguishable boxes, with no box being empty. These numbers also may be viewed as counting the number of ways to partition a set of size n into k classes; in other words, $S(n, k)$ is the number of partitions of $[n]$ into k nonempty subsets.

In Example 141, the two "boxes" were the two sounds, and with each sound we associated the distinct line numbers that ended in that sound. That is, using the language of the Stirling number of the second kind, we "distributed" the five lines (the distinguishable objects) into two sounds (the indistinguishable boxes). In our solution, we labeled the sounds as a and b for convenience only, and they should be considered as indistinguishable in their structure. Rhyming schemes (a, b, a, b, a) and (b, a, b, a, b) were considered the same: the key characteristic of the scheme is that one sound ends lines 1, 3, and 5 while the other sound ends lines 2 and 4. In the format shown in the *Further Thoughts*, this scheme was represented as $\{135, 24\}$.

The above enumeration showed that $S(5, 2) = 15$. In the following example we find $S(5, 3)$, which may be viewed as the number of 5-line rhyming patterns with three different rhyme sounds.

Example 142. Determine $S(5, 3)$ by enumerating all partitions of the set $\{1, 2, 3, 4, 5\}$ into three nonempty subsets.

Solution. Using the same abbreviated format as in the *Further Thoughts* above, we see that $S(5, 3) = 25$: $\{1, 2, 345\}, \{1, 3, 245\}, \{1, 4, 235\}, \{1, 5, 234\}$ $\{2, 3, 145\}, \{2, 4, 135\},$ $\{2, 5, 134\}, \{3, 4, 123\}, \{3, 5, 124\}, \{4, 5, 123\}, \{1, 23, 45\}, \{1, 24, 35\}, \{1, 25, 34\}, \{2, 13, 45\},$ $\{2, 14, 35\}, \{2, 15, 34\}, \{3, 12, 45\}, \{3, 14, 35\}, \{3, 15, 24\}, \{4, 12, 35\}, \{4, 13, 25\}, \{4, 15, 23\},$ $\{5, 12, 34\}, \{5, 13, 24\}, \{5, 14, 23\}.$ $\qquad \square$

Theorem 11.9. *Stirling numbers of the second kind satisfy the following equations for integers k and n, $2 \le k \le n$.*

(1) $S(1, 1) = S(n, 1) = S(n, n) = 1$.

(2) $S(n, 2) = 2^{n-1} - 1$.

(3) $S(n, n - 1) = C(n, 2) = \binom{n}{2}$.

(4) $S(n, k) = S(n - 1, k - 1) + kS(n - 1, k)$.

---

[2]There are also Stirling numbers of the first kind, but we will not address them in this book.

Proof. In our proof we assume that the set in question is $[n]$, the first n positive integers.

(1) There is only one subset of $[n]$ containing n elements and only one way to partition $[n]$ into n subsets of one element each, which proves part (1).

(2) *First Thoughts.* Notice that each partition of $[n]$ into two subsets is determined by knowing the elements in just one of those subsets. We can thus count the subsets that contain one specific element. For example, in each partition of $\{1, 2, 3, 4, 5\}$ in the solution to Example 141, the number 5 is appended to a proper subset of integers from the set $\{1, 2, 3, 4\}$. (We only consider the proper subsets, because after 5 is appended, we must have at most four elements total.) There are $2^4 = 16$ subsets of a 4-element set, all but one of them proper, so we see that $S(5, 2) = 16 - 1 = 15$.   ◇

There are 2^{n-1} proper subsets of $[n - 1]$. Since n can be joined with each such subset to form a nonempty subset of $[n]$, only one of which is not proper, there are $2^{n-1} - 1$ ways to partition $[n]$ into exactly two subsets.

(3) We leave the proof of this part to Problem 24.

(4) *First Thoughts.* Often the reasoning behind a valid recurrence can be observed by building small examples. Let's explore that here by constructing all $S(6, 3)$ partitions of $\{1, 2, 3, 4, 5, 6\}$ into three nonempty classes. We group them into two types. In the first type, $\{6\}$ will be one of the three classes; in this type, the other two (nonempty) classes must partition $[5]$, and there are $S(5, 2)$ ways to do this, as listed in the *Further Thoughts* after Example 141. In the second type of partition, when $\{6\}$ is not one of the classes, the numeral 6 must be appended to one of the three classes of a partition of $[5]$ into three classes. In Example 142, we conveniently listed all $S(5, 3) = 25$ such partitions of $[5]$. Therefore we have $3S(5, 3)$ partitions of $[6]$ into three classes of the second type. Altogether, we see that $S(6, 3) = S(5, 2) + 3S(5, 3) = 15 + 3(25) = 90$. The general proof is virtually identical.   ◇

The partitions of $[n]$ into k nonempty subsets may be divided into two types: those partitions that contain $\{n\}$ as a class and those that do not. We count the number in each type separately. On one hand, there is a one-to-one correspondence between the partitions of $[n]$ containing $\{n\}$ as a subset and the partitions of the $(n - 1)$-element set $[n - 1]$ into $k - 1$ nonempty classes, and there are $S(n - 1, k - 1)$ of each.

On the other hand, there is a k-to-one correspondence between the partitions of $[n]$ not containing $\{n\}$ as a class and the partitions of $[n - 1]$ into k nonempty classes. This is seen by noting that any partition of $[n - 1]$ into k subsets, $S_1, S_2, ..., S_k$, can be changed to a partition of $[n]$ into k subsets in k different ways by appending n to any one of the S_i. Thus, there are $kS(n-1, k)$ partitions of this type. Combining the two cases, we see that $S(n, k) = kS(n - 1, k) + S(n - 1, k - 1)$.   □

For a fixed n, if one adds the values of $S(n, k)$ for $1 \leq k \leq n$, one obtains B_n, the total number of partitions of an n-element set. The numbers B_n are called **Bell numbers**, named after E. T. Bell, who studied their behavior in the 1930s. The sequence of Bell numbers starts as $1, 1, 2, 5, 15, 52, \ldots$. The general form can be represented in several other interesting ways, including $B_{n+1} = \sum_{k=1}^{n} \binom{n}{k} B_n$ and $B_n = \frac{1}{e} \sum_{k=0}^{\infty} \frac{k^n}{k!}$.

Useful references include the online encylopedia of integer sequences (oeis.org), Herb Wilf's excellent book *generatingfunctionology* [**21**], and Stanley's *Enumerative Combinatorics* [**19**].

11.5.2 Partitions of an integer. We begin with an example before defining the relevant terminology.

Example 143. Fan is required to run eight laps around the track for her gym class over a 5-day period. She must run a positive integer number of laps each day, and each day she is required to run at least as many laps as she ran the previous day. In how many ways can she meet these requirements?

Solution. By observation, we see that there are three ways this can happen.

$$8 = 1 + 1 + 1 + 1 + 4 = 1 + 1 + 1 + 2 + 3 = 1 + 1 + 2 + 2 + 2 \qquad \square$$

For integers n and k, define a k-**partition** of n to be an expression of n as the sum of k positive integers, where the order of the terms is irrelevant. We let $p(n, k)$ denote the number of k-partitions of n. In Example 143, we determined that $p(8, 5) = 3$. For another example, notice that

$$7 = 1 + 1 + 5 = 1 + 2 + 4 = 1 + 3 + 3 = 2 + 2 + 3,$$

so $p(7, 3) = 4$. Since the addends may be placed in any order, for convenience we often choose the representation in which the terms are monotonically increasing, as was done in the presentation here.

Theorem 11.10. (1) $p(n, n) = p(n, n - 1) = p(n, 1) = p(1, 1) = 1$ *for* $n \geq 2$.
(2) $p(n, n - 2) = 2$ *for* $n \geq 4$.
(3) $p(n, 2) = \lfloor n/2 \rfloor$ *for* $n \geq 2$.
(4) *For integers* n *and* k, $2 \leq k \leq n$, $p(n, k) = p(n - 1, k - 1) + p(n - k, k)$.
(5) $p(6n, 3) = 3n^2$ *for* $n \geq 1$, *so* $p(m, 3) \approx \dfrac{m^2}{12}$ *for large* m.

Proof. (1) If there are n, $n - 1$, or 1 addends, the values of the addends are uniquely determined.

(2) With $n - 2$ terms, either there must be one 3 with the remaining terms all 1's or there must be two 2's with the remaining terms all 1's.

(3) We leave the proof of part (3) to Problem 25.

(4) The statement is true when $n = k$ by part (1), assuming $p(0, k) = 0$. Suppose $n > k$. The k-partitions of n must be one of two types—those where 1 is a term of the sum and those where 1 is not a term. The k-partitions of the former type can be put in a natural one-to-one correspondence with the $(k - 1)$-partitions of $n - 1$. The k-partitions of the latter type can be put into one-to-one correspondence with the k-partitions of $n - k$ (by adding 1 to each addend in the expression of $n - k$). Hence, $p(n, k) = p(n - 1, k - 1) + p(n - k, k)$.

(5) Use parts (3) and (4) and induction on n. The details are left to Problem 29. $\square$

Further Thoughts. Notice the technique used to prove the recurrences in Theorem 11.9(4) and Theorem 11.10(4). In each case we divided the partitions into two types, depending on how one specific element was treated in the partition. A similar idea was used to prove that $C(n, k) = C(n - 1, k) + C(n - 1, k - 1)$ (Theorem 8.10) and should also be used in Problems 31 and 35 of this section.   ◇

The integers $p(n, k)$ are related to the **partition number** of n,

$$p(n) = \sum_{k=1}^{n} p(n, k),$$

which is the number of ways of expressing n as the unordered sum of positive integers. The function $p(n)$ and its corresponding sequence of values (1, 2, 3, 5, 7, 11, 15, 22, 30, ...) are well studied, though no closed formula is known. Hardy and Ramanujan did determine in 1918 that

$$p(n) \sim \frac{1}{4n\sqrt{3}} e^{\pi\sqrt{2n/3}} \quad \text{as } n \to \infty.$$

The terms of $p(n)$ are often expressed using generating functions, which are beyond the scope of this book. More about the partition numbers can be found in [2], [3], [19], or many standard number theory books.

11.5.3 Occupancy problems.
While $S(n, k)$ and $p(n, k)$ can both be expressed recursively, neither is simple to calculate (at least, not by hand) for general values of n and k. In this subsection, we rephrase the partition problems we just studied and see how the numbers above arise in many other contexts. The so-called "occupancy problems" can all be stated in the following form:

> Given n objects and k boxes, how many ways are there to place the objects in the boxes?

Of course, this question is vague. Are the objects distinguishable? Are the boxes distinguishable? How many objects can be placed in each box? Must each box contain at least one object? This type of systematization was pioneered by Gian-Carlo Rota (1932–1999). There are four cases, depending on the answers to the first two questions, and each case has two subcases, which depend on the answer to the third question. The four cases are depicted in Figure 11.2, using circles to denote the objects. The circles and boxes are labeled when they are distinguishable and unlabeled when they are indistinguishable.

The solutions in the various scenarios will use the binomial coefficients $C(n, k)$, the Stirling numbers $S(n, k)$, and the k-partition numbers $p(n, k)$.

Case 1: The n objects and k boxes are both distinguishable.

(a) Assume each box must contain at least one object. If the k boxes are indistinguishable, then as we saw earlier in this section, there are $S(n, k)$ ways to allocate the objects. Therefore, there are $k! \, S(n, k)$ different allocations with distinguishable boxes.

(b) If some boxes may be empty, then there are k choices of a box for each of the n objects, giving k^n ways to place the objects.

Note that if each box is permitted to hold at most one object, then $n \leq k$ and the solution is a permutation. Choose some order for the n distinct objects. Then there are k choices for the first object, $k - 1$ choices for the second object, etc. Overall, there will be $P(k, n) = (k)(k - 1)(k - 2) \cdots (k - n + 1)$ ways to place the objects in the boxes.

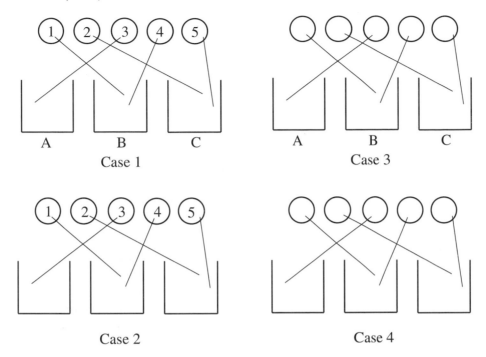

Figure 11.2

Case 2: The n objects are distinguishable but the k boxes are indistinguishable.

(a) If the k boxes must be nonempty, then these are the Stirling numbers of the second kind, so there are $S(n,k)$ ways.

(b) If some boxes may be empty, with $n \geq k$, then since there is no distinction between boxes, we can simply consider separately the number of ways to distribute the objects when the number of nonempty boxes ranges from 1 through k. The total number of allocations will be $\sum_{i=1}^{k} S(n,i)$.

Case 3: The n objects are indistinguishable but the k boxes are distinguishable.

(a) Suppose first that the k boxes must be nonempty. There are various ways to determine the solution, but none are as elegant as the one that generalizes the solution to Example 105(1). We suggest the reader go back and study it carefully. One simply views the n identical objects laid out in a row and considers the number of ways of placing $k-1$ dividers into the $n-1$ open spaces between the objects. Each such placement of the dividers will partition the objects into k groups, ordered from left to right to correspond with the k distinct boxes. This is a combination problem, and since there are $C(n-1, k-1)$ ways to place the dividers, there are $C(n-1, k-1)$ ways to place n identical objects into k distinct boxes.

(b) If we permit the boxes to be empty, then as in Example 105(2), we consider the number of arrangements of n identical objects and $k-1$ identical dividers in a row. The dividers will again partition the n objects into k groups, but since dividers can be adjacent, it will be possible to have groups (boxes) without objects. There are $C(n+k-1, k-1)$ ways to arrange these objects and dividers into $n+k-1$

positions in a row, since $k-1$ of the positions must be dividers. Thus, our answer is $C(n+k-1, k-1)$.

Case 4: The n objects and k boxes are both indistinguishable.

(a) Assume each box must be nonempty. In this case we are simply partitioning the n objects into k nonempty piles. This is essentially a k-partition of n, and there are $p(n,k)$ ways to do this.

(b) If boxes may be empty, then we could count separately the number of i-partitions of n, $1 \le i \le k$, to obtain a total of $\sum_{i=1}^{k} p(n,i)$ possibilities. However, there is a much more satisfying (and descriptive) solution! First note that there is a one-to-one correspondence between the allocations of $n+k$ indistinguishable objects into k indistinguishable boxes, where *no box is empty*, and the allocations of n indistinguishable objects into k indistinguishable boxes, where *boxes may be empty*. (This is seen either by removing an object from each box for each allocation of the first scenario or by adding an object to each box for each allocation in the second scenario.) Thus, according to part (a), there will be $p(n+k,k)$ ways to carry out either process. □

Further Thoughts. The one-to-one correspondence used to establish the value of $p(n+k,k)$ in Case 4(b) from the value of $p(n,k)$ in Case 4(a) could also have been used to derive the value of $C(n+k-1, k-1)$ in Case 3(b) from the value of $C(n-1, k-1)$ in Case 3(a). A similar argument will not work in Cases 1 and 2, because the items are distinguishable, and thus the correspondences will not be one-to-one. ◇

We summarize the various occupancy problem results in Table 11.1.

Table 11.1. Number of ways to put n items into k boxes, where in line (a) each box must be nonempty and in line (b) empty boxes are permitted.

	n Distinguishable Items	n Indistinguishable Items
k Distinguishable Boxes	(a) $k!\, S(n,k)$ (b) k^n	(a) $C(n-1, k-1)$ (b) $C(n+k-1, k-1)$
k Indistinguishable Boxes	(a) $S(n,k)$ (b) $\sum_{i=1}^{k} S(n,i)$	(a) $p(n,k)$ (b) $p(n+k,k)$

Example 144. A bus has 30 passengers and will be making six stops on its last route of the day. In how many total ways can the passengers choose the stops where they will disembark?

Solution. Let's use Table 11.1 to help choose the appropriate counting function. The $n=30$ different passengers can be interpreted as distinguishable objects to drop into $k=6$ distinguishable boxes (bus stops). Some of the boxes may be empty, as there could be bus stops where no one gets off the bus. There are 6^{30} different arrangements. If at least one passenger disembarked at each bus stop, the answer drops to $6!\, S(30,6)$. □

Example 145. A father has fifteen different tasks to assign to his three children. In how many ways can he assign these tasks to the children, assuming that each child must be given at least one task?

Solution. The $n = 15$ tasks (objects) and $k = $ children (boxes) are both distinguishable, so there are $3! \, S(15, 3)$ ways the tasks can be assigned if each child must be assigned at least one of them. □

Example 146. A physical education teacher asks her class of twelve students to run a total of 40 laps around the track. In how many ways can the laps be allocated?

Solution. The $n = 40$ laps (objects) are indistinguishable while the $k = 12$ children (boxes) are distinguishable. Since it was not required that each student must participate, by Case 3(b) there are $C(51, 11)$ ways to assign the laps to the students. □

In our final example, we revisit the opening hook problem for the chapter.

Example 147. A father has twenty identical one-dollar bills that he wants to place into four unmarked envelopes, one for each of his four children. In how many ways can he distribute the bills if the envelopes each must hold a different number of bills?

First Thoughts. This problem can be classified as one with $n = 20$ indistinguishable objects (the bills) and $k = 4$ indistinguishable parts (the envelopes). However, $p(20, 4)$ will be too large, because it includes the partitions where some envelopes contain the same number of bills. Can you think of a set of partitions of n' into four parts of not necessarily distinct sizes, for some $n' < 20$, that can be adapted to this problem? ◇

Solution. There is a one-to-one correspondence between the partitions of the integer 20 into four parts of distinct positive integer size and the partitions of the integer 14 into four positive-but-not-necessarily-distinct parts; namely, for such a partition of 14, add 0, 1, 2, and 3 to the four parts when they are listed in nondecreasing order. Thus, the answer is $p(14, 4) \, (= 23)$. See Problem 31 and subsequent problems for a generalization and more examples. □

Exercises and Problems. Unless specifically asked for, do not give numerical values as answers to these problems. When applicable, we suggest identifying the "boxes" and the "objects" in each problem before giving an expression that counts the given number of scenarios.

(1) Determine the values of (a) $S(4, 2)$ and (b) $S(6, 5)$ by listing all of the relevant partitions of [4] and [6], respectively.

(2) Determine the values of (a) $p(5, 3)$ and (b) $p(9, 4)$ by listing all of the 3-partitions of 5 and the 4-partitions of 9.

(3) Determine $p(7)$ by finding $p(7, k)$ for $1 \leq k \leq 7$ and summing the values.

(4) Let $T(n, k)$ denote the number of way of distributing n distinguishable items into k distinguishable boxes, if no boxes may be empty. We have seen that $T(n, k) = k! \, S(n, k)$. Using the fact that $T(n, 1) = 1$ and $T(n, n) = n!$ for $n \geq 1$, it turns out that there is also a recursive relationship for $n > k > 1$, namely

$$T(n, k) = k \, (T(n - 1, k - 1) + T(n - 1, k)).$$

    (a) Use this recurrence to create the first five rows of a grid of numbers $T(n, k)$. The grid should be similar to Pascal's Triangle. The first row contains the number 1, and the second row contains the numbers 1 and 2.

    (b) What is the value of $T(6, 3)$?

(5) Use the recursion in Theorem 11.9(4) to create a Pascal's Triangle-type grid for $S(n, k)$ for $1 \leq k \leq n \leq 5$.

(6) Each of 35 middle school students on a class trip must declare his or her first choice for a lunch stop. The choices are Arby's, Subway, Taco Bell, and McDonald's. In how many different ways can the preferences be made?

(7) Two dozen bars of soap of the same brand, size, and packaging must be placed into 15 unmarked equivalent boxes. If each box must contain at least one bar of soap, in how many ways can the bars be distributed?

(8) The contents of a bag of 40 Tootsie Rolls are distributed between seven trick-or-treaters cleverly costumed as the Seven Dwarves (Sneezy, Dopey, Doc, etc.). However, not all of the children like Tootsie Rolls, and some may unexpectedly refuse them with hopes of acquiring real chocolate. In how many ways can the distribution be made?

(9) A father intends to distribute 100 one-dollar bills to his four children. How many ways can he do this, assuming that each child has to get at least one dollar? What if there is no such restriction?

(10) How many positive integer solutions $(x_1, x_2, \ldots, x_9)$ are there to the equation $x_1 + x_2 + x_3 + \cdots + x_9 = 100$?

---

(11) In how many ways can a complete set of twelve identical red balls, ten identical white balls, and eight identical blue balls be placed into fifteen distinguishable boxes, assuming that at most one ball of each color can go in a given box?

(12) Prove the recursive relationship for $T(n, k)$ in Exercise 4.

(13) In how many ways can 24 be expressed as the sum of four positive even integers, where the order of the integers is not important?

(14) How many solutions $(x_1, x_2, \ldots, x_6)$ does the following equation have in each case:

$$x_1 + x_2 + \cdots + x_6 = 50?$$

    (a) All x_i's are nonnegative integers.

    (b) All x_i's are positive integers.

    (c) All x_i's are positive integers and each is at least 4.

    (d) $x_2 = 0$, $x_4 = 5$, and the other x_i are nonnegative.

    (e) All of the x_i are positive and $x_2 = 10x_1$.

(15) How many labeled graphs can be formed on n vertices? (The term "labeled" refers to the vertices.) This problem appeared in Chapter 9.

(16) How many labeled graphs with m edges can be formed on n vertices?

(17) How many different equivalence relations are there on a 6-element set? Give the numerical value.

(18) What is the number of nondecreasing sequences of length 20 with entries from the set $[8] = \{1, 2, \ldots, 8\}$, where

  (a) each element of $[8]$ must appear in the sequence?
  (b) not all elements must appear in the sequence?

(19) How many 40-character sequences can be created using the 26 letters of the alphabet? How many if each letter must be used at least once?

(20) Write $T(n, k)$ (as given in Exercise 4) as the sum of multinomial coefficients.

(21) Twenty books have been returned to a library and need to be reshelved.

  (a) In how many ways can the 20 books be placed on a temporary bookshelf with 6 shelves if there is no requirement regarding the number of books on a shelf and if the order the books appear on each shelf
    (i) is important?
    (ii) is not important?
  (b) In how many ways can the 20 books be set on 6 identical roaming carts for future shelving? Assume that the order in which they are placed on each cart does not matter.

(22) How many 6-digit integers can be formed using the digits 3, 5, and 7 if each of these digits must appear at least once?

(23) In how many ways can a class of 12 students be divided into four groups of three students each? What if the groups are distinguishable from each other, because of the order in which they must give their presentations?

(24) Prove that $S(n, n - 1) = \binom{n}{2}$ (Theorem 11.9 (3)).

(25) Prove that $p(n, 2) = \lfloor n/2 \rfloor$ for $n \geq 2$ (Theorem 11.10 (2)).

(26) How many surjective functions are there from $[7]$ to $[3]$? How many from $[n]$ to $[k]$? (We saw this problem already as Problem 8 in Section 11.1, but there the answer was expressed as a summation.)

(27) How many nondecreasing functions are there from $[n]$ to $[k]$? (A function f is nondecreasing if $f(i + 1) \geq f(i)$ for each i.)

(28) Recall that the complete bipartite graph $K_{m,n}$ is the graph with $m + n$ vertices and mn edges, where each of m vertices shares an edge with each of n vertices. A proper k-coloring of a graph is an assignment of one of k colors to each vertex such that adjacent vertices are assigned different colors. How many proper 4-colorings does $K_{10,8}$ have?

(29) Prove that $p(6n, 3) = 3n^2$ and conclude that the number of 3-partitions of an integer m is asymptotic to $m^2/12$.

---

(30) Use a counting argument to prove that $\sum_{i=1}^{k} C(k, i)S(n, i)i! = k^n$.

(31) Let $D(n, k)$ denote the number of partitions of n as the sum of k distinct positive integers. Prove the following statements for $n \geq 1$.

  (a) $D(n, 1) = 1$.
  (b) $D(n, 2) = \lfloor \frac{n-1}{2} \rfloor$.

(c) $D(n, k) = p\left(n - \binom{k}{2}, k\right)$.

(d) $D(n, k) = D(n - k, k) + D(n - k, k - 1)$.

(e) $D(6n, 3) = 6\binom{n}{2} + 1 = 3n^2 - 3n + 1$.

(32) Ralph plans to run 100 miles over an 8-day period. He will run a positive integer number of miles each day, and each day he must run more miles than he did the previous day. In how many ways can he accomplish this task?

(33) A probability instructor has 100 identical red beads, 80 identical white beads, and 5 identical shakers that can hold beads. The instructor wants to distribute the beads into the shakers in order to carry out sampling experiments. Each shaker must contain at least one bead of each color. In how many ways can this be done, assuming the numbers of red beads in the shakers are all different.

Note: While you are invited to think about how to solve the problem without this final assumption, we don't know of a solution in that case. The problem is difficult enough, even with the assumption!

(34) How many positive integer solutions $(x_1, x_2, \ldots, x_9)$, where the x_i are distinct from each other, are there to the equation

$$x_1 + x_2 + x_3 + \cdots + x_9 = 100?$$

(35) For integers n and k, define $p^*(n, k)$ to be the number of partitions of n into parts, with the largest part having size k.

(a) Prove that $p^*(n, k) = p^*(n - 1, k - 1) + p^*(n - k, k)$.

(b) Use part (a) to prove that $p^*(n, k) = p(n, k)$ for all n and k.

Remark. An alternate proof of (b) uses a configuration known as Ferrers Diagram: for a given partition of n into k parts, create the corresponding diagram of n dots in k rows, where each row has at least as many dots as the previous row. Now consider the numbers of dots in the columns. ◇

(36) How many 20-digit numbers use each of the ten digits at least once? Note that the number may not start with the digit 0.

Have a friend choose an integer between 1 and 64 and tell you the columns (labeled Red, Blue, Yellow, or Pink) of the boxes it appears in for each of the three sections below. For example, if the number is 33, then the columns, in order from the top box to the bottom box, are labeled Yellow, Red, and Red, respectively. Devise a fast method for determining the number your friend chose. Can you implement your method without having to look at the grids?

	Red				Blue				Yellow				Pink		
1	2	3	4		17	18	19	20		33	34	35	36		49	50	51	52
5	6	7	8		21	22	23	24		37	38	39	40		53	54	55	56
9	10	11	12		25	26	27	28		41	42	43	44		57	58	59	60
13	14	15	16		29	30	31	32		45	46	47	48		61	62	63	64

1	2	3	4		5	6	7	8		9	10	11	12		13	14	15	16
17	18	19	20		21	22	23	24		25	26	27	28		29	30	31	32
33	34	35	36		37	38	39	40		41	42	43	44		45	46	47	48
49	50	51	52		53	54	55	56		57	58	59	60		61	62	63	64

1	5	9	13		2	6	10	14		3	7	11	15		4	8	12	16
17	21	25	29		18	22	26	30		19	23	27	31		20	24	28	32
33	37	41	45		34	38	42	46		35	39	43	47		36	40	44	48
49	53	57	61		50	54	58	62		51	55	59	63		52	56	60	64

12

Topics in Number Theory

Mathematics is the queen of sciences,
and the theory of numbers is the queen of mathematics.
—Carl F. Gauss (1777–1855)

As the set of integers is central in discrete mathematics, the operations and relations on integers explored in Chapters 2 and 7 provides a basic foundation of number theory. Not surprisingly, the more we build upon this foundation, the better equipped we will be for problem solving in discrete mathematics. In this chapter, we revisit prime numbers, represent integers in nondecimal bases, and solve equations using modular arithmetic, including systems of congruences and nonlinear Diophantine equations. Finally, we present a modern application of number theory, Rabin's method, which uses congruences in public-key cryptography.

12.1 More on Primes

In Chapter 7, we raised a number of questions about prime numbers; some answers were given there, including a proof that there are infinitely many prime numbers. We now continue this exploration of prime numbers, including their distribution and the potential for finding them with a formula.

Is there any pattern in the distribution of primes? In a word, the answer is "No." No observable pattern of the placement of primes within a small set of integers will continue indefinitely. Of course, one cannot have consecutive primes, other than 2 and 3, because even integers are divisible by 2. In fact, the "gaps" between two consecutive primes can be as large as we wish. For example, there is a set of one billion consecutive integers with no prime among them, as the proof of the following theorem demonstrates.

Theorem 12.1. *For any $n \in \mathbb{N}$, there exist n consecutive composite integers.*

Proof. Consider the following list of n consecutive integers:

$$(n+1)! + 2, (n+1)! + 3, (n+1)! + 4, \dots, (n+1)! + (n+1).$$

Since $(n+1)! = 1 \cdot 2 \cdot 3 \cdots n \cdot (n+1)$, we see that $(n+1)! + k$ is divisible by k, but not equal to k, for $k = 2, 3, \dots, n+1$. Therefore, all n numbers are composite.   □

Further Thoughts. Note that we did not claim that the set exhibited in the proof above was the first appearance of n consecutive composite integers. The sequence

$$P + 2, P + 3, P + 4, \dots, P + (n+1),$$

where P is the product of all primes not exceeding $n + 1$, also consists of n composite numbers, and they are much smaller than the ones above. For example, suppose we want to find nine consecutive composite numbers. There are four primes whose value is at most 10, namely 2, 3, 5, and 7. Their product is $P = 210$, and indeed we see that 212, 213, ..., 220 is a sequence of nine consecutive composite numbers that are much smaller than those starting with $10! + 2$.   ◇

In some sense, the following statement runs counter to Theorem 12.1. It was formulated by J. L. F. Bertrand (1822–1900) in 1845 and proven by P. L. Chebyshev (1821–1894) in 1850. All known proofs use facts that are outside of the scope of our course, and we omit them.

Theorem 12.2. *For any $n \geq 4$, there exists at least one prime number p such that $n \leq p \leq 2n - 2$.*

Another natural pattern to look for involves the distance of primes from known composite numbers. Certainly, all odd primes can be written in the form $4n + 1$ or $4n + 3$; they are members of the arithmetic sequences 1, 5, 9, 13, 17, ... or 3, 7, 11, 15, 19, ..., respectively. Are there infinitely many primes in each of these two sequences? For each sequence, the answer is in the affirmative; the latter will be shown here, and the former will be left as a challenging problem (see Problem 15, Section 12.3). The proofs for both are reminiscent of Euclid's proof of the infinitude of the primes (see Theorem 7.4).

Proposition 12.3. *There are infinitely many primes of the form $4n + 3$ for $n \in \mathbb{N}$.*

Proof. By way of contradiction, suppose there are only finitely many primes of the form $4n + 3$, namely $p_1, p_2, \dots, p_m$. Let $N = (p_1 p_2 \cdots p_m)^2 + 2$. Since $(p_1 p_2 \cdots p_m)^2$ is the square of an odd integer, it has the form $4q + 1$ for some $q \in \mathbb{N}$. Hence, $N = 4q + 3$ for some $q \in \mathbb{N}$. Since $N > 2$, N must have a prime divisor p, and any such divisor is odd. The prime p cannot belong to $\{p_1, p_2, \dots, p_m\}$; since p would divide $N - (p_1 p_2 \cdots p_m)^2 = 2$. We conclude that all prime divisors of N have the form $4k + 1$ for some $k \in \mathbb{N}$. Since N is a product of factors that are each congruent to 1 mod 4, by Theorem 7.15(3), $N \equiv 1$ (mod 4). This contradicts N having the form $4q + 3$ for some $q \in \mathbb{N}$.   □

In 1837, a far more general question was answered by P. G. L. Dirichlet (1805–1859). We present it without proof as all proofs are beyond the scope of this book.

Theorem 12.4. *Let a and d be two relatively prime integers. Then the arithmetic sequence*

$$a, a + d, a + 2d, \dots, a + nd, \dots \ (n \in \mathbb{N})$$

contains infinitely many primes.

Thus, for example, there are infinitely many primes in the sequences

$1, 4, 7, 10, 13, \dots \ (a = 1, d = 3)$,

$1, 7, 13, 19, \dots \ (a = 1, d = 6)$,

$5, 13, 21, 29, 37, \dots \ (a = 5, d = 8)$, and

$3, 7, 11, 15, \dots \ (a = 3, d = 4)$.

The last theorem we mention with regard to this question, the Prime Number Theorem, is arguably the greatest known result about prime numbers. Various versions of the theorem were originally conjectured late in the 18th century by Adrien-Marie Legendre and Carl F. Gauss and in the middle of the 19th century by Pafnuty Chebyshev. Bernard Riemann's 1859 paper "On the Number of Primes Less than a Given Prime", made important connections between the distribution of primes and the zeros of the so-called Riemann zeta function of a complex variable. Several proofs of the Prime Number Theorem were found late in the 19th century based on Riemann's work, and in the 20th century, various other proofs were given, some elementary (i.e., not based on complex numbers) in nature. Essentially, the theorem says that for large values of n, the number of primes less than n is approximately $n/\ln(n)$. Here is the formal statement.

Theorem 12.5 (Prime Number Theorem). *Let $\pi(x)$ denote the number of primes less than x. Then*

$$\lim_{x \to \infty} \frac{\pi(x)}{x/\ln(x)} = 1.$$

Is there any formula for producing prime numbers?

For example, can one find a function f of one variable such that $f(n)$ is prime for all $n \in \mathbb{N}$? The quadratic function $y = x^2 + x + 41$ takes prime values for all $x = 1, 2, \dots, 39$ (as well as $x = -40, -39, \dots, -2, -1, 0$), but not for $x = 40$. This was observed by Leonhard Euler. One can do even better: $y = x^2 - 79x + 1601$ takes prime values for the first eighty values of x. On the other hand, it is not hard to show (see Problem 12) that no nonconstant polynomial $p(x)$ with integer coefficients can take prime values for all $x \in \mathbb{N}$. But what if f is not a polynomial of one variable?

It was noticed by Pierre de Fermat that $F_n = 2^{2^n} + 1$ is prime for $n = 0, 1, 2, 3, 4$, and he conjectured that this pattern continues. Any number of this form is called a **Fermat number**, and $F_0, F_1, \dots, F_4$ are **Fermat primes**. The conjecture was disproved by Euler, who showed that $2^{2^5} + 1 = 2^{32} + 1 = 4{,}294{,}967{,}297 = 641 \cdot 6{,}700{,}417$ is not prime. Despite extensive computer searches, no other Fermat primes have been found.

Little progress in finding a formula for primes had been made through the centuries. A breakthrough came in 1947, when William H. Mills proved the existence of a real number α, such that $\lfloor \alpha^{3^n} \rfloor$ is prime for all $n \in \mathbb{N}$.

Another deep result in this direction was obtained by Yuri Matijasevich in 1972. He proved the existence of a polynomial in 58 variables $x_1, \dots, x_{58}$ of degree 4 with the property that if it is evaluated for all $(x_1, \dots, x_{58}) \in \mathbb{Z}^{58}$, then, remarkably, the set of its positive values is precisely the set of all primes! Other examples of such polynomials have been found subsequently. For example, in 1976, J. P. Jones, D. Sato, H. Wada, and D. Wiens found a polynomial of degree 25 with 26 variables—so it may be written using all of the letters of the English alphabet.

12.1.1 Conjectures about Primes.

Many conjectures about prime numbers have been made by prominent mathematicians over the years. Some have been proven false, some have been proven true, and the veracity of others has remained elusive. Here are some of the more famous open conjectures about prime numbers.

- Twin Prime Conjecture: There are infinitely many primes p for which $p + 2$ is also prime. For example, $\{3, 5\}$, $\{11, 13\}$, $\{41, 43\}$, and $\{59, 61\}$ are examples of twin primes.[1]

- Goldbach Conjecture: Every even integer greater than 2 can be expressed as the sum of two primes.

- Waring's Prime Number Conjecture: Every odd integer exceeding 3 is either a prime number or is the sum of three prime numbers. Waring's Conjecture follows almost immediately from Goldbach's Conjecture, and it is sometimes called Goldbach's Weak Conjecture.

- Mersenne Prime Conjecture: There are infinitely many prime numbers of the form $2^p - 1$, where p is prime. Such primes are called **Mersenne primes**; the smallest examples are $2^2 - 1 = 3$, $2^3 - 1 = 7$, $2^5 - 1 = 31$, $2^7 - 1 = 127$, and $2^{13} - 1 = 8{,}191$. The primality of p is necessary, but not sufficient, for the primality of $2^p - 1$ (see Problem 13).

- There are infinitely many prime numbers of the form $n^2 + 1$.

Exercises and Problems.

(1) Determine the size of the largest known prime number using a web search. Also determine the type of prime (mentioned in this section) that has been the source of the largest known prime for several decades.

(2) Find eleven consecutive composite integers using the following:

  (a) the method in Theorem 12.1,

  (b) the alternative method to Theorem 12.1 using a product of primes,

  (c) a search of a list of prime numbers less than 150.

(3) List all twin primes less than 100. How many pairs are not of the form $6n - 1$ and $6n + 1$ for some integer n?

(4) Find a prime number p between n and $2n - 2$ for the following values of n. There are many correct answers.

  (a) $n = 4$, (b) $n = 40$, (c) $n = 400$, (d) $n = 4000$, (e) $n = 40{,}000$.

(5) Are there infinitely many primes of the form $100n + 1$? Find four prime numbers of this form. Are there infinitely many primes of the form $100n + 15$?

---

[1]This conjecture lies in contrast to Theorem 12.1. On the one hand, arbitrarily large gaps can be found between prime numbers. On the other hand, all evidence points to infinitely many pairs of primes that differ by 2.

(6) Which sequence contains more prime numbers less than 100, $(4n + 1)_{n \geq 0}$ or $(4n + 3)_{n \geq 0}$?

(7) Approximate $\pi(x)$ using $\dfrac{x}{\ln x}$ for the following values of x:

   (a) $x = 10$, (b) $x = 100$, (c) $x = 1{,}000$, (d) $x = 10{,}000$.

(8) By consulting a list of prime numbers or using some mathematical software, determine the exact value of $\pi(x)$ for each x in Exercise 7. Compare the exact values with the approximate values.

(9) For each of the formulas, determine how many primes are present for the given values of n. Give your opinion as to whether the formula appears to be a promising way to produce prime numbers.

   (a) $n! + 1$ for $1 \leq n \leq 5$

   (b) $n! - 1$ for $2 \leq n \leq 6$

   (c) $n^2 + 1$ for $1 \leq n \leq 10$

(10) Verify Goldbach's Conjecture for each even number below. Write each number as a sum of two primes in three different ways.

   (a) 42

   (b) 66

   (c) 100

---

(11) Prove that there are infinitely many integers n such that $7n^3 - 5n^2 + 6n + 11$ is composite.

(12) Prove that any nonconstant polynomial $a_n x^n + a_{n-1} x^{n-1} + \cdots + a_2 x^2 + a_1 x + a_0$ cannot be prime for all integers x.

(13) (a) Prove that if $2^p - 1$ is prime, then p is prime.

   (b) Show that the converse of this implication is false by showing that $2^{11} - 1$ is not prime.

12.2 Integers in Other Bases

Our familiar base-ten number system seems to be the most natural one to use, given that many of us learn to count to ten on our fingers. However, historically, other bases have been and continue to be used. The Babylonian number system essentially used 60 as a base, and our method of keeping time, with 60 seconds in a minute and 60 minutes in an hour, is a remnant of that system. In modern times, binary arithmetic (base 2) has become important for computer design and information storage, where switches are either on or off and locations store either 0 or 1. Thus, quantities related to digital computing are usually powers of 2. For example, a device with "one gigabyte" of memory contains $2^{10} = 1{,}024$ megabytes of memory, where a byte is a unit of storage that typically consists of eight bits (short for binary digits—0's and 1's). Each megabyte equals 2^{10} kilobytes, and a kilobyte equals 2^{10} bytes; hence, a gigabyte equals $2^{30} = 1{,}073{,}741{,}824$ bytes.

12.2.1 Introduction. Our decimal (base-ten) system uses the numerals $0, 1, 2, \ldots, 9$, and the digits of an integer tell us how many 1's, 10's, 100's, and other powers of 10 appear in the number. For example, in base 10, the number 83,406 can be expressed as

$$83,406 = 8(10,000) + 3(1,000) + 4(100) + 0(10) + 6$$
$$= 8\left(10^4\right) + 3\left(10^3\right) + 4\left(10^2\right) + 0\left(10^1\right) + 6\left(10^0\right).$$

In general, an arbitrary decimal integer $a = (a_n a_{n-1} \ldots a_2 a_1 a_0)$ can be written as

$$a = a_n 10^n + a_{n-1} 10^{n-1} + \cdots + a_2 10^2 + a_1 10 + a_0.$$

To write a number in base b for some integer $b \geq 2$, the numerals $0, 1, 2, \ldots, b-1$ are used.[2] Any nonnegative integer a can be written (uniquely) in base b in the form

$$a = a_n b^n + a_{n-1} b^{n-1} + \cdots + a_2 b^2 + a_1 b + a_0.$$

For example, the number "3 dozen", which consists of 3 twelves, is represented as 36 in base 10 but could be viewed as 30 in base twelve or 210 in base four, since

$$36 = 3(12) + 0 = 2\left(4^2\right) + 1\left(4^1\right) + 0.$$

To remove ambiguity, we will use an appropriate subscript after the parentheses around the sequence of base b digits to denote a number expressed in a base other than 10. So, $36 = (30)_{12} = (210)_4$.

Example 148. Convert $(42013)_5$ to decimal form.

Solution.

$$(42013)_5 = 4\left(5^4\right) + 2\left(5^3\right) + 0\left(5^2\right) + 1(5) + 3(1)$$
$$= 4(625) + 2(125) + 0 + 5 + 3$$
$$= 2,758. \qquad \square$$

A bit more effort is needed to convert decimal numbers to numbers in another base, although the processes are conceptually equivalent.

Example 149. Convert the decimal number 1,882 to an octal number.

Solution. Since octal is a base-8 system, we need to write 1,882 as a sum of the powers of 8: 1, 8, 64, 512, 4,096, etc. We start with the largest power less than 1,882, namely 512. It divides into 1,882 **three** times, with remainder $1,882 - 3(512) = 346$. From 346, we can form **five** groups of 64 ($5 \times 64 = 320$), with 26 left over. Finally, 26 is broken into **three** groups of 8 and **two** 1's. Summarizing,

$$1,882 = 3\left(8^3\right) + 5\left(8^2\right) + 3\left(8^2\right) + 2(1) = (3532)_8. \qquad \square$$

In Example 149 we see how the Division Theorem can be used repeatedly to convert an integer a from base 10 to base b. For the first step, the highest power of b that is less than a is divided into a; the quotient gives the digit corresponding to that power of b, and the remainder is used in the next step. At each subsequent step, the highest

---

[2]When $b > 10$, other symbols may be used for numerals greater than 9. For example, the hexadecimal system with $b = 16$ utilizes the digits $0, 1, \ldots, 8, 9, A, B, C, D, E, F$.

power of b that is less than the remainder from the previous step is divided into that remainder, and the new quotient and remainder are found. This process continues until the remainder is less than b, and the final remainder becomes the unit's digit in the base-b representation. The existence and uniqueness of the quotient and remainder in the Division Theorem ensure that for any integer a there will be a unique way to represent a in a base-b system for any $b \in \mathbb{N}$, $b \geq 2$. We prove this formally in Theorem 12.6.

12.2.2 Conversion Algorithms. In Examples 148 and 149 we utilized a basic method for converting between a decimal representation and representation in another base. However, there are more efficient algorithms for performing these conversations, two of which are presented here.

Notice that converting from base b to base 10 is actually an exercise in evaluating a polynomial at b. In Example 148, we found $f(5)$, where $f(x) = 4x^4 + 2x^3 + 0x^2 + 1x + 3$, and it is easy to see that this same process is used for converting from numbers base b to base 10. Accordingly, we seek an efficient way to evaluate a polynomial of degree n.

By direct substitution, evaluating

$$f(b) = a_n b^n + a_{n-1} b^{n-1} + \cdots + a_2 b^2 + a_1 b + a_0$$

uses $3n - 1$ operations; therefore this method has linear complexity. To see this, note that it will take $n - 1$ multiplicative operations to calculate the powers $b^2, b^3, \ldots, b^n$ (first multiply b by b to get b^2, then multiply that product by b again to get b^3, and so on). Then each product $a_i b^i$ must be found, $1 \leq i \leq n$, which requires another n operations. Finally, summing these $n + 1$ products will take n addition operations. Altogether, the total is $(n - 1) + n + n = 3n - 1$ operations.

Horner's Algorithm gives a method of evaluating $f(b)$ that requires only $2n$ operations, and while this is also linear in its complexity level, it is more efficient than using the substitution method.[3] To use Horner's Algorithm to convert $(42013)_5$ to a decimal number, by evaluating

$$f(5) = 4\left(5^4\right) + 2\left(5^3\right) + 0\left(5^2\right) + 1\left(5^1\right) + 3(1),$$

we proceed as follows.

Multiply the first digit, $a_n = 4$, by the base, $b = 5$, and add the second digit, $a_{n-1} = 2$. Now multiply by $b = 5$ again. Add this product to the third digit, and multiply by $b = 5$ again. Continue this process, moving left to right, until completing the calculation by adding $a_0 = 3$ at the last step. We get

$$4(5) + 2 = 22, \quad 22(5) + 0 = 110, \quad 110(5) + 1 = 551, \quad 551(5) + 3 = 2{,}758.$$

This value matches the one found in Example 148.

Example 150. Use Horner's Algorithm to convert $(4321)_5$ to a decimal.

Solution. Below we write a streamlined calculation employing Horner's Algorithm. Starting with the inside parentheses and working outward, it is clear the method is the

---

[3]We should clarify that it is more efficient for a computer. For humans, when n is small and the various powers of b are memorized, direct substitution may be more efficient.

same:
$$(4321)_5 = 5(5(5(4) + 3) + 2) + 1 = 5(5(23) + 2) + 1 = 5(117) + 1 = 586. \qquad \square$$

Proof of Horner's Algorithm. We use induction on n. Clearly, the algorithm works in the base case of a single digit, a_0, and no calculations are needed. Now let $n \geq 0$ and assume that the algorithm correctly evaluates any polynomial of degree n in $2n$ steps. Consider the polynomial

$$f(b) = a_{n+1}b^{n+1} + a_n b^n + \cdots + a_2 b^2 + a_1 b + a_0.$$

The polynomial

$$g(b) = a_{n+1}b^n + a_n b^{n-1} + \cdots + a_2 b + a_1$$

is a polynomial of degree n and by the inductive hypothesis can be evaluated in $2n$ steps by Horner's Algorithm. Note then that

$$f(b) = b \cdot g(b) + a_0$$

and these two calculations (multiplying $g(b)$ by b and adding a_0) are exactly the final two steps employed by Horner's Algorithm in evaluating $f(b)$. Therefore, Horner's Algorithm will correctly evaluate a polynomial of degree $n + 1$ in $2n + 2 = 2(n + 1)$ steps. This completes the proof by induction. $\qquad \square$

A drawback to the method used in Example 149 for converting a decimal integer a to its equivalent number in base b is that one needs to know the highest power of b less than a before even beginning to divide. Moreover, the division involved high powers of b. An alternative method uses the Division Theorem, repeatedly dividing by b and keeping track of remainders. The digits of the base-b number will consist of the remainders, in right-to-left order.

Before proving that this method works, we illustrate it by converting 1,882 to its equivalent number in base 8. At each stage after the first, we use the Division Theorem on the previous quotient; we stop once the quotient is zero.

$$1,882 = 8(235) + 2,$$
$$235 = 8(29) + 3,$$
$$29 = 8(3) + 5,$$
$$3 = 8(0) + 3.$$

Writing the remainders in right-to-left order, we see that $1,882 = (3532)_8$.

We have not yet proved the uniqueness of the representation of an integer in a given base b. In proving this fact, we will also prove that the above method yields the base-b representation.

Theorem 12.6. *For any integer k and integer $b \geq 2$, k has a unique base-b representation.*

Proof. We may assume $k \geq 0$, since the base-b representation of $-k$ will simply be the negation of the representation of k. By the Division Theorem, there are unique integers q_0 and r_0, with $0 \leq r_0 < b$ such that $k = q_0 b + r_0$. Continue using the Division Theorem, as follows, where the quotients are clearly decreasing and in each case the remainder r_i satisfies $0 \le r_i < b$. We stop when $q_n = 0$.

$$k = q_0 b + r_0,$$
$$q_0 = q_1 b + r_1,$$
$$q_1 = q_2 b + r_2,$$
$$\cdots$$
$$q_{n-2} = q_{n-1} b + r_{n-1},$$
$$q_{n-1} = r_n \quad \text{since } q_n = 0.$$

Beginning with the first line, we now solve for k, eliminating the quotient q_i. We obtain

$$k = (q_1 b + r_1)b + r_0$$
$$= q_1 b^2 + r_1 b + r_0$$
$$= (q_2 b + r_2)b^2 + r_1 b + r_0$$
$$= q_2 b^3 + r_2 b^2 + r_1 b + r_0$$
$$\cdots$$
$$= r_n b^n + r_{n-1} b^{n-1} + \cdots + r_2 b^2 + r_1 b + r_0.$$

We have expressed k as the sum of powers of b, and the representation is unique because of the uniqueness of the quotient and remainder in the Division Theorem.   □

Further Thoughts. Notice that our proof also shows that the algorithm above (repeatedly dividing by b and keeping track of the remainders) yields the base-b representation

$$k = (r_n r_{n-1} \ldots r_2 r_1 r_0)_b.$$

The number of operations needed to execute the algorithm will equal the number of division steps. (See Problem 12.)   ◇

Example 151. Convert the decimal 386 to its base-2 equivalent.

Solution.

$$386 = 2(193) + 0,$$
$$193 = 2(96) + 1,$$
$$96 = 2(48) + 0,$$
$$48 = 2(24) + 0,$$
$$24 = 2(12) + 0,$$
$$12 = 2(6) + 0,$$
$$6 = 2(3) + 0,$$
$$3 = 2(1) + 1,$$
$$1 = 2(0) + 1.$$

Writing the remainders in right-to-left order, we obtain $387 = (110000010)_2$. We verify that this is correct by noting that

$$0\left(2^0\right) + 1\left(2^1\right) + \cdots + 1\left(2^7\right) + 1\left(2^8\right) = 2 + 128 + 256 = 386.$$ □

Each algorithm just presented can be used to convert from base m to base n for any $m, n \geq 2$. The key is to perform the relevant computations in the appropriate base; of course, we are well practiced at performing them in base 10.

12.2.3 Base-2 Applications. As a simple application of binary arithmetic, we will describe the ancient Egyptian method of multiplying numbers. The Egyptians evidently recognized that every positive integer could be expressed uniquely as a sum of powers of 2, because they used this fact (and their ability to double numbers) to turn integer multiplication problems into addition problems with relatively few terms. Evidence that they used this method comes from examples in the Rhind Papyrus. The method is essentially the same as the Russian Peasant Algorithm utilized in the early 19th century. We illustrate the method with the product 21×81.

First we create a table listing the successive powers of 2, stopping when we reach the largest power of 2 that is smaller than 21. We place an asterisk by those powers that occur in the representation of 21 as the sum of such powers. Since $21 = 16 + 4 + 1$, the terms 1, 4, and 16 are marked. In the third column of the table, we place 81 in the first row and successively double it as we move down the column. This means the entries in the third column equal the products of 81 with the entries in the first column. (See Table 12.1.)

Table 12.1. Calculating 21×81 with Egyptian Multiplication

1	*	81
2		162
4	*	324
8		628
16	*	1,296

Then, since

$$(21 \times 81) = (1 + 4 + 16)(81) = 1(81) + 4(81) + 16(81),$$

we see that the product 21×81 equals the sum of the entries in the third column whose rows are marked with an asterisk. Hence, $21 \times 81 = 181 + 324 + 1{,}296 = 1{,}701$.

Example 152. Use the Egyptian multiplication method to calculate 26×37.

Solution. First note that $26 = (11010)_2$, and then construct the appropriate table:

1		37
2	*	74
4		148
8	*	296
16	*	592

Therefore, $26 \times 37 = 74 + 296 + 592 = 962$.   □

As our second application of binary arithmetic, we show how to find a winning strategy for a version of the game Nim. In this two-player game, there are initially three or four (though any number is permissible) piles of chips, each containing a specified

numbers of chips. The players alternate turns, where on each turn a player must remove at least one chip from exactly one of the piles. The player to take the last chip is
the winner. Before reading on, try this game yourself with relatively small numbers of
chips and see if you can notice any useful strategies for winning the game.

To explain the strategy for this game, we need to introduce a new type of addition
for binary numbers a and b, called the Nim sum, denoted $a \oplus b$.

Let integers a and b have respective binary representations $(a_k \ldots a_1 a_0)_2$ and
$(b_j \ldots b_1 b_0)_2$. Define the **Nim sum** of a and b as $c = a \oplus b$, where c is the integer
with binary representation given by $c_i = a_i + b_i \bmod 2$. (We may assume $a_i > 0$ when
$i > k$ and $b_i > 0$ when $i > j$.) For example, if $a = 20 = (10100)_2$ and $b = 15 = (1111)_2$,
then $a \oplus b = (11011)_2 = 27$, since $a_i = b_i$ only when $i = 2$.

The Nim sum has some simple, but important, properties we will use.

(1) $s \oplus s = 0$ for all integers s.

(2) Associativity: $(a \oplus b) \oplus c = a \oplus (b \oplus c)$ for all a, b, c. Thus, we may write $a \oplus b \oplus c$
    without ambiguity.

(3) If b's binary representation has b_j as its leading 1 (i.e., $b_j = 1$ and $b_i = 0$ for $i > j$)
    and $a_j = 1$, then $a \oplus b < a$.

To carry out our proposed strategy, denote the three piles as A, B, and C, with corresponding sizes a, b, and c. Prior to each move, determine the binary representations
of each size and calculate the Nim sum, $s = a \oplus b \oplus c$. If the binary representations for
a, b, and c are placed in a right-justified grid, then s_i is 0 or 1, depending on whether
the number of 1's above it in the set $\{a_i, b_i, c_i\}$ is even or odd, respectively. These steps
are shown in Table 12.2 for the initial move of a game where $a = 10$, $b = 15$, and
$c = 20$.

Table 12.2

	16	8	4	2	1
10	0	1	0	1	0
15	0	1	1	1	1
20	1	0	1	0	0
s	1	0	0	0	1

The strategy for winning this game is to remove chips in such a way that the bottom
row contains all zeros (i.e., $s = 0$) each time it is the opponent's turn to play. Of course,
this prompts two questions:

(1) Is this strategy possible to execute?

(2) If it is possible, why will it result in victory for the player who employs it?

The answer to the first question is "sometimes". If $s \neq 0$, the player is assured of
being able to execute the strategy for the duration of the game. He or she simply looks
for the leading 1 in s, say $s_m = 1$, and chooses one of the other rows for which column
m contains a 1. (There must be at least one such row, since there will be an odd number
of ones in the first three rows of column m.) Let's suppose that $a_m = 1$. The player
should then remove an appropriate number of chips from pile A in order to leave $a \oplus s$
chips there. By property (3) above, this will be possible. To illustrate this, consider
Table 12.2, in which the leading one in the bottom row is s_4. There is just one 1 above

it, corresponding to $c_4 = 1$ in row 3. The player therefore leaves $(10100)_2 \oplus (10001)_2 = (00101)_2 = 5$ chips in pile C, removing 15. The new configuration appears in Table 12.3.

Table 12.3

	16	8	4	2	1
10	0	1	0	1	0
15	0	1	1	1	1
5	0	0	1	0	1
s	0	0	0	0	0

Notice that the bottom row contains all zeros, so $s = 0$. Moreover, if the player follows the above strategy, he can ensure this happens each time it is his opponent's turn, regardless of the move made by the opponent. Since the c chips in pile C are replaced with $c \oplus s$ chips, the new Nim sum of the three rows will be

$$a \oplus b \oplus (c \oplus s) = (a \oplus b \oplus c) \oplus s = s \oplus s = 0,$$

as desired.

This fact enables us to answer the second question as to why the strategy is a winning one. If the opponent is always making a move when the Nim sum is zero, then since the number of chips decreases with each move, eventually the Nim sum will be zero when no chips are left. In the move taken prior to reaching this position, the player employing the outlined strategy will have just removed the last chip, thereby winning the game.

Remark. The strategy outlined above may be employed with any number of piles and any number of chips in those piles. One should always remove chips in such a way to create a Nim sum of zero with the remaining chips.   ◇

Example 153. It is your turn in a game of Nim, and there are four piles of chips with sizes 5, 6, 7, and 8. What move can you make that will put you on a path to victory?

Solution. The respective base-2 representations are $(101)_2$, $(110)_2$, $(111)_2$, and $(1000)_2$, so the Nim sum is $(1100)_2$. The leading one of s is s_3, in the fourth position from the right, so chips must be removed from the pile with 8 chips, as this pile is the only one with a 1 in the fourth position. Since we need to leave $(1000)_2 \oplus (1100)_2 = 4$ chips in the pile, four chips should be removed, leaving piles of sizes 5, 6, 7, and 4.   □

Exercises and Problems.

(1) Convert $(21021)_3$ to decimal form.

(2) Write 674 as a base-b number for $b = 2$, $b = 5$, and $b = 8$.

(3) Write 387 as a base-b number for $b = 2$, $b = 5$, and $b = 16$.

(4) Use Horner's Algorithm to convert $(2641)_8$ to decimal form. Check your answer.

(5) Use Horner's Algorithm to convert $(10011010)_2$ to decimal form. Check your answer.

(6) Use the Egyptian multiplication method to calculate 76×14.

(7) Use the Egyptian multiplication method to calculate 33×65.

(8) What move should you make when playing Nim if it is your turn and the pile sizes are 11, 15, and 25?

(9) What move should you make when playing Nim if it is your turn and the pile sizes are 18, 31, 54, and 29?

---

(10) Suppose you are playing Nim and the largest pile has 63 chips. If it is your turn and all other pile sizes are equally likely, what is the probability of you being able to employ a winning strategy?

(11) Based on the manner in which Egyptian multiplication utilizes binary representation of integers, doubling, and addition, discover for yourself the method of Egyptian division. (The method may only yield exact answers for certain quotients.) Apply the method to $51 \div 12$.

(12) What is the complexity of the algorithm that will find the base-b representation of an integer k? You should determine the number of division steps needed to execute the algorithm described in the section, as a function of b and k.

(13) Solve this 13th-century weighting problem posed by Fibonacci in his book *Liber abaci*: *Find a system of four weights, such that using them one by one, you can weigh any whole load of Q in the range of 1 to 40 kg in a balance. Each weight can be placed in either pan. The first pan contains the load Q and the second pan is free of load.*

(14) (See the opening hook problem for the chapter.) Consider the following simple number guessing game. Ask someone to select an integer between 1 and 64, and tell him or her that you will determine the number by asking three questions. Show three cards to that person, with each card having each of the numbers 1 to 64 color-coded in one of four colors, and ask, "What is the color of your secret number on this card?" For example, you may get the answers "Blue" for the first card, "Yellow" for the second card, and "Pink" for the third card, giving the sequence (Blue, Yellow, Pink). The number of different color-sequences of length 3 using 4 colors is $4^3 = 64$, so the secret number can be determined if the cards are created appropriately. Give a method for coding the numbers on each card to perform the trick without having to reference the cards.

12.3 More on Congruences

An elementary mathematical technique is that of solving linear equations with one unknown—i.e., solving for x in equations of the form $ax = b$, where a and b are known real numbers. The theory of such equations is very simple: if $a = 0$, but $b \neq 0$, there are no solutions; if $a = b = 0$, every real number is a solution; and if $a \neq 0$, then there exists a unique, real solution that can be found with these steps:

$$
\begin{aligned}
ax = b \iff & a^{-1}(ax) = a^{-1}b \\
\iff & (a^{-1}a)b = a^{-1}b \\
\iff & 1x = a^{-1}b \\
\iff & x = a^{-1}b.
\end{aligned}
$$

This argument uses numerous properties of real numbers given in Section 2.1, including the existence of the multiplicative inverse for every nonzero real number and the fact that multiplying both sides of an equation by a nonzero number yields an equivalent equation.

A **linear congruence** has the form $ax \equiv b \pmod{n}$, where a, b, and $n > 0$ are known integers and x denotes the unknown. A solution of a linear congruence is any integer x that satisfies the congruence. In this section, we explore similarities and differences between solving linear equations with real numbers and linear congruences with integers. While there are definite similarities, there is a richer theory for characterizing whether linear congruences have no solutions, a unique solution, or multiple solutions.

12.3.1 Solving Linear Congruences.
The variety of solutions that can occur is readily observed by experimenting with a few linear congruences. By testing small values of x, we see that $x = 4$ is a solution of the congruence $5x \equiv 8 \pmod{12}$. There are many other solutions, such as $x = -8, 16$, and 28, but each solution has remainder 4 upon division by 12. On the other hand, $4x \equiv 1 \pmod{6}$ has no solutions: if it did, then $4x - 1 = 6t$ for some integer t, or $1 = 4x - 6t$, in which case $2|1$, a contradiction. But a slightly different congruence, $4x \equiv 2 \pmod{6}$, has two classes of solutions. One class consists of $x = 2$ and all other integers congruent to 2 modulo 6, and the other class consists of all integers congruent to 5 modulo 6.

When attempting to solve $ax \equiv b \pmod{n}$ in general, it is natural to try to replicate the method used for solving linear equations, namely multiplying both sides by an integer c for which $ca \equiv 1 \pmod{n}$. This would leave both sides congruent, by Theorem 7.15(2). The following theorem specifies when this is possible.

Theorem 12.7. *Let $ax \equiv b \pmod{n}$ be a linear congruence with respect to x, and suppose $\gcd(a, n) = 1$. Then there exists an integer c such that $ca \equiv 1 \pmod{n}$, and all solutions of the congruence can be written in the form $x \equiv cb \pmod{n}$.*

Proof. According to Corollary 7.9, there exist integers u and v such that $ua + vn = 1$. Let $c = u$. Then $ca = 1 - vn \equiv 1 \pmod{n}$. Note that c and n are relatively prime by the same corollary. Therefore we can invoke Theorem 7.15 and multiply both sides of $ax \equiv b \pmod{n}$ by c to obtain $ax \equiv b \pmod{n} \iff c(ax) \equiv cb \pmod{n} \iff (ca)x \equiv cb \pmod{n} \iff 1 \cdot x \equiv cb \pmod{n} \iff x \equiv cb \pmod{n}$. $\square$

Further Thoughts. The number c in the above theorem is not determined uniquely; in fact, there are infinitely many such c. This is because there are infinitely many u satisfying $ua + vn = 1$: for every (u, v) satisfying this equality, the pair $(u - nt, v + at)$ also satisfies the equality for every t. Viewed another way, all solutions to the congruence belong to one congruence class modulo n (i.e., all solutions have the form $x = u - nt$ for $t \in \mathbb{Z}$). ◇

From the previous proof and discussion, for a given value of c, the solution set of the linear congruence is a congruence class, namely $[cb]$. Exactly one of these solutions is an element of the set $\{0, 1, \dots, n - 1\}$; specifically, it is the remainder upon dividing cb by n. This remainder is called the **least residue modulo** n of cb. The set of least residues modulo n is $\{0, 1, \dots, n - 1\}$.

When $\gcd(a, n) = 1$, the integer c for which $ca \equiv 1 \pmod{n}$ is called the **inverse** of a mod n, often denoted a^{-1} modulo n. In this case, a is said to be **invertible modulo** n. When $\gcd(a, n) = 1$, the linear congruence $ax \equiv b \pmod{n}$ has a unique least residue solution, namely the least residue of $a^{-1}b$. Therefore, the method and result of solving this linear congruence is analogous to solving the equation $ax = b$, using the language of inverses and least residues modulo n.

What can be said about the solutions of $ax \equiv b \pmod{n}$ when $d = \gcd(a, n) \neq 1$? Recall that $ax \equiv b \pmod{n}$ is equivalent to $ax = b + nt$ for $t \in \mathbb{Z}$. Applying Theorem 7.12 directly to $ax + n(-t) = b$ reveals that a solution (x, t) exists if and only if $\gcd(a, n)|b$, with solutions obtained by using backward substitution in the Euclidean Algorithm. Therefore, we have the following theorem.

Theorem 12.8. $ax \equiv b \pmod{n}$ *has a solution if and only if $d|b$, where $d = \gcd(a, n)$.*

Example 154. Solve the following congruences:

(1) $7x \equiv 5 \pmod{12}$.

(2) $6x \equiv 4 \pmod{10}$.

(3) $4x \equiv 3 \pmod{6}$.

Solution. (1) First, $\gcd(7, 12) = 1$, which guarantees a single congruence class of solutions. We notice that $7 \cdot 7 \equiv 1 \pmod{12}$, so 7 is its own multiplicative inverse modulo 12; therefore, multiplying both sides of the original congruence by 7, we obtain $x \equiv 35 \equiv 11 \pmod{12}$. The answer also can be written as

$$\{x : x = 11 + 12t, t \in \mathbb{Z}\} = \{11 + 12t : t \in \mathbb{Z}\}.$$

Instead of "noticing" above, we can also solve the equation $7x - 12y = 5$ for x algebraically. This type of problem was solved in Example 81, where the Euclidean Algorithm was employed. We use it again here, first in the forward direction and then in reverse, solving for the number 1 in terms of 7 and 12. We have

$$12 = 1 \cdot 7 + 5, \quad 7 = 1 \cdot 5 + 2, \quad 5 = 2 \cdot 2 + 1.$$

Therefore,

$$1 = 5 - 2 \cdot 2 = 5 - 2 \cdot (7 - 1 \cdot 5) = (-2) \cdot 7 + 3 \cdot 5$$
$$= (-2) \cdot 7 + 3 \cdot (12 - 1 \cdot 7) = 3 \cdot 12 + (-5) \cdot 7.$$

Then, similar to the proof of Theorem 12.7, we multiply both sides of the congruence in (1) by -5, to obtain

$$-35x \equiv -25 \pmod{12}.$$

Since $-35 \equiv 1 \pmod{12}$ and $-25 \equiv 11 \pmod{12}$, we arrive at the solution $x \equiv 11 \pmod{12}$, which is the same answer obtained before.

If the modulus n is large, proceeding with the Euclidean Algorithm can be much faster than "noticing" by using trial and error to find a value of c that works.

(2) Since $\gcd(6, 10) = 2$ and $2|4$, solutions of the congruence exist. A solution x satisfies $6x - 10t = 4$ for some $t \in \mathbb{Z}$. Using the Euclidean Algorithm in reverse, $2 = 2(6) + (-1)10$, and doubling both sides yields $4 = 4(6) + (-2)10$. Hence, $x = 4$ is a solution to the congruence.

An alternative, described in Exercise 7, is to divide both sides of $6x - 10t = 4$ by $\gcd(6, 10) = 2$, resulting in $3x - 5t = 2$. This implies $3x \equiv 2 \pmod{5}$; its solution can be obtained using Theorem 12.7.

(3) Since $\gcd(4, 6) = 2$ and 2 is not a divisor of 3, the congruence has no solutions. To confirm this fact, notice that any solution x must satisfy $4x - 6t = 3$ for some $t \in \mathbb{Z}$. But, $4x - 6t$ is a multiple of 2, while 3 is not. □

We now turn our attention to solving systems of congruences. We utilize a procedure similar to the method of substitution for solving linear equations.

Example 155. (1) Find all integers x that give remainder 2 when divided by 6 and remainder 10 when divided by 11.

(2) Solve the system of congruences $x \equiv 4 \pmod{11}$ and $x \equiv 1 \pmod{19}$.

Solution. (1) We are asked to solve simultaneously two congruences: $x \equiv 2 \pmod 6$ and $x \equiv 10 \pmod{11}$. One way to proceed is as follows. The general solution for the first congruence can be written in the form $x = 2 + 6t$, where $t \in \mathbb{Z}$. This formula describes *all* solutions of the first congruence. Therefore we can try to find those values of t for which these solutions will satisfy the second congruence as well.

By substituting this expression into the second congruence, we obtain

$$2 + 6t \equiv 10 \pmod{11} \iff 6t \equiv 8 \pmod{11}.$$

The last congruence, which has t as the unknown, can be solved easily, since $\gcd(6, 11) = 1$. Multiplying both sides by 2 (since $6 \cdot 2 = 1 \pmod{11}$), we get

$$6t \equiv 8 \iff 12t \equiv 16 \iff t \equiv 5 \pmod{11}.$$

Therefore $t = 5 + 11k$, where $k \in \mathbb{Z}$, and these are all values of t for which solutions $x = 2 + 6t$ of the first congruence are also solutions of the second one. Hence

$$x = 2 + 6t = 2 + 6 \cdot (5 + 11k) = 32 + 66k,$$

so $\{32 + 66k : k \in \mathbb{Z}\}$ is the solution set of the system of congruences. Note that $x = 32$ is the unique least residue solution modulo 66.

(2) The approach here is similar to that of (1), though in this case we will find just one solution instead of the infinite set of solutions. Since $x \equiv 4 \pmod{11}$, x can be written in the form $4 + 11k$ for some integer k. Hence, $4 + 11k \equiv 1 \pmod{19}$. Therefore,

$$11k \equiv -3 \equiv 16 \pmod{19}.$$

Since $\gcd(11, 19) = 1$, we can use the Euclidean Algorithm in reverse to find that $1(16) = 7(11)(16) - 4(19)(16)$. Therefore 7 and 11 are inverses modulo 119. Combining this with the previous congruence, we see that

$$k \equiv (7)(16) \equiv 112 \equiv 17 \pmod{19}.$$

Then, $x = 4 + 11(17) = 191$ is a solution to the original system. □

Further Thoughts. While each system of congruences in Example 154 had a solution, that will not always be the case. Criteria for when a system of congruences has a solution are given in the Chinese Remainder Theorem. See Problems 8 and 16. When there is a solution, there will be infinitely many solutions, namely the entire congruence class that contains the given solution. In part (1), this class was [32] modulo 66. The reader should verify that the infinite solution set in part (2) is [191] modulo 209.   ◇

12.3.2 Euler's Congruence Theorem. In the previous subsection, we explored the role of inverses mod n in solving congruences and showed that an integer a is invertible modulo n if and only if $\gcd(a, n) = 1$. The natural combinatorial question is to ask how many least residues are invertible modulo n, or equivalently, how many integers in $\{1, 2, \dots, n\}$ are relatively prime to n.

For $n \in \mathbb{N}$, let U_n denote the integers in $\{1, 2, \dots, n\}$ that are relatively prime to n. The **Euler phi function** (also called the **Euler totient function**), $\phi(n)$, is defined to be the number of elements in U_n. Thus, $\phi(n)$ counts the number of invertible least residues modulo n. For example, $U_5 = \{1, 2, 3, 4\}$, so $\phi(5) = 4$; $U_8 = \{1, 3, 5, 7\}$, so $\phi(8) = 4$ as well.

Key properties of the Euler phi function and the straightforward method of computing $\phi(n)$ for any $n \in \mathbb{N}$ are explored in Problems 4 and 10. We now present and prove Euler's Congruence Theorem, which is an important theoretical tool in number theory, as well as a key aspect of the inner workings of the RSA Public-Key Cryptosystem (named after Ron Rivest, Adi Shamir, and Len Adleman). Details of the latter are explored in Problem 9 of Section 12.5.

We begin with an example that illustrates an important lemma. Notice that if we multiply each element of $U_8 = \{1, 3, 5, 7\}$ by $a = 2$ and reduce modulo 8, the resulting set is $\{2, 6, 2, 6\} = \{2, 6\}$. However, if the multiplier is $a = 3$, the resulting set is $\{3, 1, 7, 5\}$, which equals U_8. In general, the two sets will be equal when a and n are relatively prime.

Lemma 12.9. *Suppose $n \geq 2$ and $a \in \mathbb{Z}$ with $\gcd(a, n) = 1$. Let $U_n = \{r_1, r_2, \dots, r_{\phi(n)}\}$. Then, the least residues modulo n of ar_1, ar_2, ..., $ar_{\phi(n)}$ are a permutation of the elements of U_n.*

Proof. For $i = 1, 2, \dots, \phi(n)$, since a and r_i are both relatively prime to n, ar_i is relatively prime to n; hence, ar_i is congruent to a member of U_n. To see that each (invertible) least residue is obtained exactly once as the remainder of ar_i for some $i \in \{1, 2, \dots, \phi(n)\}$, we show that the function $f : U_n \to U_n$ via $f(x) = ax$ is one-to-one. To wit, if $ar_i \equiv ar_j$ (mod n), then by Theorem 7.15(3), since $\gcd(a, n) = 1$, $r_i \equiv r_j$ (mod n); hence r_i and r_j are equal least residues. Therefore, f is a permutation on U_n. $\square$

Theorem 12.10 (Euler's Congruence Theorem). *Suppose $n \geq 2$ and $a \in \mathbb{Z}$ with $\gcd(a, n) = 1$. Then, $a^{\phi(n)} \equiv 1$ (mod n).*

Proof. By Lemma 12.9, $(ar_1)(ar_2) \cdots (ar_{\phi(n)}) \equiv r_1 r_2 \cdots r_{\phi(n)}$ (mod n). Collecting factors of a yields $a^{\phi(n)}(r_1 r_2 \cdots r_{\phi(n)}) \equiv r_1 r_2 \cdots r_{\phi(n)}$ (mod n). Since $\gcd(r_i, n) = 1$, for $i = 1, 2, \dots, \phi(n)$, $\gcd(r_1 r_2 \cdots r_{\phi(n)}, n) = 1$. By Theorem 7.15(3), $a^{\phi(n)} \equiv 1$ (mod n). $\square$

In general, if $a^k \equiv 1$ (mod n) for any $k \geq 2$, then a is invertible modulo n and the inverse of a mod n is a^{k-1} (or its least residue, if preferred), since $aa^{k-1} \equiv 1$ (mod n). Thus, Euler's Theorem provides a method of computing inverses mod n. For example, $3^6 \equiv 1$ (mod 91). Thus, modulo 91, the inverse of 3 is $3^5 \equiv 243 \equiv 61$.

Corollary 12.11. *Suppose $n \geq 2$ and $a \in \mathbb{Z}$ with $\gcd(a, n) = 1$. The inverse of a mod n is $a^{\phi(n)-1}$.*

In the special case when the modulus is a prime p, every nonmultiple of p is relatively prime to p; hence $\phi(p) = p - 1$. The following famous result is obtained as a corollary to Euler's Theorem.

Corollary 12.12 (Fermat's Little Theorem). *Suppose p is prime and $a \in \mathbb{Z}$ with $p \nmid a$. Then, $a^{p-1} \equiv 1 \pmod{p}$.*

The proof of the following corollary of Fermat's Little Theorem is left to Problem 11.

Corollary 12.13. *Suppose p is prime and $a \in \mathbb{Z}$. Then, $a^p \equiv a \pmod{p}$.*

Exercises and Problems.

(1) Solve each system of congruences using the technique employed in Example 154(2).

   (a) $x \equiv 7 \pmod{11}$, $x \equiv 3 \pmod{6}$.

   (b) $x \equiv 3 \pmod{12}$, $x \equiv 7 \pmod{16}$.

   (c) $x \equiv 9 \pmod{19}$, $x \equiv 3 \pmod{13}$.

   (d) $x \equiv 2 \pmod{3}$, $x \equiv 3 \pmod{5}$, $x \equiv 2 \pmod{7}$.

(2) Find the least positive integer N such that when N is divided by 3 the remainder is 2, when N is divided by 4 the remainder is 3, when N is divided by 5 the remainder is 4, and when N is divided by 7 the remainder is 6.

(3) Describe the set of all integers x satisfying the following two congruences simultaneously:

$$x \equiv 3 \pmod{7}, \quad x \equiv 6 \pmod{8}.$$

(4) In Problem 10 you are asked to prove that the Euler totient function is multiplicative (i.e., $\phi(ab) = \phi(a)\phi(b)$ if a and b are relatively prime). Use this fact to complete the following.

   (a) Compute $\phi(20)$.

   (b) Compute $\phi(180)$.

   (c) If $n \geq 3$, prove that $\phi(n)$ is even.

   (d) If n is odd, show that $\phi(2n) = \phi(n)$.

(5) Suppose $p > 2$ is prime and $a \in \mathbb{Z}$ with $p \nmid a$. Show that a^{p-2} is the inverse of a modulo p.

---

(6) Prove that for every integer $c \geq 20$, the Diophantine equation $7x + 4y = c$ has a solution (x, y) with both x and y being nonnegative integers.

(7) If $d = \gcd(a, n) \neq 1$, then the existence of a solution of a linear congruence $ax \equiv b \pmod{n}$ will depend on b. If $d \nmid b$, then no solutions exist. If $d \mid b$, let $a = da_1, b = db_1$, and $n = dn_1$. Prove that

$$ax \equiv b \pmod{n} \iff a_1 x \equiv b_1 \pmod{n_1}.$$

Note. Since $\gcd(a_1, n_1) = 1$ (Theorem 7.10(4)), we reduced the problem to the case described in Theorem 12.7. Therefore, if $d \mid b$, solutions exist and can be found effectively.

(8) Let $a, b \in \mathbb{N}$ and $a_1, b_1 \in \mathbb{Z}$. Prove that if a and b are relatively prime, then the system of congruences

$$x \equiv a_1 \;(\mathrm{mod}\; a), \quad x \equiv b_1 \;(\mathrm{mod}\; b)$$

has a solution. This statement (as well as the one in Problem 16) represents a particular case of the so-called Chinese Remainder Theorem.

(9) For odd prime p, prove that $x^2 \equiv 1 \;(\mathrm{mod}\; p)$ has exactly two congruence classes of solutions, namely $[-1]$ and $[1]$.

(10) For $n \in \mathbb{N}$, let $\phi(n)$ denote the Euler phi function.

  (a) Find $\phi(5)$, $\phi(6)$, $\phi(30)$, and $\phi(25)$.
  (b) Determine $\phi(p)$, $\phi(p^2)$, $\phi(p^3)$, and $\phi(p^m)$, if p is prime and $m \in \mathbb{N}$.
  (c) Prove that if $\gcd(a, b) = 1$, then $\phi(ab) = \phi(a)\phi(b)$.
  (d) Prove that if $n = p_1^{e_1} p_2^{e_2} \cdots p_k^{e_k}$ is the prime decomposition of n, then

$$\phi(n) = n\left(1 - \frac{1}{p_1}\right)\left(1 - \frac{1}{p_2}\right) \cdots \left(1 - \frac{1}{p_k}\right).$$

  (e) Prove that if a and n are positive integers, then $a^{\phi(n)+1} \equiv a \;(\mathrm{mod}\; n)$.

(11) Prove Corollary 12.13.

(12) The following statement is sometimes called the *Freshman's Dream*: For p prime, $(a + b)^p \equiv (a^p + b^p) \;(\mathrm{mod}\; p)$, for all $a, b \in \mathbb{Z}$.

  (a) Explain how the result follows from Fermat's Little Theorem.
  (b) Prove the result by using the Binomial Theorem.

(13) Let $a, b \in \mathbb{N}$ be relatively prime. Prove that the equation $ax + by = ab$ has no solution with $x, y \in \mathbb{N}$. Can it be solved with $x, y \in \mathbb{Z}$?

---

(14) Let a and b be two positive, relatively prime integers. Prove that $ab - a - b$ is the greatest integer that cannot be written as $ax + by$ with x and y being nonnegative integers.

(15) The goal of this problem is to develop the proof that there are infinitely many prime integers of the form $4n + 1$, $n \in \mathbb{N}$.

We make use of the following well-known result from number theory, the proof of which lies outside the focus of this book.

Let p be an odd prime. The congruence $x^2 \equiv -1 \;(\mathrm{mod}\; p)$ has a solution $x \in \mathbb{Z}$ if and only if $p \equiv 1 \;(\mathrm{mod}\; 4)$.

  (a) Use the given result to prove the following: if $N = k^2 + 1$ for some integer k, then all odd prime factors of N are congruent to 1 modulo 4.
  (b) Prove that there are infinitely many primes of the form $4n + 1$. (Consult the proof of Proposition 12.3.)

(16) Let $a, b, c \in \mathbb{N}$ and $a_1, b_1, c_1 \in \mathbb{Z}$. Prove that if any two of the integers a, b, c are relatively prime, then the system of congruences

$$x \equiv a_1 \;(\mathrm{mod}\; a), \quad x \equiv b_1 \;(\mathrm{mod}\; b), \quad x \equiv c_1 \;(\mathrm{mod}\; c)$$

has a solution. This statement (as well as the one in Problem 8) represents a particular case of the Chinese Remainder Theorem.

12.4 Nonlinear Diophantine Equations

In Section 7.3 we explained how to find all integer solutions to a linear Diophantine equation of the form $ax + by = c$ for integers a, b, and c. In this short section we consider how to find integer solutions (or show they do not exist) to certain nonlinear integer equations.

Example 156. Find all integer solutions of the equation $x^2 - y^2 = 115$.

Solution. We have $x^2 - y^2 = (x - y)(x + y) = 115$. Since both $x - y$ and $x + y$ are integers, the problem is reduced to solving the following eight systems of two equations with two unknowns, where each case corresponds to factoring 115 into two factors:

$$(x - y, x + y) \in \{(1, 115), (-1, -115), (5, 23), (-5, -23),$$
$$(115, 1), (-115, -1), (23, 5), (-23, -5)\}.$$

Solving each system, we find the solution set of the equation:

$$\{(58, 57), (-58, -57), (14, 9), (-14, -9), (58, -57),$$
$$(-58, 57), (14, -9), (-14, 9)\}.$$

The following observation could streamline the process: if (a, b) satisfies the equation, then so do $(-a, b), (a, -b)$, and $(-a, -b)$. Then it would be sufficient to consider only the systems $(x - y, x + y) = (1, 115)$ and $(x - y, x + y) = (5, 23)$, namely, solutions with $x \geq y \geq 0$. □

Example 157. Prove that the equation $x^2 - 2y^2 + 8z = 3$ has no integer solutions.

Solution. If y is even, i.e., $y = 2k$ for $k \in \mathbb{Z}$, then

$$x^2 = 3 - 8z + 2y^2 = 3 - 8z + 8k^2 \equiv 3 \pmod 8.$$

If y is odd, i.e., $y = 2k + 1$ for $k \in \mathbb{Z}$, then

$$x^2 = 3 - 8z + 2y^2 = 3 - 8z + 8k^2 + 8k + 2 \equiv 5 \pmod 8.$$

Thus $x^2 \equiv 3$ or $5 \pmod 8$. But neither case is possible; indeed, if $x \equiv 0, 1, 2, 3, 4, 5, 6,$ or $7 \pmod 8$, then $x^2 \bmod 8 = 0, 1, 4, 1, 0, 1, 4,$ and 1, respectively. □

The reader may wonder why we utilized modulus 8 in the solution above. If a Diophantine equation has a solution, then so does the corresponding congruence for an arbitrary modulus m (equal numbers are congruent for every modulus). The contrapositive to this statement is:

If there exists a positive integer m such that a congruence modulo m has no solution, then the corresponding Diophantine equation has no solution.

Therefore, a general approach of showing that a Diophantine equation has no solutions is to find a positive integer m such that the corresponding congruence modulo m has no solutions. One may start with small moduli, like 2, 3, 4, 5, etc. In Example 157, it turns out that modulus 8 is the smallest modulus to yield a congruence with no solutions. Sometimes the needed modulus can be found quickly, sometimes not. Unfortunately, the approach may not work at all. There are Diophantine equations that have no integer solutions, but for which the corresponding congruences have a solution for every modulus $n \geq 2$. One such example is given by the equation $(2x + 1)(3x + 1) = 0$, which obviously has no integer solutions. It can be shown that the corresponding congruence $(2x + 1)(3x + 1) \equiv 0 \pmod{n}$ has a solution for every $n \geq 2$ (see Problem 6).

Exercises and Problems.

(1) Prove that the following Diophantine equations have no solutions:

    (a) $x^2 - 5y = 3$.

    (b) $2x^2 - 5y^2 = 7$.

    (c) $x^4 - 2y^2 = 1$.

(2) Find at least three integer solutions for each of the following equations:

    (a) $x^2 + y^2 = z^2$.

    (b) $x^2 - 5y^2 = 1$.

    (c) $x^2 + 1 = 5y$.

---

(3) Find all integer solutions of the equation $2xy = x^2 + 2y$.

(4) Find all integer solutions of the equation $x^3 + 91 = y^3$.

(5) Solve the Diophantine equation $x^3 - 2y^3 - 4z^3 = 0$.

(6) Prove that the congruence $(2x + 1)(3x + 1) \equiv 0 \pmod{n}$ has a solution for every integer $n \geq 2$, yet the corresponding equation has no integer solutions.

12.5 Cryptography: Rabin's Method

One fascinating modern application of number theory is the area of public-key cryptosystems. A public-key cryptosystem is one in which an individual (often called "Bob" in the literature) will make certain parameters public, and any other person (often referred to as "Alice") can then use those parameters to send a message[4] to Bob. The most commonly discussed public-key system is undoubtedly the RSA scheme introduced by M.I.T. researchers Ron Rivest, Adi Shamir, and Len Adleman in 1976. RSA is a noteworthy scheme; a brief primer is given in Problem 9, and we encourage the reader to research the method further. However, in order to offer an accessible alternative, we will discuss a special case of Rabin's method, a technique published by Michael Rabin in 1979. The presentation here follows the one given in *Mathematics: A Discrete Introduction*, by Edward Scheinerman [**18**].

A key component of Rabin's method involves finding square roots modulo n. For a positive integer n and $a \in \mathbb{Z}$ with $0 < a < n$, we say a is a **quadratic residue modulo** n if there is an integer b such that $a \equiv b^2 \pmod{n}$.[5] We call b a **square root** of a modulo n. Since each square root is a representative of a congruence class modulo n, if b is a square root of a modulo n, then $b + nk$ for any $k \in \mathbb{Z}$ will be as well. Our preference is to select the representative of the congruence class containing b from the set $\{0, 1, \dots, n - 1\}$ or the negatives of these.

---

[4] Using an encoding system such as ASCII or Unicode, any printable message can be converted to a string of numbers in order to implement a cryptographic scheme. Thus, for our purposes here, all messages will be sent as numbers.

[5] Notice that if a is a quadratic residue modulo n, then $n \nmid a$. Some sources require a quadratic residue modulo n to be relatively prime to n.

For example, 3 is a quadratic residue modulo 11, since $5^2 \equiv 3 \pmod{11}$. Also notice that $6^2 \equiv 3 \pmod{11}$, where 6 can be viewed as -5 or $11 - 5$ modulo 11. It may seem very natural that we have found two square roots of 3 modulo 11, namely 5 and 6. For some values of n and a, there may be no square roots of $a \pmod{n}$. For other values, we will see that there may be two or four square roots.

Theorem 12.14. *If p is an odd prime and a is a quadratic residue modulo p, then a has exactly two square roots modulo p.*

Proof. Since a is a quadratic residue, a has a square root $x \in \mathbb{Z}$. Then $x^2 \equiv a \pmod{p}$, and since $(p - x)^2 = p^2 - 2px + x^2 \equiv x^2 \equiv a \pmod{p}$, we see that $p - x$ is another square root of a. Since $p \nmid a$, $p \nmid x$; hence $x \not\equiv 0 \pmod{p}$. Moreover, $x \not\equiv p - x \pmod{p}$, since $p \nmid 2$. Therefore, a has at least two square roots.

If y is another square root of a, then $(x - y)(x + y) = x^2 - y^2 \equiv a - a \equiv 0 \pmod{p}$. Hence $p | (x - y)(x + y)$, and since p is prime, either $p | (x - y)$ or $p | (x + y)$. By selecting x and y strictly between 0 and p, we conclude that either $x - y = 0$ (so $y = x$) or $x + y = p$ (so $y = p - x$). Thus, a has exactly two square roots modulo p. $\square$

When p is a prime congruent to 3 modulo 4, there is a straightforward method of computing the two square roots of a quadratic residue modulo p.

Theorem 12.15. *If a is a quadratic residue modulo p, where p is a prime with $p \equiv 3 \pmod{4}$, then $\pm a^{(p+1)/4} \bmod p$ are square roots of a modulo p.*

Proof. Since a is a quadratic residue modulo p, there is some integer x such that $x^2 \bmod p = a$. We will show that $\left((x^2)^{(p+1)/4}\right)^2 \bmod p = a$ as well. A key step of the proof is the fact that $a^p \equiv a \pmod{p}$ for any $a \in \mathbb{Z}$ when p is prime (Problem 17 in Section 7.1). We leave the details to the reader in Problem 10. $\square$

Notice in the example prior to the theorems above, with $p = 11$ and $a = 3$, that the square roots of a are $\pm 3^{12/4} = \pm 27 \equiv \pm 5 \pmod{11}$, namely 5 and 6. As another example, if $p = 23$ and $a = 3$, the square roots of a are $\pm 3^6 \equiv \pm 16 \pmod{23}$, namely 7 and 16.

When p is not prime, there can be more than two square roots. For example, we next show there are four square roots of 3 modulo 253.

Example 158. If b is a square root of 3 modulo 253, then $b \bmod s$ will also be a square root of 3 modulo s for any divisor s of 253.

Solution. If $253 = rs$ and $b^2 = 3 + 253t$ for some integer t, then $b^2 \equiv 3 + (rt)s \equiv 3 \pmod{s}$, as desired. $\square$

Example 159. Show that there are four square roots of 3 modulo 253.

Solution. Notice that $253 = 23 \cdot 11$, and we previously found two square roots of 3 modulo 23 (namely 7 and 16) and two square roots of 3 modulo 11 (namely 5 and 6). So, if b is a square root of 3 modulo 253, then b must be congruent to 5 or 6 (mod 11) and b must be congruent to 7 or 16 (mod 23).

Together, there are four systems of congruences to solve:

(1) $b \equiv 5 \pmod{11}$ and $b \equiv 7 \pmod{23}$,

(2) $b \equiv 5 \pmod{11}$ and $b \equiv 16 \pmod{23}$,

(3) $b \equiv 6 \pmod{11}$ and $b \equiv 7 \pmod{23}$,

(4) $b \equiv 6 \pmod{11}$ and $b \equiv 16 \pmod{23}$.

In each of the systems, a solution may be found using the method in Example 154, and the four solutions are $b = 214$, $b = 16$, $b = 237$, and $b = 39$. Notice that these solutions are respectively congruent to -39, -237, -16, and -214 modulo 253, so, as before, each root is the negative of another root.   □

The technique used above can be utilized to find the four square roots of a modulo n, where $1 \leq a \leq n = pq$ for odd primes p and q, when a is a quadratic residue both modulo p and modulo q. The primary steps are:

(i) Compute the two square roots modulo p and the two square roots modulo q. This computation is expeditious when p and q are primes congruent to 3 modulo 4.

(ii) Solve the systems of two congruences. A solution for each is guaranteed by the Chinese Remainder Theorem, because the two moduli p and q are relatively prime; see Problem 8 in Section 12.3.

The latter calculations will be computationally fast, as long as p and q are known.

Here then is a summary of the encryption and decryption schemes for Rabin's method:

(i) The number n is determined by Bob (the receiver) as the product of two large primes p and q, each congruent to 3 modulo 4. The value of n is made public, while the individual values of p and q are known only by Bob.

(ii) A message M is encrypted by Alice (the sender) by computing $S = M^2 \bmod n$, who then sends the value S to Bob. Although the number S can be seen by the public, Bob is the only one who can successfully calculate M, a square root of S, using the above method, because he alone knows the factors p and q. While S may have up to four square roots, a carefully chosen conversion method between message characters and numbers should ensure that only one of these square roots will translate into anything meaningful—otherwise Bob will have to do a bit of guessing! For instance, in Example 159, the square roots obtained were 214, 16, 237, and 39. If the usable characters had been specified to be the letters A through Z, with A corresponding to 1 up through Z corresponding to 26, then the proper interpretation would be the 16th letter of the alphabet, P.

Can this scheme be easily hacked? Suppose, for example, that we knew $n = 253$ was the product of two *unknown* primes (call them p and q) and that we calculated that $3 \equiv 214^2 \equiv 16^2 \equiv 237^2 \equiv 39^2 \pmod{253}$. Could we use this information to factor n? The answer is "yes!" Notice that

$$(214 - 16)(214 + 16) \equiv 214^2 - 16^2 \equiv 3 - 3 \equiv 0 \pmod{253}.$$

This means that $198 \cdot 230 = pqk = 253k$ for some integer k. (Alternatively, we could have used the factors 237 and 39 to find that $(237 - 39)(237 + 39) \equiv (198)(23) \equiv 0 \pmod{253}$.)

Then, one way to factor n would be to find a common divisor of 198 and 253 ($= pq$). Now, if 198 and 253 were relatively prime, then 198 would not be divisible by p or q. In this case, since $198 \cdot 230 = 253k$, it follows that 230 would be divisible by both p and q. This is impossible, because $230 < pq = 253$. Thus, 198 and $pq = 253$ share a nontrivial common divisor, and it must be p or q. We could therefore use the Euclidean Algorithm to efficiently find $\gcd(198, 253)$. This turns out to be 11, and then we calculate $253/11 = 23$ to find that n has prime factors 11 and 23.

The same method could always be used to factor n if we knew the four square roots of some number a modulo n. Moreover, the process would be efficient, because of the efficiency of the Euclidean Algorithm. This suggests that finding square roots modulo n is as difficult as factoring n, which is computationally infeasible for large n. Therefore, Rabin's method is as secure as other cryptographic methods, such as RSA, whose security relies upon the presumed difficulty of factoring integers that are products of two large primes.

Exercises and Problems.

(1) Find all quadratic residues modulo n in each case by computing all nonzero values of $a = b^2 \bmod n$, for $1 \le b \le n - 1$.

  (a) $n = 7$.   (b) $n = 8$.   (c) $n = 10$.   (d) $n = 11$.   (e) $n = 12$.

(2) Find all values of a, $1 \le a \le 20$, for which a has four square roots modulo 21.

(3) Show that if a is a perfect square (in the integers), then a is a quadratic residue modulo n for each $n \in \mathbb{N}$.

(4) As a newcomer to Rabin's cryptosystem, Bob selects small primes $p = 7$ and $q = 19$ and publishes $n = pq = 133$. It is understood that only 2-letter words (or sequences) will be encrypted, with each letter represented by a 2-digit number: 01 for "A", 02 for "B", ..., and 26 for "Z". Notice that Bob's choice of n restricts the 2-letter words that can be encrypted with his modulus, from "AA" (0101) to "AZ" (0126).

  (a) Alice wants to send information about the weapon in a murder mystery to Bob. Convert the word "AX" to numerical form; call this number M. Encrypt M by computing M^2 modulo n. The result must be an integer between 0 and $n - 1$.

  (b) Suppose Bob has received the encrypted message 130 from Alice, indicating the culprit in the mystery. Decrypt this message to reveal a 2-letter word within Bob's range of possibilities.

  (c) Suppose Bob has received the encrypted message 42 from Alice. Decrypt this message to reveal a 2-letter word within Bob's range of possibilities.

  (d) Explain how an observant eavesdropper on the communication between Alice and Bob could observe that the first letter in any message sent through this particular cryptosystem must be "A".

(5) Explain why any message M encrypted with Rabin's method should satisfy $0 < M < n$, where n is the modulus used in the encryption.

---

(6) Find all of the square roots of 100 (mod 247). Use the method shown in Example 159.

(7) Find all of the square roots of 6 (mod 1380).

(8) Suppose Bob is now more experienced in implementing Rabin's method than he was in Exercise 4. He perceives the need to encrypt all 2-letter sequences with his modulus, where each letter is represented by a 2-digit number: 01 for A, 02 for B,..., 26 for Z.

   (a) Find all pairs of primes p and q that are suitable for Rabin's method, with p and q between 40 and 70, for which Bob's modulus $n = pq$ can encrypt all 2-letter sequences (see Exercise 5).

   (b) Suppose Bob has chosen $n = 43 \cdot 71 = 3{,}053$. If Alice wishes to securely say "Hi" to Bob using Rabin's method, determine the numerical form M of the message, and encrypt the message.

   (c) Suppose Bob receives the encrypted message 2417 from Alice, using the modulus $n = 43 \cdot 71 = 3{,}053$. Decrypt and determine the 2-letter word that indicates Alice's favorite soft drink.

   (d) Suppose Bob receives the encrypted message 2921 from Alice, using the modulus $n = 43 \cdot 71 = 3{,}053$. Decrypt and determine the 2-letter word that indicates Alice's favorite eatery.

   (e) Suppose it is public knowledge that Bob's modulus is 3,953. We will use the method given in this section, involving known square roots of a quadratic residue, to determine p and q for which $pq = 3{,}953$. It is easy to check that 2,094 is a quadratic residue modulo 3,953, $100^2 \equiv 2{,}094 \pmod{3{,}953}$, and $1{,}575^2 \equiv 2{,}094 \pmod{3{,}953}$. Find p by computing the greatest common divisor of 3,953 and one of the factors of $(1{,}575 + 100)(1{,}575 - 100)$ and then find q.

(9) In the RSA Public-Key Cryptosystem, each user obtains two large prime numbers p and q (each is typically larger than 10^{100}). He keeps p and q secret, but the number $n = pq$, called the *enciphering modulus*, is made public. He computes $\phi(n)$ and keeps its value secret.

   (a) For $n = pq$, with p and q distinct primes, compute $\phi(n)$.

   The user then selects an *enciphering exponent* e between 1 and $\phi(n)$ that is relatively prime to $\phi(n)$. The inverse of e mod $\phi(n)$ is denoted by d and is called the *deciphering exponent*. He keeps d secret, but e is public knowledge.

   (b) Explain why $ed = 1 + k\phi(n)$ for some $k \in \mathbb{Z}$.

   In order for another user to send a secure message to him, she looks up his values for n and e in a public directory. For a numerical message P, where $\gcd(P, n) = 1$, she computes P^e mod n. This is openly sent to him, but only he can easily recover the original message, since only he has access to d.

   (c) If $ed \equiv 1 \pmod{\phi(n)}$ and $\gcd(P, n) = 1$, use Euler's Theorem to show $(P^e)^d \equiv P \pmod{n}$.

(10) Complete the details of the proof of Theorem 12.15.

(11) Suppose p is an odd prime. Prove the following statements about quadratic residues and square roots modulo p. Use Exercise 1 to confirm your belief that the statements are true.

(a) The square roots of 1 modulo p are 1 and $p - 1 \equiv -1 \pmod{p}$, as well as members of their congruence classes modulo p.

(b) The integer -1 is a quadratic residue modulo p if and only if $p \equiv 1 \pmod 4$.

(c) The number of quadratic residues modulo p between 1 and $p - 1$, inclusive, is $\dfrac{p - 1}{2}$.

(12) Using data gathered from Exercise 1 and additional examples, explore the question of whether the facts in Problem 11 hold for composite moduli. Do some of these facts appear to hold for some composite n, but not others? Can you find alternative results that hold for certain collections of composite numbers?

Your Discrete Math professor has students put solutions to homework problems on the board. Below is a table denoting twelve students and twelve problems. For each student, a Y is placed under the problems that the student has completed. The goal is to assign a problem to each student in such a way that each student has completed the one assigned. Will that be possible in this case? If so, assign an appropriate problem to each student.

	1	2	3	4	5	6	7	8	9	10	11	12
Al	Y	Y	Y		Y	Y	Y	Y	Y	Y		Y
Bo		Y	Y	Y			Y	Y	Y	Y	Y	
Cal	Y	Y					Y				Y	
Dee	Y	Y		Y	Y		Y				Y	
Ed	Y		Y	Y	Y	Y	Y	Y		Y	Y	Y
Fay	Y		Y	Y		Y	Y	Y	Y	Y		Y
Gad	Y				Y						Y	
Hal		Y		Y	Y		Y				Y	
Ian	Y	Y		Y			Y					
Jo	Y			Y	Y						Y	
Kay	Y		Y	Y		Y	Y	Y	Y	Y		Y
Len	Y	Y		Y	Y		Y				Y	

13

Topics in Graph Theory

Read Euler, read Euler, he is the master of us all.
—Pierre-Simon Laplace (1749–1827)

13.1 Planar Graphs

In this section we return to planar graphs. We first encountered them in Chapter 9 in the context of graph colorings and the Four Color Theorem. Here we will prove a few simple properties that will lead to a surprising connection to the Platonic solids.

We begin by reviewing some terminology. Recall that a **planar graph** is one that can be drawn on the plane so that none of its edges cross. A manifestation of a planar graph in which the edges do not cross is called a **plane graph**. In a plane graph, the edges divide the plane into regions, where a finite region is defined to be the interior of the closed curve formed by a cycle of the graph in which no edges between vertices of the cycle pass through this interior. Note that in a plane graph there will also always be one infinite region, but the infinite region can be replaced by a finite region in an isomorphic plane graph simply by drawing the edges differently. See Figure 13.1 where the region B bordered by the cycle $(1, 2, 3, 1)$ is finite in the left depiction but infinite in the right depiction.

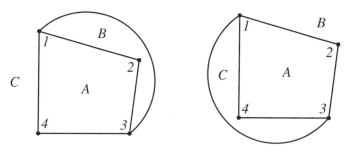

Figure 13.1

One of the best known results in planar graph theory (along with the Four Color Theorem) is due to Euler. He showed that all plane graphs with a fixed number of vertices and edges divide the plane into the same number of regions. Thus there is a linear relationship between the number of vertices, edges, and regions (often called "faces" in this theorem) in all plane graphs. Before reading on, we invite the reader to discover this relationship by creating a few small examples.

Theorem 13.1 (Euler's formula). *In a connected plane graph with e edges, v vertices, and f faces, $v - e + f = 2$.*

Proof. We first prove this result for tree graphs. In Section 9.4, we proved that a tree on v vertices has exactly $v - 1$ edges. Since a tree has no cycles, it has only the infinite face, so $f = 1$. Therefore, $v - e + f = v - (v - 1) + 1 = 2$, as desired.

We now prove the general result by using induction on the number of edges. Assume that the result holds for any connected plane graph with v vertices, e edges, and f faces, and let G be a connected plane graph with $e + 1$ edges. The statement is clearly true in the base case for an isolated vertex, where there are no edges and one (infinite) face. Now assume that the result holds for any connected plane graph with $e - 1$ edges, for some fixed $e \geq 1$, and let G be a connected plane graph with v vertices, f faces, and e edges. Let G' be the graph obtained from G by removing any edge from that cycle. Clearly, G will still be a connected plane graph. Moreover, G' will have v vertices but one fewer edge and one fewer face than G. Therefore, applying the inductive hypothesis to G', we obtain $v - (e - 1) + (f - 1) = 2$, whence it follows that $v - e + f = 2$ for G. This proves the result by induction. □

Euler's formula yields two inequalities relating the number of edges and vertices in planar graphs.

Corollary 13.2. *Let G be a connected plane graph with $v \geq 3$ vertices and e edges. Then*

(1) $e \leq 3v - 6$ *and*

(2) *if G has no triangles (3-cycles), $e \leq 2v - 4$.*

Proof. (1) For each face of G, form all ordered pairs (face, edge) for which the specified edge borders the given face. Let N denote the total number of all such pairs, over all faces of G. Since each face is bordered by at least three edges (unless $v = 3$ and $e \leq 2$, in which cases we already have $e \leq 3v - 6$), we see that $N \geq 3f$. Furthermore, no edge can border more than two faces, so $N \leq 2e$. Summarizing, $2e \geq N \geq 3f$. Then, by Euler's formula, since $f = e - v + 2$, we have

$$2e \geq 3f = 3(e - v + 2) = 3e - 3v + 6,$$

from which it follows that $e \leq 3v - 6$.

(2) Proof left to the reader. See Problem 6 in Section 13.2. □

Corollary 13.3. *If G is a planar graph, then G has at least one vertex with degree at most 5.*

Proof. The sum of the vertex degrees in any graph is $2e$. On the other hand, if every vertex of G has degree 6 or greater, then the sum of the vertex degrees will be at least

$6v$, which means that $2e \geq 6v$ and $e \geq 3v$. Since we may assume that G is connected (we could make our current argument on any connected component of G), we have contradicted the previous corollary. Hence G must have a vertex of degree 5 or less. $\square$

These corollaries can be used to prove that the complete graph K_5 and the complete bipartite graph $K_{3,3}$ are not planar. (See Problem 7 in Section 13.2.) Surprisingly, this observation actually helps characterize *all* nonplanar graphs! Of course, if a graph G has K_5 or $K_{3,3}$ as a subgraph, then G will not be planar, but it turns out that the converse is essentially true as well. However, it first takes some work to explain what we mean by "essentially".

A **subdivision** of a single edge $\{u, w\}$ consists of adding a (new) vertex to the edge; more precisely, $\{u, w\}$ is replaced by the edges $\{u, v\}$ and $\{v, w\}$ where v is a new vertex. A **subdivision** of a graph G is a graph that can be built from G by replacing some edges with subdivisions. We then say graphs G and H are **homeomorphic** if there is an isomorphism from a subdivision of G to a subdivision of H. Roughly speaking, G and H are homeomorphic if one can place new vertices on existing edges of G or H (or both) to form two identical graphs. For example, the two graphs in Figure 13.2 are homeomorphic. A graph is planar if and only if all graphs homeomorphic to it are planar, because any additional vertices of degree 2 in no way affect edge crossings of either graph.

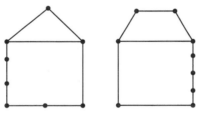

Figure 13.2

We omit the proof of the following theorem, published by Kazimierz Kuratowski in 1930. A proof can be found in standard texts on advanced graph theory ([**5**], for example).

Theorem 13.4 (Kuratowski's Theorem, 1930). *A graph is planar if and only if it has no subgraph homeomorphic to K_5 or $K_{3,3}$.*

Example 160. Show that the graph in Figure 13.3 is nonplanar. This regular graph with ten vertices and fifteen edges is called the Petersen graph. It is known for being a counterexample to a variety of properties that hold in many graphs.

Solution. From Kuratowski's Theorem, it is sufficient to find a subgraph of G that is a subdivision of K_5 or $K_{3,3}$. We leave the details to the reader in Problem 18 in Section 13.2. $\square$

We now turn to some applications of these corollaries to other areas of mathematics. We begin with a discussion of regular polyhedra in space. A convex polyhedron is called **regular** if all of its faces are congruent regular polygons and the same number

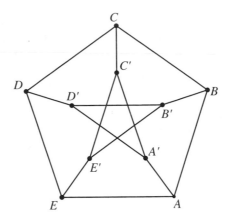

Figure 13.3

of faces (and thus, edges) meet at each vertex. One can associate a planar graph with a convex polyhedron by envisioning a hole in one face and looking through the hole to see a configuration of vertices and edges. The face containing this hole can serve as the infinite face of the corresponding planar graph, and the vertices of this face will form a cycle, within which lie all of the other vertices and edges. See Figure 13.4 for planar graphs of a cube (eight vertices with six congruent squares as faces, including the infinite face (A, B, C, D)) and tetrahedron (four vertices with four congruent triangles as faces, including the infinite face (A, B, C)). The reader should pause and convince himself that a graph obtained in this way from a convex polyhedron will indeed be planar.

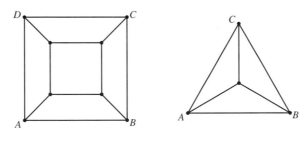

Figure 13.4

Theorem 13.5. *There are exactly five regular polyhedra.*

Proof. For the graph G associated with a given regular polyhedron, let d denote the (common) degree of each vertex, n the (common) number of edges bordering each face, v the number of vertices, e the number of edges, and f the number of faces. We make the following observations about G.

(1) From Euler's formula, $v + f - 2 = e$.

(2) Since the sum of the vertex degrees is twice the number of edges, $vd = 2e$, yielding
$$v = \frac{2e}{d}.$$

(3) Each face is bordered by $n \geq 3$ edges and each edge borders two faces; thus $fn = 2e$, yielding $f = \frac{2e}{n}$ (by counting the number of face-edge pairs where the edge borders the given face).

Combining these results, we obtain

$$\frac{2e}{d} + \frac{2e}{n} = e + 2.$$

Dividing through by $2e$ yields

$$\frac{1}{d} + \frac{1}{n} = \frac{1}{2} + \frac{1}{e} > \frac{1}{2}. \tag{13.1}$$

Obviously $d \geq 3$, since each vertex of a polyhedron lies on at least three edges. Therefore, $\frac{1}{n} > \frac{1}{2} - \frac{1}{d} \geq \frac{1}{2} - \frac{1}{3} = \frac{1}{6}$, showing $n < 6$.

Since n is an integer, it follows that n equals 3, 4, or 5. Because of the symmetry in (13.1), by a similar argument, d equals 3, 4, or 5. Now, using the just-proven formulas for v, f, and e, we examine all $3 \cdot 3 = 9$ possibilities in Table 13.1. The reader should complete the table, thereby verifying that only five of these nine possibilities satisfy the equality in (13.1). Since each one of the five indeed corresponds with an actual regular polyhedron, the proof is complete. The common names for these five polyhedra are given in the table.   □

Table 13.1. Possible Regular Polyhedra

d	n	e	v	f	Name
3	3	6	4	4	Tetrahedron
3	4	12	8	6	Cube
3	5	30	20	12	Dodecahedron
4	3	12	6	8	Octahedron
4	4	–	–	–	
4	5	–	–	–	
5	3	30	12	20	Icosahedron
5	4	–	–	–	
5	5	–	–	–	

13.2 Chromatic Polynomials

In Section 9.3, we introduced vertex colorings and saw some of their applications. Recall that in a proper vertex coloring, each vertex is assigned a specific color, subject to the requirement that adjacent vertices are given different colors. The **chromatic number** of G, denoted $\chi(G)$, is the fewest number of colors needed to properly color all of the vertices.

The number of proper vertex colorings of a graph is of great theoretical interest to graph theorists and has applications to other areas of mathematics and the study of recreational puzzles [13]. In this section we explore the properties of the chromatic polynomial (to be defined shortly) of a labeled graph and its relation to proper vertex colorings.

Consider the number of ways one could properly color the three labeled vertices of triangle ABC using five colors. There are 5 choices for vertex A, 4 choices for vertex B, and 3 choices for vertex C. Each vertex is adjacent to the other two vertices, so three colors must be used, resulting in a total of $5 \cdot 4 \cdot 3 = 60$ proper colorings. It is not hard to see that if x colors are available, then there will be $x(x-1)(x-2)$ proper colorings. Notice the impact of the vertices being assigned labels. Coloring the vertices red, yellow, and blue clockwise, starting at a specified vertex, is considered a different coloring than coloring them yellow, blue, and red, respectively.

The number of proper colorings of a graph G in x colors is a function of G and x, and we will denote this "chromatic function" as $P(G, x)$. Using this notation, we just showed

$$P(C_3, x) = x(x-1)(x-2) = x^3 - 3x^2 + 2x.$$

Example 161. Determine $P(\overline{K}_n, x)$.

Solution. Because the null graph on n vertices has no edges, each vertex may be colored in any one of the x colors available. Therefore $P(\overline{K}_n, x) = x^n$. □

Example 162. Determine $P(C_4, x)$.

Figure 13.5

Solution. See Figure 13.5. When $x = 1$, there are no proper colorings. When $x = 2$, there are two choices of colors for vertex 1, and then the color of each of the other three vertices is determined. Therefore, $P(C_4, 2) = 2$. For larger values of x, there are two cases:

(1) When vertices 1 and 3 are assigned the same color (x ways to do this), then vertices 2 and 4 can each be assigned any of the remaining $x - 1$ colors. Thus, in this case, there are $x(x-1)^2$ possible ways to color the vertices.

(2) When vertices 1 and 3 are assigned different colors (with $x(x-1)$ ways to do this), then vertices 2 and 4 can each be assigned any of the remaining $x - 2$ colors. In this case, there are $x(x-1)(x-2)^2$ possible ways to color the vertices.

Therefore, combining the two cases,

$$P(C_4, x) = x(x-1)^2 + x(x-1)(x-2)^2$$
$$= x(x-1)(x^2 - 3x + 3) = x^4 - 4x^3 + 6x^2 - 3x.$$

Notice that this formula holds when $x = 1$ and $x = 2$ as well. □

In general, the chromatic function can be computed recursively by using an **edge contraction**. Given a graph G on n vertices with m edges, including an edge $e = \{u, v\}$, define the graphs $G - e$ and G/e as follows:

$G - e$: This graph is obtained from G by deleting edge e. It will have the same vertex set as G, with $m - 1$ edges.

G/e: This graph is obtained from G by **contracting** edge e. This means we delete edge e and merge vertices u and v into one vertex, w; in other words, edges of the form $\{y, u\}$ or $\{y, v\}$ are replaced with the single edge $\{y, w\}$. Notice that G/e will have $n - 1$ vertices and $m - d - 1$ edges, where $d = |N_G(u) \cap N_G(v)|$, the number of common neighbors of vertices u and v.

An example with graphs G, $G - e$, and G/e is shown in Figure 13.6.

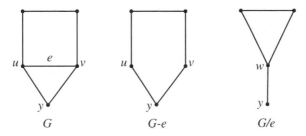

Figure 13.6

The proper colorings of G are closely related to the proper colorings of $G - e$ and G/e, as follows. The colorings of $G - e$ are of two types—those where u and v are colored the same and those where they are colored differently. The colorings where u and v are assigned the same color coincide exactly with the proper vertex colorings of graph G/e, and the colorings where u and v are assigned different colors coincide with the proper vertex colorings of G. Therefore the number of proper colorings of $G - e$ is the sum of the number of proper colorings of G and G/e. This proves the following theorem.

Theorem 13.6 (Graph Reduction Theorem). $P(G, x) = P(G - e, x) - P(G/e, x)$.

In our examples thus far, the chromatic function $P(G, x)$ has been a monic polynomial of degree n, and the Graph Reduction Theorem gives us an easy way to prove that this is always the case.

Theorem 13.7. *Let G be a graph with n vertices, m edges, and c connected components, $G_1, G_2, \ldots, G_c$. Let x be a nonnegative integer.*

(1) *$P(G, x)$ is a monic polynomial of degree n.*

(2) *For a nonnegative integer k, $x - k$ is a factor of $P(G, x)$ if and only if $k < \chi(G)$.*

(3) *$P(G, x) = P(G_1, x) \cdot P(G_2, x) \cdots P(G_c, x)$.*

(4) *$P(G, x)$ has 0 as a root of multiplicity c.*

(5) *The coefficient of x^{n-1} in $P(G, x)$ is $-m$.*

(6) *G is a tree if and only if $P(G, x) = x(x - 1)^{n-1}$.*

Proof. We prove (1) and (5) simultaneously by using strong induction on m. When $m = 0$, G is the null graph $\overline{K}_n$, and we have seen that $P(\overline{K}_n, x) = x^n$, the quintessential monic polynomial of degree n. Now assume the statement is true for all graphs with at most m edges, and let G be a graph with n vertices and $m + 1$ edges. Let $e = \{u, v\}$ be an edge of G. By the reduction theorem, $P(G, x) = P(G - e, x) - P(G/e, x)$. Then $G - e$ is a graph on n vertices and m edges; by the inductive hypothesis, $P(G - e, x) = x^n - mx^{n-1} + p(x)$, where $p(x)$ is a polynomial of degree at most $n-2$. On the other hand, it does not matter exactly how many edges G/e has (it will be at most m), but G/e has $n-1$ vertices, since u and v merged into one vertex. Therefore, $P(G/e, x) = x^{n-1} + q(x)$, where $q(x)$ is a polynomial of degree at most $n - 2$. Thus,

$$P(G, x) = x^n - mx^{n-1} + p(x) - x^{n-1} - q(x) = x^n - (m + 1)x^{n-1} + p(x) - q(x),$$

as desired.

To prove (3), note that the vertices of each component may be colored separately, so by the Multiplication Rule, $P(G, x)$ will simply be the product of the individual $P(G_i, x)$, $1 \le i \le c$.

Statement (4) is clearly true when there are no edges or one edge. Proceeding by strong induction on the number of edges, assume that any graph with c components and fewer than $m \ge 2$ edges will have $x^c g(x)$ proper vertex colorings, where $g(x)$ is a monic polynomial of degree $n - c$ (based on part (1) of this theorem) and $g(0) \ne 0$. We consider two cases.

If G has at least two components with an edge, then we can apply the inductive hypothesis to each component and conclude that $P(G_i, x) = xg_i(x)$, where $g_i(0) \ne 0$, for $1 \le i \le c$. Since

$$P(G, x) = \prod_{i=1}^{c} P(G_i, x) = x^c \prod_{i=1}^{c} g_i(x),$$

it follows that $P(G, x)$ has 0 as a root of multiplicity c.

If G has only one component with edges, then we may assume $c = 1$. If G is a tree, then by part (6) of this theorem, $P(G, x) = x(x - 1)^{n-1}$, and 0 is a root of multiplicity 1. On the other hand, if G has a cycle with an edge e, then $G - e$ and G/e are both connected. Using the Graph Reduction Theorem and the inductive hypothesis, we see that

$$P(G, x) = P(G - e, x) - P(G/e, x) = xf(x) - xh(x) = x(f(x) - h(x)),$$

where f and h are polynomials of degree $n-1$ and $n-2$, respectively, and 0 is not a root of either one. Moreover, $f(0)$ and $h(0)$ are the coefficients of the x terms in $P(G - e, x)$ and $P(G/e, x)$, respectively, so by Problem 22, $f(0)$ and $h(0)$ will have different signs. Therefore, $f(0) - h(0) \ne 0$, proving that 0 is a root of $P(G, x)$ of multiplicity 1.

We leave the proofs of (2) and (6) to the problem set. We suggest using induction to determine the chromatic polynomial of a tree on n vertices. For the converse, note that if the chromatic polynomial of G is $x(x - 1)^{n-1}$, the number of vertices, edges, and components can be determined using other parts of this theorem. $\square$

The above theorem allows us to call $P(G, x)$ the **chromatic polynomial** of G. It was introduced by David Birkhoff in 1912 to study colorings of planar graphs in an effort to prove the Four Color Theorem. Hassler Whitney generalized the notion in 1928 and the chromatic polynomial has been much studied since that time.

Exercises and Problems.

(1) A planar graph G has 12 vertices and 18 edges.

   (a) How many faces does G have if G is connected?
   (b) How many does it have if G has two components?

(2) Draw a plane graph for the octahedron.

(3) Draw a plane graph for the icosahedron.

(4) Draw a plane graph for the dodecahedron.

(5) For each graph G in Exercises 2 through 4, construct a new graph G' (drawn in a different color so as not to be confused with G) as follows.

   • Place a vertex in each face, including the infinite face.
   • Place a new edge between any two new vertices that reside in adjacent faces.

   How many vertices and edges are in each graph G'? How do these values compare with the numbers of vertices, faces, and edges in plane graph representations of the cube, octahedron, icosahedron, and dodecahedron? (Euler's formula may give some insight here.)

(6) Prove Corollary 13.2(2).

(7) Prove that K_5 and $K_{3,3}$ are not planar graphs.

(8) Let G be the graph on five vertices and six edges, consisting of a 4-cycle and a 3-cycle sharing an edge. Determine $P(G, x)$ and use it to find $\chi(G)$.

(9) Let G be the graph on six vertices and seven edges, consisting of a two 4-cycles sharing an edge. Determine $P(G, x)$ and use it to find $\chi(G)$.

(10) Find and draw two nonisomorphic graphs that have chromatic polynomial $x(x - 1)^5$.

(11) Let G be a regular, planar graph with 3 components, 14 faces, and 30 edges. What is the degree of each vertex?

---

(12) A certain graph has chromatic polynomial $p(x) = x^{10} - 15x^9 + 105x^8 - 455x^7 + 1{,}353x^6 - 2{,}861x^5 + 4{,}275x^4 - 4{,}305x^3 + 2{,}606x^2 - 704x$. Based on this polynomial, how many vertices, edges, and components does the graph have, and what is its chromatic number? (Based on this information, can you determine the graph? It is a famous one.)

(13) (a) Prove that if G is 2-colorable, then G is bipartite.
     (b) Prove that if G is bipartite, then G is 2-colorable.
     (c) Recall that a graph is k-**partite** if its vertices can be partitioned into k subsets in such a way that no edges exist between vertices of any subset. Prove that a graph G is k-colorable if and only if it is k-partite.

(14) Find a generalization of Euler's formula for planar graphs with c components.

(15) Explain why $x^2(x-1)(x-3)$ cannot be the chromatic polynomial of any graph.

(16) Explain why $x^2(x-1)^2(x-2)(x-3)(x-4)$ cannot be the chromatic polynomial of a planar graph.

(17) For a graph G, let $\Delta(G)$ denote the maximum degree of a vertex in G. Prove that $\chi(G) \leq 1 + \Delta(G)$.

(18) Show that the Petersen graph is nonplanar. See Example 160.

(19) Prove Theorem 13.7(2) by proving the following statements.

(a) If $x-k$ is a factor of the chromatic polynomial $P(G,x)$ then $k < \chi(G)$.

(b) If $k < \chi(G)$, then $x-k$ is a factor of $P(G,x)$.

(20) Prove that $P(C_n, x) = (x-1)^n + (-1)^n(x-1)$.

(21) (a) Determine $P(K_{4,4}, 3)$.

(b) Determine $P(K_{n,n}, 3)$ for general n.

(22) Prove that if G is a connected graph on n vertices, then the coefficient of x in $P(G,x)$ is positive if n is odd and negative if n is even.

Remark. This result is actually a special case of the following fact about graphs with c components and n vertices: if $P(G,x) = \sum_{i=1}^{n} a_i x^i$, then for $i \geq c$, $a_i > 0$ if $n-i$ is even and $a_i < 0$ if $n-i$ is odd. In other words, the coefficients of a chromatic polynomial alternate in sign. ◇

(23) Prove Theorem 13.7(6) by proving the following two statements about a connected graph G on n vertices.

(a) If G is a tree, then $P(G,x) = x(x-1)^{n-1}$.

(b) If $P(G,x) = x(x-1)^{n-1}$, then G is a tree.

(24) Prove the "Six Color Theorem": the vertices of any planar graph can be colored properly with six colors.

(25) In this problem, we use Euler's formula to find a solution to Example 60 from Chapter 5. Consider $n \geq 1$ points on a circle, with every two of them joined by a chord in such a way that no three chords intersect at a point.

(a) Show that the chords intersect at $\binom{n}{4}$ points inside the circle.

(b) Let G be the planar graph whose vertex set consists of the original n points on the circle and $\binom{n}{4}$ intersection points in its interior and whose edge set consists of pairs of vertices that are consecutive endpoints of arcs of the circle or segments on the chords. (See Figure 13.7, where $n = 6$.) Determine the number of edges of G as a function of n by finding the sum of the degrees of the vertices.

(c) Show that the chords divide the disc into $\binom{n}{4} + \binom{n}{2} + 1$ regions ("faces").

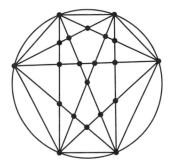

Figure 13.7

13.3 Spanning Tree Algorithms

*Nothing takes place in the world whose meaning
is not that of some maximum or minimum.*

—Leonhard Euler (1707–1783)

Recall that in Section 9.4, a **tree** was defined as a graph that is connected and acyclic; thus, a tree is a graph in which there is a unique path between any two vertices. A **spanning tree** for a connected graph G was defined to be a subgraph of G that is a tree and contains all the vertices of G. By Theorem 9.12, a tree with n vertices must have $n - 1$ edges; thus if T is a spanning tree of connected graph G, then $|V(T)| = |V(G)|$ and $|E(T)| = |V(G)| - 1$.

Clearly, it is possible for a given graph to have more than one spanning tree. If the graph is weighted, the sum of the weights of the edges of a spanning tree is the **weight** of the spanning tree. For a connected, weighted graph G, a **minimum spanning tree** (**MST**) of G is a spanning tree for which the sum of the weights of the edges in the tree is the smallest possible. Examples of situations in which finding an MST is desirable are plentiful—laying cable in a neighborhood of new houses, designing an efficient navigation system, creating a telecommunications network, etc.

Example 163. Use trial and error to search for a minimum spanning tree for the graphs A (left) and B (right) in Figure 13.8.

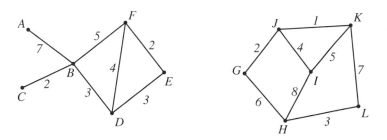

Figure 13.8

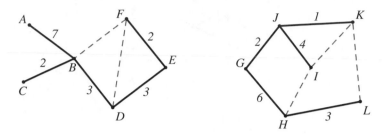

Figure 13.9. Minimum spanning trees for A and B

Solution. The minimum spanning tree of each graph is represented by the bold edges in Figure 13.9.

The weight of the MST for graph A is 17; the weight of the MST for graph B is 16. The MST for each is unique, although in general this need not be the case. □

Example 164. Graphs A and B in Example 163 were small enough to find a minimum spanning tree by inspection. Use those results to find an MST for graph C, shown in Figure 13.10.

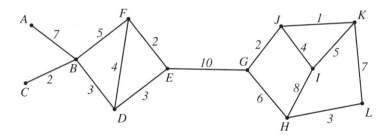

Figure 13.10. Graph C

Solution. Since graph C consists of A and B joined by a cut edge, it is not difficult to see that an MST for C is comprised of the MST for A, the MST for B, and the cut edge, $\{E, G\}$. See Figure 13.11.

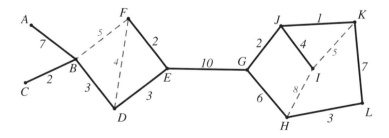

Figure 13.11. Minimal spanning tree for C

The weight of the (unique) MST for C is 43. □

Finding an MST for a given graph falls within the broad category of *optimization problems*. As is true with many concepts in discrete mathematics, the objective is easy to understand, and, as above, solutions for small cases can be obtained by brute force (checking all possibilities). However, many meaningful situations in which a minimum spanning tree is desired are too large to be calculated by hand. Furthermore, although as humans we may be able to look at graph C and immediately notice that the cut edge allows us to find a spanning tree by joining together spanning trees for two subgraphs of C, such an observation is difficult to program into a computer. Therefore, in order to implement (program) a procedure that will work for any graph, we need to delineate specific step-by-step instructions in an algorithm.

We now introduce two algorithms that can be used to obtain an MST for a given connected, undirected, weighted graph. Both of these are **greedy algorithms**. A greedy algorithm is one that makes the "best" choice at each stage of a multistep process, based on an identified requirement or condition, rather than conducting a comprehensive evaluation of the entire situation to find a solution that is better overall. This somewhat simplistic approach is easy to implement, but depending on the algorithm, its solution can range anywhere from poor to optimal.

It is somewhat surprising that though each of the algorithms we present are greedy, both turn out to be optimal in the sense that they find spanning trees with the smallest weight possible. We omit the proofs of this result; they can easily be found online or in a standard book on graph theory ([5] or [17]).

13.3.1 Kruskal's Algorithm. Given a connected, weighted, undirected graph G, the general idea of Kruskal's Algorithm is to start with an empty tree and add edges from $E(G)$ one a time, always choosing an edge of smallest possible weight, without creating a cycle. Once all elements of $V(G)$ are included, a spanning tree has been created.

Kruskal's Algorithm

(1) Sort the elements of $E(G)$ into nondecreasing order.

(2) Let T be an empty subgraph of G.

(3) Select an element of $E(G)$ of smallest weight.

(4) If this edge does not create a cycle in T, add the edge and its endpoints to T.

(5) Repeat steps (3) and (4) until T contains $|V(G)| - 1$ edges.

The subgraph, T, that results from implementing Kruskal's Algorithm is a minimum spanning tree of G.

Example 165. Use Kruskal's Algorithm to find a minimum spanning tree for the graph shown in Figure 13.12.

Solution. We begin by sorting the edges of the graph into nondecreasing order, as shown in the first column of Table 13.2. We then add edges according to the

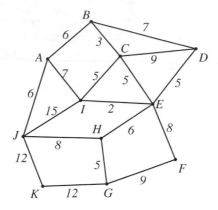

Figure 13.12

algorithm, making sure not to create a cycle. The first edge that is not added is edge $\{C, I\}$, for it would create a 3-cycle. Note that an MST for this graph must contain ten of the edges in the edge set and all eleven elements of the vertex set. The table itemizes the steps of Kruskal's Algorithm when applied to this graph.

Table 13.2. Implementing Kruskal's Algorithm

Edge	Weight	Added to T?	Vertices in T
$\{E, I\}$	2	**Added**	$\{E, I\}$
$\{B, C\}$	3	**Added**	$\{B, C, E, I\}$
$\{C, E\}$	5	**Added**	$\{B, C, E, I\}$
$\{C, I\}$	5	Not Added	$\{B, C, E, I\}$
$\{G, H\}$	5	**Added**	$\{B, C, E, G, H, I\}$
$\{D, E\}$	5	**Added**	$\{B, C, D, E, G, H, I\}$
$\{A, B\}$	6	**Added**	$\{A, B, C, D, E, G, H, I\}$
$\{A, J\}$	6	**Added**	$\{A, B, C, D, E, G, H, I, J\}$
$\{E, H\}$	6	**Added**	$\{A, B, C, D, E, G, H, I, J\}$
$\{A, I\}$	7	Not Added	$\{A, B, C, D, E, G, H, I, J\}$
$\{B, D\}$	7	Not Added	$\{A, B, C, D, E, G, H, I, J\}$
$\{E, F\}$	8	**Added**	$\{A, B, C, D, E, F, G, H, I, J\}$
$\{H, J\}$	8	Not Added	$\{A, B, C, D, E, F, G, H, I, J\}$
$\{C, D\}$	9	Not Added	$\{A, B, C, D, E, F, G, H, I, J\}$
$\{F, G\}$	9	Not Added	$\{A, B, C, D, E, F, G, H, I, J\}$
$\{J, K\}$	12	**Added**	$\{A, B, C, D, E, F, G, H, I, J, K\}$
$\{G, K\}$	12	Not Added	$\{A, B, C, D, E, F, G, H, I, J, K\}$
$\{I, J\}$	15	Not Added	$\{A, B, C, D, E, F, G, H, I, J, K\}$

The resulting spanning tree, having weight 58, is shown in Figure 13.13. This solution is not unique; a different sorting of the edges in $E(G)$ into nondecreasing order could result in a different spanning tree (still of weight 58) when the algorithm is implemented.   □

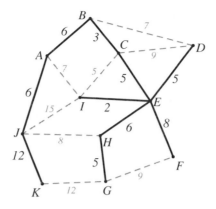

Figure 13.13

13.3.2 Prim's Algorithm. Like Kruskal's Algorithm, Prim's Algorithm begins with an empty graph and builds it up to become a spanning tree. However, whereas Kruskal's Algorithm creates a minimum spanning tree by successively seeking and adding cheap *edges*, Prim's Algorithm successively seeks and adds close *vertices* (with accompanying edges). In doing so, Kruskal's Algorithm creates a sequence of forests (preserving *acyclicity*), with the final forest ultimately being connected once $n - 1$ edges are added, while Prim's Algorithm creates a sequence of subtrees (preserving *connectedness*) of the original graph, with the final subtree ultimately spanning the graph once n vertices are added.

Prim's Algorithm

(1) Let T be an empty subgraph of G.

(2) Choose any vertex, v_0, of $V(G)$ and put v_0 in T.

(3) From among the elements of $E(G)$ having one endpoint in $V(T)$ and one endpoint in $V(G) \setminus V(T)$, select one of smallest weight. Put this edge and its other endpoint in T.

(4) Repeat step (3) until $V(T) = V(G)$.

The subgraph, T, that results from implementing Prim's Algorithm is a minimum spanning tree of G.

Example 166. Use Prim's Algorithm to find a minimum spanning tree for the graph shown in Figure 13.12.

Solution. We begin by choosing any one of the vertices to put into $V(T)$. In this solution, we have selected A as the initial vertex. After the cheapest edge with A as an endpoint is selected ($\{A, B\}$), three edges are considered ($\{A, J\}, \{A, J\}$, and $\{B, D\}$) before edge $\{B, C\}$ is next added to T. Notice that when the third edge is added, an arbitrary choice must be made between edge $\{C, E\}$ and edge $\{C, I\}$. Table 13.3 itemizes the steps of Prim's Algorithm as applied to this graph.

The resulting MST is indicated by the bold edges in Figure 13.14. As in Example 165, any spanning tree for this graph must contain ten of the edges in the edge set and all eleven elements of the vertex set. Note that although the MST again has weight

Table 13.3. Implementing Prim's Algorithm

Edge Added to T	Weight	Vertices in T
$\varnothing$		$\{A\}$
$\{A,B\}$	6	$\{A,B\}$
$\{B,C\}$	3	$\{A,B,C\}$
$\{C,I\}$	5	$\{A,B,C,I\}$
$\{I,E\}$	2	$\{A,B,C,E,I\}$
$\{E,D\}$	5	$\{A,B,C,D,E,I\}$
$\{A,J\}$	6	$\{A,B,C,D,E,I,J\}$
$\{E,H\}$	6	$\{A,B,C,D,E,H,I,J\}$
$\{H,G\}$	5	$\{A,B,C,D,E,G,H,I,J\}$
$\{E,F\}$	8	$\{A,B,C,D,E,F,G,H,I,J\}$
$\{G,K\}$	12	$\{A,B,C,D,E,F,G,H,I,J,K\}$

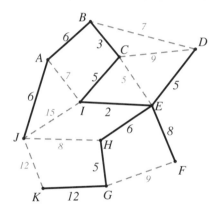

Figure 13.14

58, the tree is not identical to the one found in Example 165. As was the case with the implementation of Kruskal's Algorithm, this solution is not unique; a different choice of v_0, or a different choice of smallest-weight edge to connect a vertex from $V(T)$ with a vertex from $V(G) \setminus V(T)$ in some of the iterations of step (3) of the algorithm, could result in a different MST. $\square$

13.4 Path and Circuit Algorithms

13.4.1 Traveling Salesperson Problem. A classic and much studied problem within graph theory, previously introduced in Section 9.2, is the so-called *Traveling Salesperson Problem*, or TSP. Traditionally, the context for this problem is one in which a salesperson needs to visit each of the cities in a particular set and the optimization goal is to minimize the total weight of the circuit. In other words, the desired outcome is a minimum weight Hamilton circuit. Usually, we assume that $G = K_n$; edges that represent routes that are inconvenient to travel can be given large weights, assuring that they won't be chosen for the circuit. The weights may represent, for example, distances between cities or time required to travel between locations or the cost of plane or train travel between cities.

Example 167. Consider the weighted graph shown in Figure 13.15. By inspection, solve the Traveling Salesperson Problem. That is, use "brute force" to find a minimum weight Hamilton circuit.

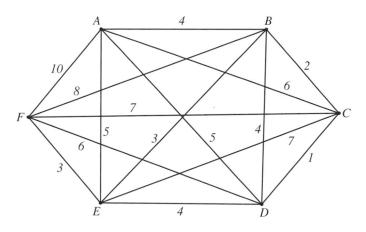

Figure 13.15

Solution. We admit to not actually examining all $5!/2 = 60$ circuits[1] starting and ending at vertex A, but we are confident that circuit $\{A, B, C, D, F, E, A\}$ in Figure 13.16 forms a circuit of minimum total weight, 21. □

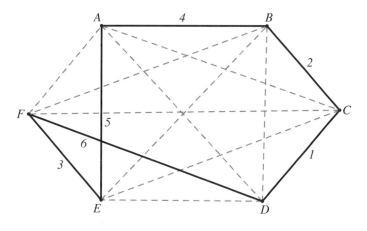

Figure 13.16

When n is large, using brute force to consider all possible $(n-1)!/2$ routes that a "traveling salesperson" might take is impractical, even with the aid of computing technology. However, because of the widespread applications of the TSP, considerable effort has been invested in producing algorithms that find or approximate a solution.

---

[1]More generally, K_n has $(n-1)!/2$ different circuits to examine starting and ending at each vertex, by considering all permutations of the other vertices and noting that a circuit traversed in the opposite direction will have the same weight as the original circuit in an undirected graph.

The **Nearest Neighbor Algorithm** is an example of a greedy algorithm; the traveler constructs a path by choosing the "nearest" unvisited location at each step.

Nearest Neighbor Algorithm

(1) Let Y be an empty subgraph of G.

(2) Let v_0 be the element of $V(G)$ that is designated as the vertex of origination for the circuit. Put v_0 in Y, and set $v_0 = v'$.

(3) Identify the vertex w such that the edge $\{v', w\}$ has minimum weight in the set
$$\{w \in V(G) \setminus V(Y) : \{v', w\} \in E(G)\}.$$
Put w and $\{v', w\}$ in Y.

(4) Set $w = v'$.

(5) Repeat steps (3) and (4) until $V(Y) = V(G)$.

The Nearest Neighbor Algorithm is a fast algorithm that performs reasonably well, but the result is usually not optimal. However, since the weight of the resulting circuit may vary depending on the choice of vertex of origination, one way to try to find a cheaper circuit is to repeat the algorithm with a different choice for the first vertex.

Example 168. Consider the weighted graph G shown in Figure 13.15 again. Letting A be the vertex of origination, apply the Nearest Neighbor Algorithm to G.

Solution. We begin by putting A into $V(Y)$. Table 13.4 details the steps of the Nearest Neighbor Algorithm as applied to G.

Table 13.4. Implementing the Nearest Neighbor Algorithm (starting at Vertex A)

Edge Added to Y	Weight of New Edge	$V(Y)$
$\varnothing$		$\{A\}$
$\{A, B\}$	4	$\{A, B\}$
$\{B, C\}$	2	$\{A, B, C\}$
$\{C, D\}$	1	$\{A, B, C, D\}$
$\{D, E\}$	4	$\{A, B, C, D, E\}$
$\{E, F\}$	3	$\{A, B, C, D, E, F\}$
$\{F, A\}$	10	$\{A, B, C, D, E, F\}$

As it turns out, with this graph, the cheapest available edge (one that doesn't create a circuit prematurely) at each step occurs by adding the vertices in alphabetical order around the outer edges of the graph. The resulting circuit, $\{A, B, C, D, E, F, A\}$, has weight 24 and is shown with bold edges in Figure 13.17. □

Further Thoughts. Notice that the weight of the resulting circuit, 24, exceeds the value of 21 found in Example 167. It also exceeds the value found by using the nearest neighbor method starting at some of the other vertices. See Exercise 9. ◇

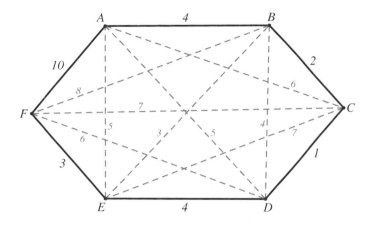

Figure 13.17

The **Sorted Edge Algorithm** is another example of a greedy algorithm, in which the traveler constructs a path by sequentially choosing the "cheapest" edge available that doesn't prematurely complete a cycle. Note that the approach is similar to that of Kruskal's Algorithm for finding a minimal spanning tree, because the graph that is being formed is not necessarily connected until the final edge is added.

Sorted Edge Algorithm

(1) Sort the elements of $E(G)$ into nondecreasing order.

(2) Let C be an empty subgraph of G.

(3) Select an element of $E(G)$ of smallest weight—that is, a "cheapest" edge.

(4) Add this edge and its endpoints to C, *unless* doing so would

    (a) complete a cycle that does not include all elements of $V(G)$ or

    (b) cause an element of $V(C)$ to have degree greater than 2.

(5) Repeat steps (3) and (4) until $V(C) = V(G)$.

The sorting required to carry out the Sorted Edge Algorithm makes this algorithm slower than the Nearest Neighbor Algorithm. However, even though it is slower than the Nearest Neighbor Algorithm, it isn't necessarily better, and it certainly need not result in an optimal solution to the Traveling Salesperson Problem. See Problem 15.

Example 169. Once again, consider the weighted graph G shown in Figure 13.15. Use the Sorted Edge Algorithm to approximate a solution to the Traveling Salesperson Problem applied to G.

Solution. We begin by sorting the edges of G into nondecreasing order (with ties broken arbitrarily), as shown in the first column of Table 13.5. The table also itemizes the remaining steps of the algorithm. The edges are selected and added according to their order in the list, unless the given edge either completes a circuit prematurely (as edge $\{B, D\}$ would) or yields a vertex of degree greater than 2 (as edge $\{A, B\}$ would). The resulting circuit, having total weight 24, is shown in bold edges in Figure 13.18. □

Table 13.5. Implementing the Sorted Edge Algorithm

Edge	Weight	Added to C?	Vertices in C
$\{C,D\}$	1	**Added**	$\{C,D\}$
$\{B,C\}$	2	**Added**	$\{B,C,D\}$
$\{E,F\}$	3	**Added**	$\{B,C,D,E,F\}$
$\{B,E\}$	3	**Added**	$\{B,C,D,E,F\}$
$\{A,B\}$	4	Not Added	$\{B,C,D,E,F\}$
$\{B,D\}$	4	Not Added	$\{B,C,D,E,F\}$
$\{D,E\}$	4	Not Added	$\{B,C,D,E,F\}$
$\{A,D\}$	5	**Added**	$\{A,B,C,D,E,F\}$
$\{A,E\}$	5	Not Added	$\{A,B,C,D,E,F\}$
$\{A,C\}$	6	Not Added	$\{A,B,C,D,E,F\}$
$\{D,F\}$	6	Not Added	$\{A,B,C,D,E,F\}$
$\{C,E\}$	7	Not Added	$\{A,B,C,D,E,F\}$
$\{C,F\}$	7	Not Added	$\{A,B,C,D,E,F\}$
$\{B,F\}$	8	Not Added	$\{A,B,C,D,E,F\}$
$\{A,F\}$	10	**Added**	$\{A,B,C,D,E,F\}$

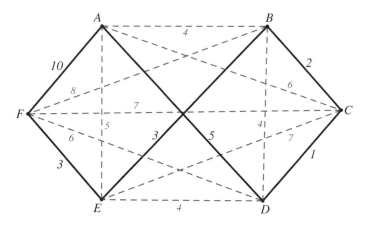

Figure 13.18

13.4.2 Shortest-Path Algorithm. Another type of optimization problem within graph theory involves finding the "cheapest" or "shortest" path—that is, the path of smallest total weight—between two specified vertices of a graph G. Unlike the process of finding a minimum spanning tree or solving a Traveling Salesperson Problem, finding a shortest path does not require visiting all vertices.

In 1959, Edsger Dijkstra published an algorithm for finding a cheapest path between two vertices in a simple, connected graph. The variant of Dijkstra's Algorithm that we present below actually finds a cheapest path between a specified vertex v_0, the origin, and every other vertex in the graph, resulting in a **shortest-path spanning tree originating at v_0.** (There is a different shortest-path spanning tree for each choice of the origin.) Versions of the algorithm exist for both directed and undirected graphs, but the essential components are the same; we present the algorithm for an undirected graph.

Dijkstra's Algorithm

(1) Let T be an empty subgraph of a connected graph G.

(2) Let v_0 be the element of $V(G)$ designated as the origin, and put v_0 in T.

(3) For each edge in $E(G)$ having one endpoint, v, in $V(T)$ and one endpoint, w, in $V(G) \setminus V(T)$, calculate the distance from v_0 to w. Select a path of smallest distance. Put the edge associated with this path, together with its other endpoint, in T.

(4) Repeat step (3) until $V(T) = V(G)$.

The subgraph, T, that results from implementing Dijkstra's Algorithm is a shortest-path spanning tree of G.

Note the similarity in process between Dijkstra's Algorithm and Prim's Algorithm. Note also, however, that the repeated step (3) in Dijkstra's Algorithm might require considerably more work than the corresponding step in Prim's Algorithm.

In one sense, Dijkstra's Algorithm may be viewed as a greedy algorithm, because at each step it adds an edge (to the shortest-path tree) which will result in the cheapest path from the starting vertex, v_0, to any available vertices. In another sense, it employs a brute force approach in that all paths between v_0 and each other vertex are considered, albeit indirectly. (We say "indirectly" because if the cheapest path from v_0 to some vertex z follows the route v_0wxyz, then necessarily the cheapest path from v_0 to y is v_0wxy, the cheapest path from v_0 to x is v_0wx, and so on.) Thus, Dijkstra's Algorithm is probably best categorized as what is called a *dynamic programming algorithm*: once the subproblem of finding the cheapest path from v_0 to another vertex is found, that solution is stored and reused later in solving the subproblem of finding the cheapest path to the next vertex. This feature of using previously found values to help make optimal choices later was not present in the greedy algorithms we encountered previously (Prim, Kruskal, Sorted Edge, and Nearest Neighbor).

Example 170. Use Dijkstra's Algorithm to find a shortest-path spanning tree, using A as the vertex of origination, for the graph shown in Figure 13.19.

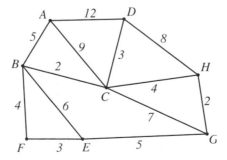

Figure 13.19

Solution. We begin by putting A into $V(T)$. Table 13.6 itemizes the steps of Dijkstra's Algorithm as applied to this graph, and the resulting spanning tree is shown in Figure 13.20.   □

Table 13.6. Implementing Dijkstra's Algorithm

Edge Added	New Path (with Distance)	$V(T)$
∅		$\{A\}$
$\{A, B\}$	A, B (5)	$\{A, B\}$
$\{B, C\}$	A, B, C (7)	$\{A, B, C\}$
$\{B, F\}$	A, B, F (9)	$\{A, B, C, F\}$
$\{C, D\}$	A, B, C, D (10)	$\{A, B, C, D, F\}$
$\{C, H\}$	A, B, C, H (11)	$\{A, B, C, D, F, H\}$
$\{B, E\}$	A, B, E (11)	$\{A, B, C, D, E, F, H\}$
$\{H, G\}$	A, B, C, H, G (13)	$\{A, B, C, D, E, F, G, H\} = V(G)$

Notice that each of the seven paths in the second column is a cheapest path from A to the last vertex listed in the path.

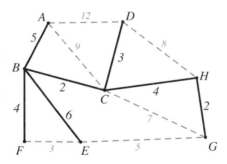

Figure 13.20

13.4.3 General comparison of algorithms. Our goal for the sections on algorithms was to provide a glimpse into the assortment of algorithms one can use on a graph to solve problems that have real-world applications. Though we did not give rigorous attention to their levels of efficiency, hopefully you noticed that algorithms can vary widely in both their performance and their runtimes (i.e., the number of steps, up to some constant factor, needed to implement the algorithm as a function of the number of vertices, n, or number of edges, m). The analysis of the tradeoffs between an algorithm that runs quickly, but may not give an optimal solution, and a slower-running algorithm that yields a better solution lies in the intersection of mathematics and computer science.

The problem of finding a minimum Hamilton circuit in a weighted graph G is not an easy one. In fact, as we saw, there is not even an efficient way to determine whether a graph has a Hamilton circuit, let alone find them. When $G = K_n$, as is usually the case for the Traveling Salesperson Problem, there are of course many Hamilton circuits, but the process of finding the length of each one is also very time consuming. Nevertheless, this brute force approach is the only one that is able to guarantee that an optimal solution is found, in general. The Sorted Edge and Nearest Neighbor Algorithms are heuristic (relatively fast, but not necessarily optimal) algorithms that usually give fairly good solutions. The Nearest Neighbor Algorithm can be improved by implementing it with each vertex as the starting vertex; however, such an approach multiplies the runtime by a factor of n, and an optimal solution still is not guaranteed.

In contrast with the Traveling Salesperson Problem, the problems of finding a minimal spanning tree and finding the cheapest path between two specified vertices can be solved quickly and optimally. Kruskal's, Prim's, and Dijkstra's Algorithms each may be implemented in a polynomial (as a function of n) number of steps, and though each is somewhat greedy, they are guaranteed to find an optimal solution for the problem they solve.

Below we summarize the algorithms encountered in this chapter. For each one, we list the magnitude of its runtime and an assessment of its level of optimality. The runtimes are not meant to be precise, as they can vary depending on the way the graph information is stored and accessed and whether the given graph is typical or extreme. In particular, there are various implementations of Dijkstra's Algorithm that can improve its efficiency. See, for example, [8].

Problem	Algorithm	Runtime	Optimal?
Traveling Salesperson	Brute Force	$(n-1)!/2$	Yes
	Sorted Edge	n^2	No
	Nearest Neighbor	n^2	No
Minimal Spanning Tree	Kruskal	$m \log n$	Yes
	Prim	n^2 or $m \log n$	Yes
Shortest-Path	Dijkstra	n^2 or $m + n \log n$	Yes

Exercises and Problems.

(1) How many different spanning trees exist for the following?

    (a) P_n

    (b) C_n

    (c) K_n

(2) Apply (a) Kruskal's Algorithm and (b) Prim's Algorithm to find a minimum spanning tree for the graph in Figure 13.21.

(3) Apply (a) Kruskal's Algorithm and (b) Prim's Algorithm to find a minimum spanning tree for the graph in Figure 13.22.

(4) Apply (a) Kruskal's Algorithm and (b) Prim's Algorithm to find a minimum spanning tree for the graph in Figure 13.23.

(5) Apply (a) Kruskal's Algorithm and (b) Prim's Algorithm to find a minimum spanning tree for the graph in Figure 13.24.

(6) Suppose you are a city planner tasked with finding a minimum spanning tree to model the laying of cable to various parts of the city. For political reasons, you are told you must include certain "edges" in the final spanning tree, even though they may prevent you from obtaining a minimum spanning tree. How would you

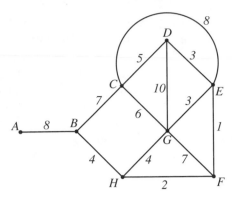

Figure 13.21

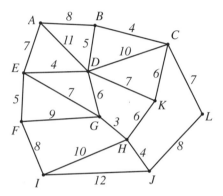

Figure 13.22

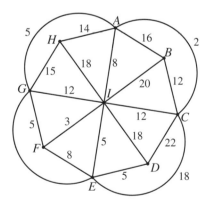

Figure 13.23

modify (i) Prim's Algorithm and (ii) Kruskal's Algorithm to accommodate such a request in each case below? Is one of these easier than the other to modify?

(a) You are given one edge that must be included.

(b) You are given a set of four edges that must be included.

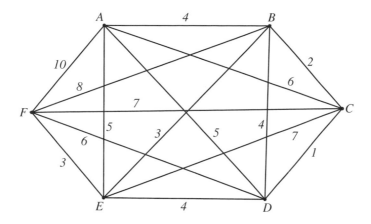

Figure 13.24

(7) How could Kruskal's Algorithm be adapted to find the *maximum spanning tree* for a graph G? Explain.

(8) Use the Nearest Neighbor Algorithm twice to find the cost of a Hamilton circuit in Figure 13.25, once using A as the vertex of origination and once using C. Is either of the circuits found minimal? (Recall that you found the cost of all $(4 - 1)!/2 = 3$ distinct circuits by a brute-force examination in Exercise 2 of Section 9.2.)

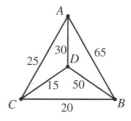

Figure 13.25

(9) Repeat Example 168 five times, using the Nearest Neighbor Algorithm to find the cost of a Hamilton circuit in Figure 13.15, with each of the vertices B, C, D, E, and F serving as the vertex of origination. Which of the five routes is the cheapest?

(10) Use the Nearest Neighbor Algorithm (try a couple of different origination vertices) and the Sorted Edge Algorithm to find the cost of a Hamilton circuit in Figure 13.26. Can you find a cheaper Hamilton circuit than the ones from these algorithms?

(11) Consider the weighted K_6 graph shown in Figure 13.27.

(a) How many circuits must be examined to guarantee finding a minimum Hamilton circuit?

(b) Use the Nearest Neighbor Algorithm to find a Hamilton circuit beginning at vertex A.

(c) Use the Nearest Neighbor Algorithm to find a Hamilton circuit beginning at vertex E.

(d) Find the Hamilton circuit determined by the Sorted Edge Algorithm.

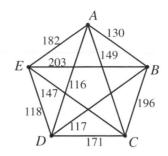

Figure 13.26

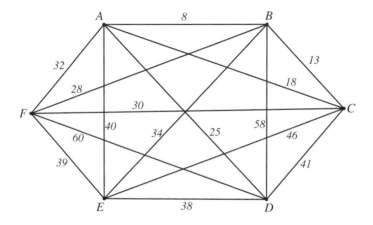

Figure 13.27

(12) Use Dijkstra's Algorithm to create a shortest-path spanning tree, with vertex of origination A, for the graph in Figure 13.22. Create a table like Table 13.6 in Example 170. What is the cheapest path from A to J?

(13) Use Dijkstra's Algorithm to create a shortest-path spanning tree, with vertex of origination I, for the graph in Figure 13.23. Create a table like Table 13.6 in Example 170. Which vertex is the farthest from I?

(14) Prove that if all edge weights of a connected, undirected graph are distinct, the minimum spanning tree is unique.

(15) Create a weighted graph for which the Sorted Edge Algorithm yields a Hamilton circuit with a higher weight than that obtained by using the Nearest Neighbor Algorithm.

---

(16) For a complete graph with n vertices, how many simple paths originate at vertex v_1 and terminate at v_n? This is the number of paths that would be checked in the least efficient, shortest-path algorithm. Leave your answer in the form of a sum.

(17) Given a graph $G = (V, E)$, a **matching** of G is a subset of its edges in which no two edges share a common vertex. A **perfect matching** is one in which all vertices appear in the matching. It is not too difficult to see that a bipartite graph containing a Hamilton circuit will contain a perfect matching (place every other

edge of the circuit into the matching), but while this is a sufficient condition for a perfect matching, it is not necessary. Hall's Theorem (see [17]) states that a bipartite graph with partition classes X and Y contains a perfect matching if and only if $|N(S)| \geq |S|$ for each subset S of X, where $N(S)$ consists of the neighbors of all vertices in S.

(a) Construct a bipartite graph that models the situation given in the opening hook problem of the chapter and explain how a perfect matching would solve the problem.

(b) Try to find a perfect matching for the graph you created in (a) or use Hall's Theorem to show that one does not exist.

Hints

Section 2.2

1. (a) 1; (b) −2; (c) 1; (d) −4; (e) 0.
2. When n is even, write $n = 2k$, and when n is odd, write $n = 2k + 1$.
3. (a) 2; (b) −1; (c) 2; (d) −3; (e) 0.
5. Make sure that $0 \leq q < |b|$.
6. Three of them are true.
7. In (b), 344 div 6 = 57 and 344 mod 6 = 2.
8. The number 2 should be used in your notation.
9. Numbers are even if and only if they are divisible by 2.
10. If m is odd, $m = 2j + 1$ for some integer j.
11. If a divides b, then $b = ja$ for some integer j.
13. Both converses are false.
14. Experiment with various values of x. Make sure to try both integers and nonintegers.
15. Use Theorem 2.5(3).
16. Denote the first number of the sum as n and the last number as $n + 3$.
17. The gcd of the coefficients is important.
18. Use the floor function.
19. In (a), one of the three integers is a multiple of 3.
20. Write n as $9k + 5$.
21. Write the integers in the form $2n - 1$ and $2n + 1$.
22. Expand $(3q + r)^2$ for appropriate values of r.
23. Examine why the given method fails for the integer 62, but notice that $62^2 = 37(100) + 12^2 = 3{,}844$.
24. Write $N = 10a + 5$.
25. Use Problem 22.
26. Simplify the sum $k + (k + 1) + \cdots + (k + (n - 1))$. Look up the formula for the sum of the integers from 1 to $n - 1$.
27. Note that integral squares have remainders 0, 1, or 4 when divided by 5.
28. If the last two digits of an integer are a and b, consider $10a + b$.

Section 2.3

1. (a) $D + E = \begin{bmatrix} 3 & 14 & 13 \\ -14 & 15 & -11 \\ 4 & 9 & 1 \end{bmatrix}$.

   (b) $4E - 3D = \begin{bmatrix} -23 & -14 & 17 \\ -7 & -3 & 26 \\ 16 & 8 & 18 \end{bmatrix}$

   (c) $DE = \begin{bmatrix} -60 & 150 & 45 \\ -89 & -24 & -95 \\ -36 & 14 & -10 \end{bmatrix}$.

   (d) $ED = \begin{bmatrix} -38 & 48 & -66 \\ -77 & -20 & -93 \\ -15 & 97 & -36 \end{bmatrix}$.

   (e) $DB = \begin{bmatrix} -60 & 0 & 125 & 30 \\ -158 & 93 & 14 & -13 \\ -48 & 28 & 24 & 8 \end{bmatrix}$.

   (f) $C^T = \begin{bmatrix} 0 & -3 \\ -8 & 4 \\ 5 & -6 \end{bmatrix}$.

   (g) $A + C^T = \begin{bmatrix} 6 & -13 \\ -6 & 1 \\ 9 & -6 \end{bmatrix}$.

   (h) $B^T E = \begin{bmatrix} 78 & -24 & 54 \\ -57 & -14 & -11 \\ -41 & 88 & 17 \\ -27 & 20 & 41 \end{bmatrix}$.

2. $E^2 = \begin{bmatrix} 8 & 56 & 4 \\ -32 & 3 & -65 \\ -31 & 61 & 36 \end{bmatrix}$.

3. (b) $CB = \begin{bmatrix} 100 & -64 & -42 & -18 \\ -76 & 54 & -3 & -2 \end{bmatrix}$.

   (d) $B^T A = \begin{bmatrix} 20 & -10 \\ -14 & -29 \\ 48 & -37 \\ 30 & -63 \end{bmatrix}$.

   (e) $A^T C^T = \begin{bmatrix} 4 & -34 \\ 24 & 18 \end{bmatrix}$.

4. (c) $(A + B) + C = \begin{bmatrix} (a_1 + b_1) + c_1 & (a_2 + b_2) + c_2 \\ (a_3 + b_3) + c_3 & (a_4 + b_4) + c_4 \end{bmatrix}$
   $$= \begin{bmatrix} a_1 + (b_1 + c_1) & a_2 + (b_2 + c_2) \\ a_3 + (b_3 + c_3) & a_4 + (b_4 + c_4) \end{bmatrix} = A + (B + C).$$

   (f) $AI_2 = \begin{bmatrix} a_1 & a_2 \\ a_3 & a_4 \end{bmatrix} \begin{bmatrix} 1 & 0 \\ 0 & 1 \end{bmatrix} = \begin{bmatrix} a_1(1) + a_2(0) & a_1(0) + a_2(1) \\ a_3(1) + a_4(0) & a_3(0) + a_4(1) \end{bmatrix} = A.$

5. $(A+B)^T = \begin{bmatrix} a_1+b_1 & a_3+b_3 \\ a_2+b_2 & a_4+b_4 \end{bmatrix} = \begin{bmatrix} a_1 & a_3 \\ a_2 & a_4 \end{bmatrix} + \begin{bmatrix} b_1 & b_3 \\ b_2 & b_4 \end{bmatrix} = A^T + B^T.$

6. The "necessary equation" referred to is $M^{-1}X = I$, where X is the purported inverse of M.

7. $C+D = \begin{bmatrix} c_1 & 0 \\ 0 & c_2 \end{bmatrix} + \begin{bmatrix} d_1 & 0 \\ 0 & d_2 \end{bmatrix} = \begin{bmatrix} c_1+d_1 & 0 \\ 0 & c_2+d_2 \end{bmatrix} \in S.$

8. D is invertible if and only if $d_1 d_2 \neq 0$.

9. $D^2 = \begin{bmatrix} d_1^2 & 0 \\ 0 & d_2^2 \end{bmatrix}.$

11. Suppose one column of A is a multiple of the other. Then $A = \begin{bmatrix} a & ka \\ c & kc \end{bmatrix}$, for some $k \in \mathbb{R}$. Now use the given hint.

Sections 3.1 and 3.2

1. All but two of these sentences are propositions.

2. Four or five of these sentences are propositions. One of them is ambiguous.

3. There are only three elements in the set described in (d).

4. $A \subseteq B$ if the following statement is true: if $a \in A$, then $a \in B$. In general, the statement $p \Rightarrow q$ is a true statement any time that p is false.

5. (a) Two of these are true.
   (b) One of these is true.
   (c) Two of these are true.
   (d) Three of these are true.

6. (a) Two of these are true.
   (b) One of these is true.
   (c) One of these is true.

8. (c) $\{n : n = k^2 \text{ for some } k \in \mathbb{Z}, k \neq 0\}$.

9. $|A| = 4, |B| = 6, |C| = 10, |S| = 2$.

10. Only of them is true.

11. (c) $\{z : z = a^3 + b^3 + c^3 \text{ for } a,b,c \in \mathbb{Z}\}$.

12. The given condition, $x^2 + 6x \geq a$, is equivalent to the condition $(x+3)^2 \geq a + 9$.

13. Use the quadratic formula to determine when there is a solution to the given equation.

Section 3.3

1. Two rows are needed in the table for (a) and four rows are needed for (b).

2. (d) $m < 3$ or $n \geq 10$.

4. The final column of the truth table should have T precisely when p and q have different truth values.

5. Three of these are true, including (a), which reads: *If m is divisible by 5, then m is divisible by 10.*

6. Five of these are true.

7. (b) $(p \wedge q) \Rightarrow r$. In general, an implication and its contrapositive will always have the same truth value.

8. Four of them are true.

Section 3.4

1. 15.

2. 5.

3. 47; 69.

4. 2,070.

5. (c) $\{1, 4\}$; (e) $\{1, 4, 6\}$.

7. (c) $(A \cap B) \cup (A \cap C) \cup (B \cap C) \setminus (A \cap B \cap C)$.

8. (c) $(0, 2, 1)$.

9. (b) 520; (c) $\{1, 2, 4\}$.

10. The answers, in no particular order, are X, Y, and $\varnothing$.

11. (d) $(5, 10]$.

12. (d) $\{\varnothing, \{x\}, \{\{x\}\}, \{x, \{x\}\}\}$.

14. For any fixed $r > 0$, $C(O, r)$ will be the set of points lying on a circle of radius r, centered at O. The union of all such sets will consist of concentric circles, and they will fill the plane, other than point O. The answer is therefore the whole plane with point O removed.

15. (a) $0 \leq |A \cap B \cap C| \leq \min\{m, n, p\}$.

16. No.

17. No.

18. Yes. If $B \neq C$, explain how $A \times B$ cannot equal $A \times C$.

19. Try finding the answer for a specific value of n, say $n = 3$.

20. (b) $\{1, 2, 3, 4, 7\}$; (d) No.

21. Each set equals $A \triangle B$.

22. The set has eleven elements.

23. Each answer is a circle with its interior.

24. Two circles and two empty sets.

Sections 3.5 and 3.6

1. (a) False. (b) True. (c) True. (d) True.

3. Nothing can be said about two of them—they could be either true or false.

4. 2^n rows are needed for n statements.

5. Let $ABCD$ be a square, and show it is a quadrilateral of the desired type.

6. Let $n \in B$. Then $n = 10k - 1$ for some integer k. Show that n is of the proper form to be in A.

7. Can you find an element that is in either set?

8. Let $A = [a_{ij}] \in S$. Then A is a 4×4 matrix for which ... (complete the statement). Now show that A meets the requirements to be in T.

9. The set $(A \setminus B) \cup (B \setminus A)$ consists of elements that are either in A but not B or are in B but not A.

10. The first line of the proof should be as follows: "Let $X \in \mathcal{P}(A) \cap \mathcal{P}(B)$."

12. Use a membership table. Alternatively, show that each set is a subset of the other. Such a proof would start with the following: "Let $x \in A \cap (B \cup C)$. Then $x \in A$ and $x \in (B \cup C)$."

13. These can be proven directly from the appropriate definitions. Membership tables are not necessary.

15. Two of them are true. Demonstrate the false ones with small counterexamples.

16. Propositions can be found that will yield any number of T's in the last column.

Section 3.7

1. A natural choice of $f : \mathbb{N} \to \mathbb{N}_0$ will send 1 to 0.

2. Doubling each number in $\mathbb{Z}$ will yield $\mathbb{E}$.

3. $(0, 1, -1, \dots)$.

5. Two of the sets are uncountable; the rest are countable.

6. Create a one-to-one onto function from (a, b) to (c, d) that maps a to c and b to d.

7. One possibility is to define $f(1/2) = 0$, $f(1/4) = 1$, and $f(1/8) = 1/2$. Of course, this is just a start.

8. In "most" cases, let $f(x) = x$.

9. Each real number in $(0, 1)$ has a unique binary representation. Use this fact to demonstrate a one-to-one correspondence between elements of these sets: the power set of $\mathbb{N}$; the set of infinite 0-1 sequences; and $[0, 1]$. For example, the infinite sequence of all zeros can correspond to both the empty set and the number 0.

10. Count the number of words in the given phrase.

Section 4.1

1. Simplify $\neg(b \Rightarrow a)$.

2. (d) Converse: If I need to walk to work, then my car is out of gas and the bus is not running. Inverse: If my car isn't out of gas or the bus is running, then I don't need to walk to work. Contrapositive: If I don't need to walk to work, then my car isn't out of gas or the bus is running.

3. (a) If it ain't broke, then don't fix it. Converse: If you don't fix it, then it must not be broken. Contrapositive: If you fix it, it must have been broken.

4. (c) The given statement is an exclusive or.

5. Each of the final columns should have exactly one False.

6. Each of the final columns should have exactly one False.

7. Show that each statement is true only when p and q have different truth values.

Sections 4.2 and 4.3

1. Two of these are predicates.

3. Two of these are true.

4. (b) False; Negation: $\forall a \in A, a^2 \neq 50$.

5. Only one is false. The negation of (c): For any distinct integers x and y, $x^2 \neq y^3$. Alternatively, there are not distinct integers x and y for which $x^2 = y^3$.

6. Two of these are true.

7. (a) We translate this as, "There exists a boy in the class and there exists a girl in the class such that the boy admires the girl." A shorter version is, "One of the boys admires one of the girls."

8. (a) There is a student who took Math 170 in the fall semester and received a B in the class.

9. (a) It starts $\forall x \in \mathbb{R} \, (x \neq 0) \Rightarrow \dots$.

    (c) The sentence includes the word "closed".

    (d) $\exists a \in \mathbb{N} \dots$.

    (e) If n is prime, $\dots$.

    (f) There is no $\dots$.

10. (2) $\forall g \, \exists b, \, \neg P(b, g)$. This may be translated as *For each girl in the class, there is a boy who doesn't admire her.*

11. In (b), each statement is true if and only if $P(x)$ and $Q(x)$ are true for all real numbers x.

12. (a) True. (b) False. (c) False (not for even x). (d) False.

13. Three of these are true.

14. One of these is true.

15. Two of these are true.

16. Interchange $\forall$ and $\exists$ and negate the inequalities.

17. One of these is true.

18. It is true. Let $x = n + 1$ and let $y = n$.

19. One of these is true.

20. There are many possibilities.

21. Factor $a^3 + b^3$.

22. Three quantifiers are needed for each of (a) and (b).

23. Four quantifiers are needed. The negation has symbolic form

$$\exists x_0 \in (a, b), \exists \epsilon > 0, \forall \delta > 0, \exists x, (|x - x_0| < \delta \wedge |f(x) - f(x_0)| \geq \epsilon).$$

Section 4.4

1. Valid.

2. Not valid.

3. Three are valid.

4. Three are valid.

5. In order to show that a conclusion is not valid, it suffices to show that the conclusion is false for at least one set of truth values of the six propositions shown below. In order to show that the conclusion is valid (always true), you must show that it is true regardless of the truth values of the six propositions. In other words, some of the statements could be true or false, depending on the values of p, q, r, s, t, and u, while others are always true, regardless of these values.

   • p: Charles takes the bus.
   • q: Charles misses the appointment.
   • r: The bus is late.

- s: Charles should go home.
- t: Charles feels downcast.
- u: Charles gets the job.

Using the above propositions, we can restate the given premises in symbolic form.

- P_1 : $p \Rightarrow (r \Rightarrow q)$, which is equivalent to $(\neg p) \vee (\neg r) \vee q$.
- P_2 : $(q \wedge t) \Rightarrow (\neg s)$, which is equivalent to $(\neg q) \vee (\neg t) \vee (\neg s)$.
- P_3 : $(\neg u) \Rightarrow (s \wedge t)$, which is equivalent to $u \vee (s \wedge t)$.

Overall, five of the statements are always true. Here is a solution to (a). The statement can be written as $p \Rightarrow (r \Rightarrow u)$, which is equivalent to $(\neg p) \vee (\neg r) \vee u$. We will show this statement is always true, so the conclusion is valid. We consider two cases, depending on the truth value of u. If $u = T$, we are done. If $u = F$, then by P_3, $s \wedge t = T$. In this case, by P_2, $q = F$. Then, by P_1, $(\neg p) \vee (\neg r)$ must be true (since q is assumed false). Hence, we have show that $(\neg p) \vee (\neg r) \vee u$ is true; therefore, $p \Rightarrow (r \Rightarrow u)$.

6. Write all propositions in symbolic form and use rules of logic.

Section 4.5

1. Use the definitions of odd and even to form expressions for n. Use a different variable in each case.

2. Start the proof by letting a be an arbitrary element of A. Then show $a \in C$.

3. Let $n = 2k$ be an even integer and let m be any integer.

4. Four of these are true.

5. Write $a = q_1 n + r$ for some q_1 and r.

6. Call each small square a cell and each set of 3×3 cells a block. In (a), there must be a 9 in the middle block of 9 cells. It can't be in the 4th row or the 6th column, so the cell in position $(5, 5)$, i.e., row 5, column 5, is the only one available for a 9 in the middle block.

8. Sets A and B are disjoint if and only if no elements of B are elements of A if and only if $A \setminus B = A$.

9. If x is odd, $x = 2n + 1$ for some integer n.

10. What must happen in order for the vowel "I" to be replaced?

11. For any x, $x - 3$ is odd if and only if $x - 3 = 2k + 1$ for some k.

12. Suppose by way of contradiction that $x + y = c/d$ is rational.

13. If $a < b$, then there is some positive number z such that $b = a + z$.

14. Prove the contrapositive. Use Problem 22 in Section 2.2.

15. Show that $r \Rightarrow (p \Rightarrow q)$ and $p \Rightarrow (r \rightarrow q)$ are logically equivalent.

16. This was proven in Chapter 3.

Sections 5.1 and 5.2

1. In (b) and (c), make sure to note the restricted set of x-values used in the graph.

2. (c) The entry in row i and column j is a 1 if and only if i is a divisor of j.

3. There are directed edges from Rock to Scissors, from Scissors to Paper, and from Paper to Rock.

4. There are three equivalence classes.

5. (e) This relation is reflexive, but the usual cousin relationship is not. Siblings are related to each other in κ, but they are not cousins.

6. Each equivalence contains all of the individuals with a particular blood type.

7. There are two equivalence classes.

8. There is some ambiguity as to the quadrant to which certain points belong.

9. Consider the number 1.

10. Two of these are partitions of S.

11. Consider $n \bmod 2$ and $n \bmod 3$.

12. There are three equivalence classes.

13. $(1, 1), (1, 3), (3, 1), (3, 3), (2, 2), (4, 4), (4, 5), (5, 4), (5, 5)$.

14. The number 2 must be in every pair.

16. R has three of these properties.

17. The relation α is not an equivalence relation and it is not a partial ordering.

18. It is neither antisymmetric nor symmetric.

19. Study the definitions!

21. Transitivity fails.

22. (a) The in-degree of any vertex is the sum of the entries in the column associated with that vertex.

23. Since $b \in [a] = \{x \in A : aRx\}$, then aRb. Why does this mean bRa? Show that $[a] \subseteq [b]$ and $[b] \subseteq [a]$.

24. Each equivalence class contains the set of words that can all be converted to each other by a sequence of valid moves.

25. Each directed edge of the digraph contributes one to the sum of the in-degrees and one to the sum of the out-degrees.

26. Yes. Consider a relation containing the single element (a, a). In general, the "=" relation is both symmetric and antisymmetric.

27. (a) $(2, 3) \sim (4, 6)$, so $((2,3), (4,6))$ is one element in $\sim$.   (c) $(6, 8)$. (f) $(0, 0)$ plays a role.

28. Lines in $[m]$ are parallel to each other.

Section 5.3

1. (b) The domain is $\{1, 2, 3\}$ and the range is $\{1, 2\}$.

2. Two of these are bijections.

3. $f_3^{-1} = \{(1, 3), (2, 2), (3, 1)\}$.

4. $f_2 \circ f_3 = \{(1, 2), (2, 1), (3, 1)\}$.

5. Find the domain of each.

6. Rationalize the denominator of g.

7. Different rules can define equal functions.

8. Two of them are injective. (d) The domain is $(-1, 0) \cup (0, \infty)$ and the range is all real numbers except for an interval of length one.

9. (a) Range(f) must have fewer than 3 elements.

13. (a) True. (d) False; let $A = [3]$ and $B = [2]$, with $f(x) = 1$ for all x.

14. One of them represents the fuel cost of driving x miles.

15. (a) $\{0, 1, 2, \dots, 9\}$; (b) $\mathbb{N}$; (c) $\{0, 1, 2, \dots, 9\}$; (d) $\{1\}$.

16. For (b) and (c), solve $f(ax + b) = af(x) + b$ for a and b.

17. Start by assuming $a \in \mathrm{Range}(R^{-1})$. This means $(a, b) \in R$ for some ....

18. $(a, b) \in R \iff (b, a) \in R^{-1}$.

19. To show f is one-to-one, start by assuming $f(a) = f(b)$ for some nonnegative real numbers a and b.

20. The domain of h^{-1} is $(-\infty, 0) \cup [1, \infty)$.

22. One answer is "yes" and one answer is "no".

23. (c) Translate and compress the tangent function. It is also possible to use a combination of logarithm functions.

24. Define $\phi : \mathbb{N} \to \mathbb{Z}$ to be $n/2$ if n is even and to be a different expression if n is odd.

25. Since $f : A \to B$ is a surjective function, each element of B is the image of at least one element from A.

26. Suppose that $(g \circ f)(x) = (g \circ f)(y)$ for some elements $x, y \in A$, and prove $x = y$.

27. Find a counterexample.

28. Yes. To prove f is onto, let $b \in B$ and find some $a \in A$ such that $f(a) = b$.

29. Consider the table following Example 20.

30. Bijective functions have inverses and the composition of bijections is also a bijection.

31. Place the circle on a horizontal line with the missing point directly above the point of tangency and find a correspondence.

32. Four of them hold.

Section 5.4

1. Three are arithmetic and two are geometric.

3. For $N = 75$, the first two terms are 1.

4. $a_n = 5n - 2$.

5. $b_4 = -2$. In the given example, the terms of (a_n) can be viewed as the coefficients of a degree 4 polynomial.

6. In (b), if d is the common difference between terms, recall that $a_k = a_1 + d(k - 1)$.

7. (a) $(17, 52, 26, 13, 40, 20, 10, 5, 16, 8, 4, 2, 1, 4, 2)$.

8. There is a recursive formula and a closed formula.

9. Use that $a_n = a + d(n - 1)$ to derive the result in (b). For (c), pair the terms up.

11. For (c), simplify $S_n - rS_n$.

12. Three consecutive terms of a geometric sequence with common ratio r will take the form cr^{n-1}, cr^n, and cr^{n+1} for some number c.

13. $f_1 = f_2 = 1$, $f_3 = 2$, and $f_{10} = 55$.

14. For the arithmetic sums, use that the sum of the first m terms equals $\frac{m}{2}(a_1 + a_m)$, with $m = (a_m - a_1)/d + 1$. (You should explain these facts.)

15. Associate a sequence of 1's and 2's with each viable way of climbing the stairs, where the sum of the terms in each sequence must be 15. We must count the number of possible sequences. Break the sequences into two types—those that end in a 1 and those that end in a 2—and consider the number of ways the beginning part of the sequence could be formed to make the sum of the terms 15. The answer is a Fibonacci number.

16. Break the sequences into two types—those that end in a 0 and those that end in a 1. The answer is a Fibonacci number.

17. Create 0-1 sequences of length 26 where the ith digit is a 1 if and only if the ith letter of the alphabet is included. Now count the number of such sequences without consecutive 1's. This is the same type of question posed in Problem 16.

18. Look for a pattern for small values of n.

Section 6.2

1. Show that $P(1)$ is true and then suppose there is some $k \geq 1$ such that $P(n)$ is true for $1 \leq n \leq k$ but fails when $n = k + 1$.

2. At some point you should reach the conclusion that $(k + 1)^3 + 2(k + 1) = k^3 + 3k^2 + 5k + 3$ is not divisible by 3.

4. $P(n)$ is the statement to prove: $F_{n+1} = F_0 \cdot F_1 \cdot F_2 \cdots F_n + 2$.

Section 6.3

2. At some point you should use the fact that $f_{2m+2} = f_{2m+1} + f_{2m}$.

3. At some point you should use the fact that $2^{k+1} + 2^{k+1} = 2^{k+2}$.

4. $P(k + 1)$ is the statement

$$1^3 + 2^3 + \cdots + k^3 + (k + 1)^3 = \frac{(k + 1)^2 (k + 2)^2}{4}.$$

5. If a convex polygon P with $n + 1$ vertices has consecutive vertices A, B, and C, then the polygon resulting from replacing sides AB and BC with side AC will be a convex polygon with n sides.

6. Depending upon how you do the problem, you may need to prove several base cases here in order for the inequality to be true for all k after using the inductive hypothesis.

7. This is a likely line of a good proof: $2^{k+1} = 2 \cdot 2^k \geq 2(k^3)$.

8. $(k + 1)! = (k + 1)(k!)$.

9. $3^{k+1} = 3 \cdot 3^k$.

10. A key step is that $(1 + x)(1 + x)^k > (1 + x)(1 + kx)$.

11. The quotient is $1 + x + x^2 + \cdots + x^{n-1}$.

12. Use the proof of Problem 11.

13. Choose one specific element x in a set A with $n + 1$ elements and use the inductive hypothesis on the set $A \setminus \{x\}$.

14. $P(n) : \overline{A_1 \cup A_2 \cup \cdots \cup A_n} = \overline{A_1} \cap \overline{A_2} \cap \cdots \cap \overline{A_n}$.

15. $P(n) : (A_1 A_2 \cdots A_n)^T = A_n^T \cdots A_2^T A_1^T$.

16. For (a), suppose for some fixed $n \geq 1$ that $n(n^2 + 5) = 6q$ for some integer q.

17. $f_{2n+2} = f_{2n+1} + f_{2n} = 2f_{2n} + f_{2n-1}$.

18. The given argument that $P(k + 1)$ follows from $P(k)$ does not work for all values of k.

19. See Problem 16.

20. $P(n + 1) : \prod_{k=2}^{n+1}(1 - 1/k^2) = \frac{n+2}{2n+2}$.

21. To move $n + 1$ disks to the second post, first move the top n discs to the third post. Then move the biggest disk to the ....

22. First show that $f(n) = 3f(n - 1) + 2$ for $n \geq 2$, with $f(1) = 2$.

23. The proof is very close to the one used in the traditional Tower of Hanoi problem.

24. You will need to use that $f_{n+1} = f_n + f_{n-1}$.

25. Try the problem for very small values of n and extend your reasoning.

26. Use the identities $\cos(A + B) = \cos(A)\cos(B) - \sin(A)\sin(B)$ and $\sin(A + B) = \sin(A)\cos(B) + \sin(B)\cos(A)$.

27. Use induction on n. By the inductive hypothesis, you can replace most of what is under the square root with something simpler.

28. Use the fact that $\cos(2\alpha) = 2\cos^2(\alpha) - 1$.

29. Use induction on m.

31. For the given number of coins, put a third of them on each side of the scale for the first weighing. Use Example 72.

32. The number $(k^2 + k)!$ can be written as the product of $k^2 + k$ integers in such a way that the first k of them are greater than or equal to 1, the next $k^2 - k$ factors are each greater than k, and the final k factors are each greater than k^2.

Section 6.4

1. In each instance, there is a smallest integer n for which ....

3. If you can form postage of n cents using the given stamps, then it is possible to form $n + 4$ stamps.

4. If $(n + 1)^2/b^2 = 2$, argue that $n + 1$ is even, and thus 2 can be written as the quotient of two squares with a smaller numerator than before. For the inductive step, assume there is some integer k such that $\frac{n^2}{b^2} \neq 2$ for any integers b and n, $1 \leq n \leq k$.

5. Use the recursive definition and the inductive hypothesis three times.

6. $a^{n+1} - b^{n+1} = a(a^n - b^n) + ab^n + b(a^n - b^n) - ba^n$.

8. Use induction on n and the fact that any polygon with at least 4 sides has a diagonal that lies completely within the polygon.

7. Induct on j, the number of digits in s_1. Use a computer to check that the statement holds for $1 \leq j \leq 3$.

9. To find S_{k+1}, note that each term (sum of squares) present in the calculation of S_k will be a term in S_{k+1}. In these terms, $k + 1$ will not be an element of the subset. In order for $k + 1$ to be in a subset, k cannot be used, since adjacent elements are not permitted. Thus, each term of the sum used to calculate S_{k-1} can be multiplied by $(k + 1)^2$, which is the effect of including $k + 1$ as a factor in the product of elements of each subset. The other new term will be $(k + 1)^2$, obtained when $\{k + 1\}$ is the only subset chosen. Add these together. To obtain a (very nice!) closed expression, look for a pattern.

Section 7.1

1. Only 127 and 1,987 are prime.

2. For (c), argue that one of $2^n - 1$ and $2^n + 1$ must be divisible by 3.

3. Use the Prime Factorization Theorem.

4. Denote the smallest number as n.

5. Each of the factors 3, 6, 12, 15, 21, 24, and 30 is divisible by 3 but not by 9.

6. (a) Prove the contrapositive.

   (b) If n is not a power of 2, then $n = pk$ for some integers p and k, with p odd.

7. Eventually, consecutive primes differ by at least 2.

8. Consider the possible remainders when n is divided by 3.

9. For each divisor k of n, necessarily n/k will be a divisor of n.

10. The proofs are similar to the one given in Example 77.

11. $n^4 + 4$ is the product of two quadratics.

12. First note that $p^2 - 1 = (p+1)(p-1)$. Argue that one of these numbers is divisible by 3.

13. Use Problem 19.

14. A given locker will be addressed one time for each of its positive divisors.

15. Of the n factors in $n!$, $\lfloor n/p \rfloor$ of them are divisible by p.

16. If there are only k positive primes of the form $4n + 3$, consider the number $N = a + p_1 p_2 \cdots p_k$ for appropriate values of a.

17. Show that the binomial coefficient $\binom{p}{k}$ is divisible by p for all k, $1 < k < p$.

18. The answer to (a), part (iii), is 8/9.

Sections 7.2 and 7.3

3. Use the fact that $\gcd(5{,}005, 33) = 11$.

4. Use the fact that $\gcd(112, 356) = 4$.

5. In (a), one solution is $x = -1, y = 2$; in all other solutions, x will increase (decrease) by a multiple of 10 while y will decrease (increase) by the same multiple of 17. For example, $(9, -15)$ is another solution. In (b), it helps to divide all numbers by the gcd first. (See Example 81.)

6. Use the fact that $\gcd(m, n) = \gcd(m - n, n)$.

7. $\gcd(x, y) = \gcd(x, x - y) = \gcd(x, 4)$. Since all divisors of x must be odd (x is odd) and 1 is the only odd divisor of 4, we see that $\gcd(x, y) = 1$.

8. One of the two converses is true.

9. Use the definition of gcd. Corollary 7.9(2) may be helpful. Remember that you must prove two implications for a biconditional statement.

10. First find the smallest positive integer that is a multiple of each of the four numbers.

11. You will need to use the formal definition of the gcd of two numbers.

12. In order, the points covered are $8, 16, 6, 14, 4, 12, \ldots$.

13. Experiment. Work backwards from the conclusion of the Euclidean Algorithm.

14. Use each definition and show that the requirements of the other definition are satisfied. One direction is very easy.

15. Let $d = \gcd(a, b)$ and prove that $\mathrm{lcm}(a, b) = ab/d$.

16. First prove that $F_{n+1} = F_0 \cdot F_1 \cdot F_2 \cdots F_n + 2$ (Problem 4 in Section 6.2) and use the fact that any common divisor of F_n and F_m must be odd.

17. Relatively prime numbers do not share any prime factors.

Section 7.4

1. Look for patterns.

2. $37^{2,018} \equiv 3^{2,018} \pmod{17}$.

4. Show that the last two digits of 37^{20} are 01.

6. For (a), write $n = 6k + t, 0 \le t \le 5$, and do all calculations modulo 6.

7. Write $n = 3q + r$, and show that $4^n + 15n - 1 \equiv 4^r + 6r - 1 \pmod 9$ and evaluate it for all pertinent values of r.

8. Show that $3^{16} \equiv 1 \pmod{64}$ and write $n = 8q + r$. Check all cases.

10. Find $x^2 \pmod 8$ for all values of x from 0 to 7.

11. The generalization is that an integer n will be divisible by 2^k if and only if the number $N = (a_k \cdots a_3 a_2 a_1)$ formed by the last k digits of $n \ldots$.

12. Given $n + 1$ natural numbers, reduce each one modulo n.

13. The 6-digit number $n = (abcabc)$ can be written as $1{,}000 \cdot (abc) + (abc)$.

14. $(ab) + c/2$ can be expressed as $10a + b + c/2$.

15. One possibility is to extend the result of Problem 14. The result is similar when c is even, but an additional modification is needed for the case when c is odd.

16. Note that $12^{2n+1} = 12 \cdot 144^n$ and $144 \equiv 11$.

17. Create a one-to-one correspondence between the integers $0, 1, 2, 3, 4, 5, 6$ and the given seven colors. The first prisoner to guess plays a very important role. There is a modulo 7 strategy that can be successful with 100% certainty.

18. Write $N = (a_{n-1} a_{n-2} \ldots a_2 a_1 a_0)$, where $a_{n-1} \le a_{n-2} \le \cdots \le a_2 \le a_1 < a_0$ are the (decimal) digits. Then

$$9N = (10 - 1)(10^{n-1} a_{n-1} + 10^{n-2} a_{n-2} + \cdots + 10^2 a_2 + 10 a_1 + a_0).$$

Distribute the product.

19. Let $x, 12x$, and $720x$ denote the angle, in degrees, that the hour hand, minute hand, and second hand have moved at the moment the angle between any two of them is 120 degrees. Assume for now that the hands are in the clockwise order hour, minute, second. Form three equations (one of them is $12x - 1x = 120 + 360m$) with four unknowns and try to solve it.

15. Within the long division algorithm used to calculate a/n, the Division Algorithm is repeatedly employed. Each time it is used, another digit in the quotient of a/n is obtained. Once the algorithm has moved past the decimal point with regard to the number that is "brought down" for the next use of the Division Theorem, that number will be a 0. From that point on, the value of the next digit of the quotient will therefore completely depend on the remainder r found in the Division Theorem, $0 \le r \le n - 1$. Once a remainder repeats, the digits in the quotient will repeat as well, with a period equal in length to the number of distinct remainders found between the two repeating remainders. Hence, the period has length at most n.

20. $1{,}234{,}567{,}894 \equiv 4 \pmod 9$. However, the possible values of the cube of an integer, modulo 9, are 1 and 8 only, and one cannot obtain a multiple of 9 by adding three numbers from among $\{1, 8\}$. So, no x, y, and z exist.

21. By the divisibility rules, a must be divisible by 9. Therefore b is divisible by n, in which case c is also divisible by 9. Moreover, since $a \le 9(1{,}972) = 17{,}748$, a 5-digit number, $b \le 9(5) = 45$. Therefore, $b \in \{9, 18, 27, 36, 45\}$, so $c = 9$.

Section 7.5

1. 7.

2. 2.

3. 8.

4. 6.

5. 4.

9. If digits a_i and a_{i+1} are transposed, then the terms $(10 - i + 1)a_i + (10 - j + 1)a_j$ will be replaced by the terms $(10 - i + 1)a_j + (10 - j + 1)a_i$.

10. If a single digit is changed from k to j, its contribution to the sum will change by either $|k - j|$ or $3|k - j|$.

Section 7.6

1. $403 + 80 + 16 + 3 = 502$.

2. Use the Euclidean Algorithm.

4. Prove that $a^3 + b^3 + c^3$ is divisible by both 2 and 3 by using casework on the parities of a, b, and c. It may help to expand $(a + b + c)^3$ first.

5. Substitute $n = 11q + r$ and determine which remainders can yield a multiple of 11 in this expression. Of these, determine if a multiple of 121 can be obtained.

6. Multiply out the product of the two sums and then factor in a different way by completing the squares.

7. There are several questions you could ask yourself that could lead to an answer. One is, "Is the number divisible by 3?" Another is, "What is the remainder when you divide the number by 4?"

8. What are the possible values of $x^2 \bmod 3$ for integers x?

9. Find a simple expression for x as a function of n for which $nx + 1$ factors.

10. Prove the contrapositive. Note that if $m = rs$, then $2^m = (2^r)^s$.

11. Write $M = 1 + 1/2 + 1/3 + \cdots + 1/n$ and multiply both sides by $2^{\lfloor \log_2 n \rfloor}$. Show this is a contradiction if M is an integer.

12. Determine the possible values of $x^2 \bmod 3$ and $x^2 \bmod 7$ for integers x.

14. Use the recursive relation repeatedly to write f_{5n} as the sum of lower Fibonacci numbers, including $f_{5(n-1)}$.

15. Consider the remainders that can occur when dividing by n at each step of the long division process.

18. Consider the n values of $(a_1 + a_2 + \cdots + a_i) \bmod n$ for $1 \le i \le n$.

Section 8.2

1. Use Proposition 3.1.

2. Use the Multiplication Rule.

4. Use a complement argument for (b) and (c).

5. For (c), a nonempty subset of A gets mapped to one element of B, and its complement gets mapped to the other.

6. Order matters here.

7. The Multiplication Rule will be used for most of these. Use a complement argument for (d).

8. For any positive integer k, the number of integers in $[1,000]$ divisible by k is $\lfloor 1,000/k \rfloor$.

9. The integers in question will be the complement of the set of multiples of 3 in $[3^{100}]$.

10. Find the number of integers in the set that are divisible by neither 2 nor 5.

11. Let M_3 be the multiples of 3 in A and let M_5 be the multiples of 5 in A. Find $|A - (M_3 \cup M_5)|$.

12. There is a unique winning row associated with each outer square/cube of the configuration that has $n + 2$ squares/cubes per side.

13. In (d), it is helpful to use the Multiplication Rule to count from the right-hand side.

14. In (d), each divisor must be of the form $p^i q^j$.

15. In (a), form the lines without Peter, and then place him in front of Noga in each one. For the other parts, first count the number of ways Noga and Peter can be placed and then multiply by 10!.

16. In this case, a palindrome is determined by the first four characters.

17. Consider two cases, those with a 237 subsequence and those without.

18. Count the number of 8-digit integers and compare to the number of them that do not contain a "1".

19. Explain the connection to the subsets of $[9]$.

20. (a) Explain why it is sufficient to count the number of ways of placing numbers in the first $m - 1$ rows and $n - 1$ columns.

Section 8.3

1. Any two points uniquely determine a chord.

2. 2,520.

3. (c) $C(59,5) + C(59,4) \cdot 41 + C(59,3) \cdot C(41,2)$.

4. There are $C(22,4)$ ways to choose the easy ones. Proceed similarly.

5. First write in terms of factorials.

6. The greatest number of planes that can be obtained will occur when no plane contains more than three of the given points.

7. See Example 102.

9. Try with very small n and look for a pattern.

10. First choose the groups and then order their meeting times.

11. The points are distinct, so this is a permutation problem.

12. The greatest number of lines will occur when no line contains three or more of the points. Experiment to find the different numbers of lines that are possible.

13. The teams are unlabeled.

14. One of the answers is $\dfrac{15!}{(5!)^3}$.

15. Consider rotations around the table and also reflections (reversals of the orderings).

17. The answer to (a) is more than 34 billion. In (b)(ii), there are five scenarios for each operator of a given city.

18. There are $\binom{100}{3}$ ways to form triples (a, b, c) with distinct elements from X such that $a < b < c$.

19. If k is the largest element in a subset B of A, then the other elements of B must be chosen from the set $\{1, 2, 3, \ldots, k-1\}$.

20. See Example 105.

21. Place the white balls in a row and determine the positions where black balls can be placed between them.

22. In (a), each valid walk consists of eight steps East and four steps North, in some order.

23. In (a) and (b) you are choosing distinct digits from the set [9]. In (c), zero can also be used.

24. There is a one-to-one correspondence between choosing six integers from 1 to 40, no two of them adjacent, and the number of 0-1 sequences of length 40 with exactly six 1's, none of them adjacent.

25. The solution is similar to Problem 24, with a different number of open spaces.

26. This is similar to Problem 24(a).

28. See Example 105.

29. First consider the number of 5-digit sequences with three even digits. Now place an odd digit in front of each one.

30. When two intersecting chords meet in the interior of the circle, the two chords have four distinct endpoints.

Section 8.4

2. Use the Binomial Theorem, with $a = 5x$ and $b = y$.

3. Use the Binomial Theorem, with $a = 2x$ and $b = -3y$.

5. See Example 103.

6. Use that $m! = m(m-1)!$.

7. Let $b = 1$ in the Binomial Theorem.

8. Use that $(a+b)^{n+1} = (a+b)^n (a+b)$ and apply the inductive hypothesis.

9. The kth term has coefficient $a_k = C(100, k) \cdot 2^k 3^{100-k}$.

10. Make an appropriate substitution in the Binomial Theorem.

11. Separate the terms in the binomial expansion of $(1-1)^n$ according to the sign of the term of each coefficient.

12. Let $a_k = \binom{2n}{k}$, and determine when $a_{k+1}/a_k \geq 1$.

Section 8.5

1. (a) 70; (b) 525; (c) 945.

2. There are three cases: one of them is when each die roll is the same.

3. (a) 12; (b) 360.

4. There are two routes associated with each set of chords that are used. Thus there are $2 \cdot 2^{10}$ routes.

5. The answers for even and odd n are different.

6. (a) Each relation is a subset of $A \times B$.

   (b) First count the number of ways to form the image of a specific element from A.

   (c) First determine the number of ways to form the domain.

   (d) First determine the number of one-to-one functions from A to B if the range has k elements.

   (e) If the domain has k elements, there will be $P(t, k)$ bijections.

   (f) There are t choices for the element from B, but that is not the answer to the given question.

7. The first book can be placed in 5 ways. Then what? Alternatively, try with a much smaller number of books and shelves.

8. By symmetry, each digit will be used equally often in the different 4-digit numbers formed.

9. In order to maximize the number of regions obtained, each new line should cross each previous line (which will happen if the line is not parallel to any of them). This leads to a recurrence relation that has a quadratic solution in n.

11. There will be a region for each possible configuration of the intersection of the n circles.

12. Consider the coins and sticks argument in Example 105.

13. 126 and 504.

Section 9.1

1. In each case, two vertices have degree 2 and the others have degree 1.

2. Two.

3. If the edge incident to the vertex of degree 1 is removed, the resulting graph will be disconnected.

4. $n = 2$.

5. All vertices of graphs in the families K_n and C_n have degrees $n - 1$ and 2, respectively.

6. (a) There are $n - 1$ edges coming from each of n vertices.

   (b) Each edge corresponds to the selection of two vertices.

   (c) Add the numbers in the upper triangular part of the matrix, so as to not double-count.

   (d) Each person can be represented by a vertex, with an edge placed between two persons who shake hands.

7. $\overline{G}$ has the same vertex set as G, but nonedges become edges and vice versa.

8. Lots of 1's!

9. (a) 4; (b) 6.

10. The first row is 01001.

11. Exactly m of the vertices have degree n and n of the vertices have degree m.

12. $\overline{K}_{3,3}$ has two components, each with three edges.

13. Each vertex is adjacent to each of the other $n-1$ vertices.

14. Order and label the vertices of C_4 and label the vertices of $K_{2,2}$ as $a_1, a_2, b_1,$ and b_2. Create a bijection between the two sets and show that incidence is preserved.

16. All vertices of cycles have degree 2.

17. Count the number of labeled subgraphs with i edges, $0 \le i \le 3$.

18. Start with K_5.

19. Note that $\overline{C}_5$ can be be viewed as a cycle.

20. This happens when $n(n-1)/2 = 2n$.

22. $C(n, 2)$ edges can be placed on a set of n vertices.

23. Use the result of Problem 22 and express your answer as a sum.

24. Start the proof by letting ϕ be an isomorphism from $V(G_1)$ to $V(G_2)$. Then, $\{a, b\}$ is an edge of G_1 if and only if $\{\phi(a), \phi(b)\}$ is an edge of G_2.

25. Note that $C(10, 2) = 45$.

Section 9.2

2. List all Hamilton cycles starting at A in the weighted graph in Figure 9.23. There are three different ones, as you don't need to consider the cycle formed by reversing the order of vertices in a previous cycle. Which of the three is a minimum Hamilton cycle?

3. All Euler circuits have the same weight.

4. Each vertex of K_n has degree $n-1$.

7. Find the degree of each vertex.

10. By way of contradiction, suppose every degree is greater than 1. Show there is a path for which the last vertex has degree 1.

11. Use the theorems.

12. Start with a clique and add vertices and edges.

13. Show that $m = \binom{n-1}{2} + 1$ and describe a graph with n vertices and m edges that has no Hamilton cycle.

14. By way of contradiction, if G is disconnected, then it has at least two components; say these components have n_1 and n_2 vertices, where $n_1 + n_2 \le n$. Any vertex v_1 in the first component has degree at most....

Section 9.3

1. Three of them have chromatic number 3, and three of them have chromatic number 2.

2. K_8 does not meet the hypotheses of the Four Color Theorem.

3. The chromatic number will be at most ... (complete the sentence).

4. Replace each letter with a vertex and adjoin vertices of neighboring regions.

5. Compare with the chromatic number of C_n.

6. Find a graph with chromatic number 3 that does not have any 3-cliques.

8. Create a conflict graph with authors as vertices. Put in appropriate edges and properly color the vertices with as few colors as possible.

9. Create a conflict graph with species as vertices.

10. Create a conflict graph with radio stations as vertices.

11. $k = 3$ suffices.

12. See the solution to Problem 14.

13. One requirement is that vertices $v_{i,j}$ and $v_{i,k}$ must be colored differently for $1 \leq i \leq 9$.

14. Create a clique on $n - 1$ vertices numbered from 1 to n. Add vertices $v_1, v_2, \ldots, v_{n-1}$ and place edges appropriately.

    Start by creating a clique of size $n - 1$, with vertices numbered from 1 to n, and adding vertices $v_1, v_2, \ldots, v_{n-1}$, where vertex v_i is adjacent to all vertices in the clique except vertex i.

Section 9.4

1. A tree is connected and has no cycles.

2. (b) 16.

4. (c) A cut edge cannot be part of a cycle. (d) The extreme graphs are paths and stars.

5. The number of edges in a tree is one less than the number of vertices.

6. (a) 6; (b) $C(n - 1, 2)$.

7. $n - k$.

8. Each graph should be connected, without cycles.

9. Experiment by drawing trees for small values of n.

10. From the vertex of largest degree, one can follow a path through each of its children to eventually reach a leaf.

11. If there is only one vertex of degree 1 and each of the other $n - 1$ vertices has degree at least 2, then the sum of the degrees is at least ....

12. Use combinatorics with multiple cases or simply sketch all of them. Two edges must be removed to create a spanning tree.

13. Three of the properties are met.

14. (b) No. Find a counterexample.

18. One possibility is to guess the "half-way" point each time, as closely as is possible.

19. Count the number of vertices at each level, with one vertex (the root) at the top level.

20. All vertices except the leaves have k children.

21. Use Problems 19 and 20.

22. The graph formed by deleting a vertex v and all edges incident with v can have up to $p = \deg(v)$ components; the inductive hypothesis can be applied to each.

23. Show that since all d_i are positive, if their sum is $2(n - 1)$, at least two of them must equal 1. Then use induction, removing one of those numbers from the sequence and adjusting another number as well. There are some subtleties involved.

24. Use the Multinomial Theorem for the last step.

25. Use Theorem 11.7. For (b), view K_n as a forest on n trees, where each tree is a single vertex. There are $n(n-1)$ choices for the first directed edge, and after it is placed, the resulting graph is a forest with $n-1$ trees, each with a root. The next directed edge can start at any of the n vertices and must terminate at the root of any one of the $n-1$ trees other than the root of the tree containing the vertex just chosen. The resulting graph is a forest with $n-2$ trees, each with a root. Continue this process, and explain what happens in the kth step.

26. Set each a_i equal to 1 in the Multinomial Theorem (Theorem 11.7).

Section 10.1

1. Experiment, and keep track of the number of right-side up glasses after each move. What do you notice?

2. $a \bmod n + b \bmod n = (a+b) \bmod n$.

3. Find the product of the sixteen entries.

4. Find the difference between the number of red squares and black squares covered at any point in time during the placing of the tiles.

5. Label the squares with numbers in such a way that the sum of the numbers covered by a tile will be constant, regardless of its placement.

6. Consider what happens to the number of components each time an edge is added during the building of the graph.

7. Use inversions in each row and compare the parities.

Section 10.2

1. Let $f(n)$ denote the difference between the greatest and fewest number of candies after round n. Show that $f(n)$ will eventually decrease to 0.

2. Randomly divide the members into two houses and let n denote the number of members who have at least two enemies in their house. Consider how to decrease n.

3. Label the red points from R_1 to R_n and the blue points from B_1 to B_n, and join R_i to B_i for $1 \le i \le n$. If the number of intersections is zero, we are done. Otherwise, suppose segment $R_i B_i$ intersects $R_j B_j$ for some i and j, and make an appropriate change.

4. Randomly drop a perpendicular from each point to a different line. If two perpendicular segments cross, swap the line to which the perpendicular is dropped for each of the two points.

5. Show that the sum $ab + bc + cd$ decreases with each move, and try to construct a monovariant based on this observation.

6. If G has nonadjacent vertices x and y with $\deg(x) > \deg(y)$ and if y has a neighbor z that is not a neighbor of x, then swap two appropriate edges.

7. Notice that sometimes, when you take an arc of knights and reflect their positions along that arc, you decrease the number of pairs of enemies which are seated next to one another.

8. See Problem 7.

9. For (a), divide the vertices $v_1, v_2, \ldots, v_n$ into two groups, A and B, and let d_i denote the number of neighbors that v_i has in the group for which it doesn't belong. Choose an arbitrary vertex v_i. If v_i has more neighbors in its own group, do something to increase d_i.

10. Show that if v_1 is not adjacent to a vertex of degree d_i with $1 < i \le d_1 + 1$, then it is adjacent to a vertex v_j, with $j > d_1 + 1$, and that $d_i > d_j$. Then show that there exists a vertex t distinct from v_1, v_i, and v_j such that $\{t, v_i\}$ is an edge and $\{t, v_j\}$ is not an edge. Delete edges $\{v_1, v_j\}$ and $\{t, v_i\}$, and add edges $\{v_1, v_i\}$ and $\{t, v_j\}$.

Section 11.1

1. (a) $5(26)^4$ words begin with a normal vowel.

   (b) $5(25)(24)(23)(22)$ words begin with a normal vowel and have distinct letters.

   (d) There are two cases to consider, depending on whether the first letter is a consonant.

2. 1,334,961.

3. Think derangements.

4. For $1 \le i \le 6$, let A_i denote the set of all rolls in which i does not appear on any of the eight dice.

5. For $A \subseteq [3]$, define S_A to be the number of functions from $[5]$ to $[3] \setminus A$.

6. Let A_p denote the multiples of p in $[1{,}000]$, for $p = 2, 3, 5, 7$, and calculate $|A_2 \cup A_3 \cup A_5 \cup A_7|$.

7. Let C, D, H, and S represent the sets of hands that contain a void in clubs, diamonds, hearts, or spades, respectively.

8. See Problem 5 and sum over the possibilities.

Section 11.2

2. It is impossible for every team to have a win if some team is undefeated.

3. The students are the pigeons and the letter grades are the pigeonholes.

4. Determine the possible numbers of hands that can be shaken by each person.

5. At least two of the seven integers must be congruent modulo 6.

6. Calculate the average of the number of integers having each last digit.

7. If one person knows everyone, then it is not possible for someone to know no one.

8. Divide the square into four congruent squares (pigeonholes).

9. The solution is similar to the one in Problem 8.

10. For there to be a different number of red cells in each row, there must be at least $0 + 1 + 2 + \cdots + (n - 1)$ red cells.

11. Choose one point, x_0, as a common vertex of ten triangles with nonoverlapping interiors.

12. Define s_k to be the kth partial sum modulo n; i.e., $s_k = x_1 + x_2 + \cdots + x_k \pmod{n}$.

13. Pick a point O and consider the points on the circle centered at O with radius one.

14. At least one row must contain at least three rooks. Argue that one of the remaining rows contains at least two rooks and that yet another must contain at least one rook.

15. Argue that each element of A can be written in the form $m2^k$ for some odd m and some $k \ge 0$.

Section 11.3

1. (a) 495; (b) 15,840; (c) 840; (d) 720; (e) 90.

4. $\binom{14}{5,2,7} = 72{,}072.$

6. 816,480.

7. $\binom{14}{2,1,1,2,3,2,2,1} = \dfrac{14!}{2!1!1!2!3!2!2!1!}.$

11. This can be written as either a product of k-permutation numbers or as one multi-nomial coefficient.

12. $\binom{10}{2,3,5}.$

13. Make an appropriate substitution in the Multinomial Theorem.

14. Make an appropriate substitution in the Multinomial Theorem.

15. Do not forget to take into account the number of ways of assigning suits.

16. First determine the number of ways of distributing the 48 nonaces to four distinct players.

17. Show that for $0 \le i \le n - 1$, $\dfrac{((n-i)k)!}{((n-1-i)k)!\,k!}$ is divisible by $n - i$. Also, see Problem 3 in Section 7.6.

Section 11.4

1. Consider two squares appropriately placed inside another square.

2. In the combinatorial argument, use the formulation of the identities without fractions as given in Theorem 11.8 for parts (a) and (b), since those formulation are equivalent to the ones shown here.

4. Every subset has a complement. Also, consider the mean size of a subset of $[n]$.

6. Partition all 1-step and 2-step climbing sequences of n stairs according to k, the number of 2's that start the sequence.

7. Count the number of ways to form sequences on Monday and Tuesday that differ for the first time after a specific number of stairs.

8. How many $(k + 1)$-element subsets does $[n + 1]$ have?

9. Count the number of ways of forming sequences that are identical for a specified number of positions.

10. Count the number of labeled subgraphs of K_n with three edges by breaking into cases according to the number of vertices used.

11. The given expression counts the number of ways to partition the $[nk]$ into n labeled subsets, each of size k.

12. Consider n white balls and n black balls.

13. Count certain subsets of $[n]$.

14. See Problem 13.

Section 11.5

1. $S(4, 2) = 7$ and $S(6, 5) = 15$.

2. $p(5, 3) = 2$ and $p(9, 4) = 6$.

3. $p(7) = 15$.

6. Use Table 11.1 for this and subsequent questions.

11. This is the product of three binomial coefficients.

12. Consider the two cases where object n is in a box by itself and where it is not in a box by itself; count the number of distributions in each case.

13. The sums can be put in one-to-one correspondence with the 4-partitions of 12.

14. Use the fact that the number of integer solutions to $x_1 + x_2 + \cdots + x_k = n$ is either $\binom{n+k-1}{k-1}$ or $\binom{n-1}{k-1}$, depending on whether the variables are required to be nonnegative or positive, respectively.

15. It is a power of 2.

16. It is a binomial coefficient.

17. Notice the connection between equivalence relations and partitions. The answer is a Bell number.

18. This the number of integer solutions to $x_1 + x_2 + \cdots + x_8 = 20$, where (a) $x_i \geq 1$ and (b) $x_i \geq 0$.

21. In(a)(i), if the books were indistinguishable, there would be $C(25, 5)$ arrangements.

22. 1,080.

23. This is a multinomial question.

24. Since we are forming $n - 1$ nonempty subsets using n objects, one subset has two objects and each other subset has one object. There are $C(n, 2)$ ways to choose the two objects for the 2-element subset, so $S(n, n - 1) = \binom{n}{2}$.

25. In writing $n = a + b$, the smaller of a and b can take on all integral values from 1 to $n/2$, inclusive.

27. This generalizes Problem 18.

29. Use induction on n and the recursive formula for $p(n, k)$.

30. Count the number of functions from $[n]$ to $[k]$.

31. In (c), find a one-to one-correspondence between the partitions of $n - \binom{k}{2}$ into k parts and the partitions of n into k distinct parts. In (d), consider the two cases where 1 is and is not a term in the sum. In (e), use induction on n and the recursive formula.

32. Use Problem 31.

33. 76,396,035,914.

34. Use Problem 31.

36. In order to be a 20-digit number, the first digit cannot be a zero. By symmetry, this will happen in one-tenth of the arrangements.

Section 12.1

3. There are 8 pairs of twin primes less than 100.

5. Yes (Dirichlet's Prime Number Theorem) and no.

7. $10/(\ln 10) \approx 4.34$, and $\pi(10) = 4$.

9. For a majority of the suggested values of n, it turns out that $n! + 1$ and $n! - 1$ are prime. However, upon further examination, these expressions are decreasingly likely to be prime as n increases.

10. (a) $42 = 5 + 37$; (b) $66 = 5 + 61$; (c) $100 = 7 + 93$.

11. Find values of n that will make the sum divisible by 11.

12. Consider the two cases where a_0 is prime and when it is not.

13. Prove the contrapositive using a polynomial factorization formula.

Section 12.2

1. It is $1 \cdot 3^0 + 2 \cdot 3^1 + \cdots$.

2. $674 = 1 \cdot 5^4 + 0 \cdot 5^3 + 1 \cdot 5^2 + 4 \cdot 5 + 4 \cdot 1 \Rightarrow 674_{10} = 10144_5$.

4. 1,441.

5. 154.

8. Compute the Nim sum of 1011_2, 1111_2, and ....

9. Compute the Nim sum of 10010, 11111, 110110, and ....

10. As long as the Nim sum is not at zero, you can win.

11. $51 = 48 + 3 = 12(4 + 1/4)$.

13. There are three possible locations for each of the four weights when weighing a given load: in the first pan, the second pan, or neither pan.

14. Consider a 3-digit representation in base 4, where each color represents a digit 0, 1, 2, or 3.

Section 12.3

1. For (a), you should arrive at $x = 7 + 11t \equiv 3 \pmod 6$, in which case you will need to solve the Diophantine equation $6s - 11t = 4$.

3. See Example 155.

4. $\phi(20) = \phi(4) \cdot \phi(5)$. For part (c), if $n \geq 3$, n is divisible by an odd prime.

6. Find solutions when $20 \leq c \leq 23$ and prove it from there.

7. If $ax \equiv b \pmod n$, then substitute for a and b to show that $xa_1 - b_1$ is a multiple of n_1.

8. Turn the system of congruences into a Diophantine equation and use Theorem 7.12.

9. By definition, $p|(x^2 - 1)$. Factor $x^2 - 1$ and apply Lemma 7.11.

10. In calculating $\phi(p^m)$, note that the only integers in $[p^m]$ that are not relatively prime to p^m are the positive multiples of p.

10. If p is prime, the only integers in $[\phi(p^k)]$ that are not relatively prime with p are those that are not multiples of p. How many of those are there?

12. Show that every term other than a^p and b^p is a multiple of p.

13. Show that $x \equiv 0 \pmod b$ and $y \equiv 0 \pmod a$; let $x = bs$ and $y = at$ and see what solutions in s and t are possible.

14. Note that $ab - a - b = (a-1)(b-1) - 1$, and show that if $m \geq (a-1)(b-1)$, then m can be written in the desired form. To this end, show that $m, m - b, m - 2b, \ldots, m - (a-1)b$ are all distinct modulo a.

15. In (b), assume by way of contradiction that there are only finitely many primes of the form $4n + 1$. Consider the integer N formed by adding 1 to the product of these primes squared. Then use part (a) to obtain a contradiction.

Section 12.4

1. In (a), consider a congruence modulo 5. In (c), factor $x^4 - 2y^2 - 1$ and explain why it has no integer roots.

2. All of them have solutions with single-digit values.

3. Show that $y^2 - 2y$ must be a perfect square. When is this possible?

4. Factor $x^3 - y^3$ and choose values of x and y that cause the two factors to have a product of 91. Note that 91 does not have many factors.

6. If $n \equiv r \pmod 6$ for $1 \le r \le 5$, then show directly that either $2x + 1 \equiv 0 \pmod n$ or $3x + 1 \equiv 0 \pmod n$ has a solution. If $6|n$, write $n = (2^s)(3^t)(k)$ and use the Chinese Remainder Theorem to prove that the following system of congruences has a solution:

$$2x + 1 \equiv 0 \pmod{3^t k}, \quad 3x + 1 \equiv 0 \pmod{2^s}.$$

Section 12.5

1. (b) 1 and 4.

2. Compute b^2 mod 21 for $b = 1, \ldots, 10$ and notice how this provides the values of b^2 mod 21 for $b = 11, \ldots, 20$.

3. Write $m = a^2$ and $a = k + nt$.

4. As a newcomer to Rabin's cryptosystem, Bob selects small primes $p = 7$ and $q = 19$ and publishes $n = pq = 133$. It is understood that only 2-letter words (or sequences) will be encrypted, with each letter represented by a 2-digit number: 01 for A, 02 for B,..., 26 for Z. Notice that Bob's choice of n restricts the 2-letter words that can be encrypted with his modulus, from AA (0101) to AZ (0126).

   (a) AX has the form 0124.

   (b) Show that $b \equiv \pm 2 \pmod 7$ and $b \equiv \pm 4 \pmod{19}$ and check all cases.

   (c) In this case, M^2 is a multiple of the prime $p = 7$; hence M is a multiple of 7. (Why?) This reduces the number of systems of congruences to solve, and each system is of similar difficulty to those in part (b).

6. $247 = (19)(13)$.

7. $1{,}380 = (23)(15)$.

8. (a) p and q must be congruent to 3 modulo 4.

   (b) Find the square of 809 for the appropriate modulus.

   (c) Note that $1 = (20)71 + (-33)43$.

   (d) You should find that $b \equiv \pm 13 \pmod{43}$ and $b \equiv \pm 9 \pmod{71}$.

   (e) $p = \gcd(1{,}475, 3{,}953)$.

11. If $p = 4k + 1$ for some k, then $(-1)^2 \equiv (p-1)^2 \equiv 0 \pmod 4$.

Sections 13.1 and 13.2

1. Use Euler's Theorem on each component.

2. The graph has six vertices of degree 3, with seven finite faces and an infinite face.

5. The values of f and v are switched.

6. Use the same approach as the one in the proof of Corollary 13.2(2). Explain why $N \ge 4f$.

7. Use the corollaries to Euler's Theorem.

8. One way to find $P(G, x)$ is with the Graph Reduction Theorem.

9. Use the Graph Reduction Theorem. Even without finding $P(G, x)$, it is easy to see that the graph can be properly colored with two colors.

10. Both must be trees.

11. All vertices of a regular graph have the same degree.

12. Use Theorem 13.7.

13. Vertices in the same color class do not have edges between them.

14. All components will share the infinite face.

15. Use Theorem 13.7.

16. What would be the chromatic number for a graph with this polynomial?

17. Use induction on the number of vertices.

19. In polynomial theory, $x - a$ is a factor of a polynomial $f(x)$ if and only if $f(a) = 0$.

20. Use induction and the Graph Reduction Theorem.

21. Consider separately the cases where two and three colors are used.

22. Use induction.

23. Use induction on n and the fact that all trees have a vertex of degree 1.

24. Use Corollary 13.3 and induction.

Sections 13.3 and 13.4

1. A path is a tree, a cycle is almost a tree, and every labeled tree on n vertices corresponds to a unique labeled subgraph of K_n.

2. The MSTs have weight 26.

3. The MSTs have weight 59.

6. Assign new weights to the required edges.

8. Neither is optimal.

10. The Nearest Neighbor Algorithm starting at A has total weight 661, which is less than the total found with the Sorted Edge Algorithm, which has a total weight higher than 700.

11. Recall $(n - 1)! / 2$. In (b), the circuit starts with $A - B - C$.

14. Suppose there are two different spanning trees of minimum weight and consider an edge that is in one tree but not the other.

15. Weight the edges in such a way that one is forced to choose a very expensive edge in the last step of the Sorted Edge Algorithm.

16. The number of paths of length k, $k = 1, \ldots, n - 1$, can be calculated as a permutation.

17. Let the students and the problems be vertices of two partition classes of a bipartite graph, with an edge between a student and a problem if the student can solve that problem. Use Hall's Theorem.

List of Names

The first names of the following individuals, all of whom have made notable contributions to discrete mathematics, appear in examples and exercises throughout the book. It is virtually impossible to create an exhaustive list, and we make no claims that we have done so.

(1) Abraham Fraenkel (1891–1965), a German-born Israeli mathematician, known for his contributions to axiomatic set theory

(2) Ada Lovelace (1815–1852), an English mathematician and writer, believed to have written instructions for the first computer program

(3) Béla Bollobás (1943–), a Hungarian-born British mathematician and a researcher in combinatorics and graph theory who was heavily influenced by Paul Erdős

(4) Brahmagupta Bhillamalacarya (589–668), an Indian mathematician, credited with significant advances in arithmetic, algebra, number theory, and geometry

(5) Charles Dodgson (1832–1898), an English author and logician, better known by his pen name Lewis Carroll and for authoring *Alice's Adventures in Wonderland*

(6) Donald Knuth (1938–), an American computer scientist and mathematician, considered by some to be the "father of the analysis of algorithms"

(7) Endre Szemerédi (1940–), a Hungarian-American mathematician who has published in the fields of combinatorics and theoretical computer science

(8) Ernst Zermelo (1871–1953), a German logician known for axiomatizing set theory

(9) Fan Chung (1946–), a Chinese-born American mathematician who worked in various aspects of graph theory

(10) Felix Lazebnik (1953–), a USSR-born American mathematician who works in extremal graph theory and combinatorics using algebraic methods, the Ph.D. advisor of author Owen Byer, and the original source of portions of this book

(11) Frank Harary (1921–2005), an American mathematician, considered to be one of the founders of modern graph theory

(12) Frank Plumpton Ramsey (1903–1930), a British logician and economist, known for proving Ramsey's Theorem, an important result in extremal graph theory

(13) Gary Ebert (1947–), an American combinatorialist who specializes in finite geometries, a pioneer of inductive methods in discrete mathematics based on computer models, and the Ph.D. advisor of author Kenneth Wantz

(14) Georg Cantor (1845–1918), a German mathematician who founded set theory and the first to prove that the real numbers cannot be put in one-to-one correspondence with the natural numbers

(15) Godfrey Hardy (1877–1947), an English number theorist who authored *A Mathematician's Apology* and a mentor of Ramanujan

(16) Gottlob Frege (1848–1925), a German philosopher and mathematician and one of the founders of modern logic

(17) Harold N. (Thann) Ward (1936–), an American mathematician who specializes in algebraic coding theory and its connections to designs and geometries and the Ph.D. advisor of author Deirdre Longacher Smeltzer

(18) Herb Wilf (1931–2012), an American mathematician who specialized in combinatorics and graph theory and author of *generatingfunctionology* [21]

(19) Jacob Bernoulli (1654–1705), Swiss mathematician after whom Bernoulli numbers are named and author of the probability text *Ars Conjectandi*

(20) Joel Spencer (1946–), an American combinatorialist, known for his work on probabilistic methods in combinatorics

(21) Kurt Gödel (1906–1978), an Austrian-born American logician, best known for his two Incompleteness Theorems

(22) László Lovász (1948–), a Hungarian combinatorialist who wrote books on combinatorial optimization and matching theory

(23) Leopold Kronecker (1823–1891), a German mathematician who worked in number theory and algebra and was critical of Georg Cantor's work on set theory

(24) Leonardo Bonacci, also known as Leonardo of Pisa and Fibonacci (c. 1170–c. 1250), an Italian mathematician, considered to be the best mathematician of the Middle Ages, known for popularizing the sequence of Fibonacci numbers in his book *Liber Abaci*

(25) Noga Alon (1960–), an Israeli mathematician, known for his extensive contributions to combinatorics and theoretical computer science

(26) Paul Erdős (1913–1996), a Hungarian-born mathematician considered to be the "father of discrete mathematics", one of the most prolific mathematicians of the 20th century, and known for traveling the world and collaborating with more than 500 mathematicians on problems in various areas of mathematics

(27) Peter Cameron (1947–), an Australian mathematician who has written books on algebra, combinatorics, and coding theory

(28) Peter Winkler (1946–), an American mathematician who specializes in combinatorics and probability, known for his love of mathematical puzzles

(29) Ron Graham (1935–), an American mathematician who has worked in a variety of areas of discrete mathematics and a close friend of Paul Erdős

(30) Srinivasa Ramanujan (1887–1920), an Indian mathematician who was largely self-taught but made extraordinary contributions to various areas of mathematics, most notably number theory

(31) Susanna Epp (1943–), an American mathematician who is the author of the popular text *Discrete Mathematics with Applications*

(32) Terence Tao (1975–), an American mathematician and 2006 Fields Medal winner who co-discovered the Green-Tao Theorem, which states that there are arbitrarily long arithmetic progressions of prime numbers

Bibliography

[1] Martin Aigner and Günter M. Ziegler, *Proofs from The Book*, 4th ed., Springer-Verlag, Berlin, 2010. MR2569612

[2] George E. Andrews, *The theory of partitions*, reprint of the 1976 original, Cambridge Mathematical Library, Cambridge University Press, Cambridge, 1998. MR1634067

[3] George E. Andrews and Kimmo Eriksson, *Integer partitions*, Cambridge University Press, Cambridge, 2004. MR2122332

[4] Arthur T. Benjamin and Jennifer J. Quinn, *Proofs that really count*, The art of combinatorial proof, The Dolciani Mathematical Expositions, vol. 27, Mathematical Association of America, Washington, DC, 2003. MR1997773

[5] J. A. Bondy and U. S. R. Murty, *Graph theory*, Graduate Texts in Mathematics, vol. 244, Springer, New York, 2008. MR2368647

[6] David M. Burton, *The history of mathematics: An introduction*, Allyn and Bacon, Inc., Boston, MA, 1985, pages 463 and 515. MR890956

[7] D. Fomin and L. Kurlyandchik. Light at the end of the tunnel. *Quantum*, March/April 1994.

[8] Michael L. Fredman and Robert Endre Tarjan, *Fibonacci heaps and their uses in improved network optimization algorithms*, J. Assoc. Comput. Mach. **34** (1987), no. 3, 596–615, DOI 10.1145/28869.28874. MR904195

[9] Dmitri Fomin, Sergey Genkin, and Ilia Itenberg, *Mathematical circles (Russian experience)*, Mathematical World, vol. 7, American Mathematical Society, Providence, RI, 1996. Translated from the Russian and with a foreword by Mark Saul. MR1400887

[10] Andrew Granville and Greg Martin, *Prime number races*, Amer. Math. Monthly **113** (2006), no. 1, 1–33, DOI 10.2307/27641834. MR2202918

[11] P. J. Heawood. Map-colour theorem. *Quarterly Journal of Mathematics*, 1890.

[12] I. N. Herstein and I. Kaplansky, *Matters mathematical*, Chelsea Publishing Co., New York, 1978. MR0497399

[13] Agnes M. Herzberg and M. Ram Murty, *Sudoku squares and chromatic polynomials*, Notices Amer. Math. Soc. **54** (2007), no. 6, 708–717. MR2327972

[14] Y. Ionin and L. Kurlyandchik. Some things never change. *Quantum*, September/October 1993.

[15] Stasys Jukna, *Extremal combinatorics*, with applications in computer science, Texts in Theoretical Computer Science. An EATCS Series, Springer-Verlag, Berlin, 2001. MR1931142

[16] J. E. Littlewood, *Distribution des nombres premiers*, C. R. Acad. Sci. Paris **158** (1914), 1869–1872.

[17] Fred S. Roberts, *Applied combinatorics*, Prentice Hall, Inc., Englewood Cliffs, NJ, 1984. MR735619

[18] Edward Scheinerman, *Mathematics: A discrete introduction*, 3rd ed., Cengage Learning.

[19] Richard P. Stanley, *Enumerative combinatorics. Vol. 1*, with a foreword by Gian-Carlo Rota; corrected reprint of the 1986 original, Cambridge Studies in Advanced Mathematics, vol. 49, Cambridge University Press, Cambridge, 1997. MR1442260

[20] H. Weber, *Leopold Kronecker* (German), Math. Ann. **43** (1893), no. 1, 1–25, DOI 10.1007/BF01446613. MR1510799

[21] Herbert S. Wilf, *generatingfunctionology*, 3rd ed., A K Peters, Ltd., Wellesley, MA, 2006. MR2172781

Index